Undergraduate Lecture Notes in Physics

More information about this series at http://www.springer.com/series/8917

Undergraduate Lecture Notes in Physics (ULNP) publishes authoritative texts covering topics throughout pure and applied physics. Each title in the series is suitable as a basis for undergraduate instruction, typically containing practice problems, worked examples, chapter summaries, and suggestions for further reading.

ULNP titles must provide at least one of the following:

- An exceptionally clear and concise treatment of a standard undergraduate subject.
- A solid undergraduate-level introduction to a graduate, advanced, or non-standard subject.
- A novel perspective or an unusual approach to teaching a subject.

ULNP especially encourages new, original, and idiosyncratic approaches to physics teaching at the undergraduate level.

The purpose of ULNP is to provide intriguing, absorbing books that will continue to be the reader's preferred reference throughout their academic career.

Series editors

Neil Ashby
Professor Emeritus, University of Colorado Boulder, CO, USA

William Brantley
Professor, Furman University, Greenville, SC, USA

Michael Fowler
Professor, University of Virginia, Charlottesville, VA, USA

Michael Inglis
Professor, SUNY Suffolk County Community College, Selden, NY, USA

Heinz Klose
Professor Emeritus, Humboldt University Berlin, Germany

Helmy Sherif
Professor, University of Alberta, Edmonton, AB, Canada

Mark A. Cunningham

Neoclassical Physics

 Springer

Mark A. Cunningham
Katy, TX, USA

Wolfram *Mathematica*® is a registered trademark of Wolfram Research, Inc.

ISSN 2192-4791   ISSN 2192-4805 (electronic)
ISBN 978-3-319-10646-5   ISBN 978-3-319-10647-2 (eBook)
DOI 10.1007/978-3-319-10647-2
Springer Cham Heidelberg New York Dordrecht London

Library of Congress Control Number: 2014953382

Printed on acid-free paper

Springer is part of Springer Science+Business Media (www.springer.com)

For Elizabeth Ann

Preface

Neoclassical Physics evolved from my growing sense of frustration that introductory physics texts seem designed to drive students away from the subject. While physicists take pride in describing hosts of phenomena from a few guiding principles, introductory textbooks introduce too many subjects and require too much memorization of disconnected facts. Worse, because introductory physics courses are treated as tool-building exercises for subsequent studies in physics, topics like special relativity that hold great student interest are not covered. Instead, a disproportionate amount of the material presented, like bricks sliding on the floor, is simply uninspiring. I have, in fact, had students remark "You seem to be a really smart guy, how can you *do* this stuff?"

So, I have written this book with the determination that I will spend more time covering the wonderful and that I will make better use of modern technology. It has been a constant source of frustration to me that students lose sight of the physical principles as they struggle with the mathematics. Tools now exist to assist students with algebraic manipulations; I think that it is time to incorporate them in introductory classes. I recognize that using a tool like Wolfram *Mathematica*® software brings with it another set of problems but, in my experience, the advantages outweigh the disadvantages. I note also that translators exist to convert between *Mathematica* notebook files and *Maple*® software formats. For those who have made the intellectual (and/or financial) investment in an alternative software package, it should be possible to find equivalents for the *Mathematica* functions identified in the text.

A particular benefit of using sophisticated software is that we can now routinely incorporate visualization into the course. Rather than stopping at the derivation of some result, we can ask students to examine the shape of the resulting function and ask how does that shape change if we alter any of the parameters? As I mention later in the text, one of my former colleagues was fond of stating that humans evolved to recognize tigers—those who couldn't were removed from the gene pool. Providing students with an opportunity to visualize the results of their labors will add to their understanding of the material and provide another, geometrical perspective.

In selecting the title for this text, I thought that neoclassical was a word that sounded pithy and captured my intent. While wandering through a museum, it struck me that, because the word was already in use in architectural and artistic settings, my appropriation of the word might be in conflict with accepted usage. After a bit of research, I concluded that the neoclassical movement was founded in reaction to the Baroque and Rococo schools in an attempt to return to the "purity" found in classical Greek and Roman arts and architecture. I am relieved to find that my original choice of title does, indeed, reflect my purpose and intent: I am departing from the Baroque treatment of physics found in present books and focussing on how the process of physics allows us to construct models of physical systems.

It is my expectation that Chaps. 1–5 will constitute a semester's worth of material. Each chapter will require two to three weeks to cover; this certainly marks a significant departure from the current chapter per week methodology employed in most courses. The text has been written to support a deliberate pace and will work well with a guided-discovery approach to instruction. Additionally, initial progress will be slowed by a necessary introduction to the *Mathematica* program. I have made no attempt to provide any *Mathematica* tutorials as part of this text, as a number of exceptionally detailed tutorials exist on the Wolfram company website. As a practical matter, the *Mathematica* documentation can be cut and pasted into an active notebook, so if you can find an example of something that is interesting or relevant, you can experiment on a working script immediately. Scripts in the Wolfram *Mathematica* Language® are identified in the text with a typewriter-like font.

The fastest means for learning a programming language, in my experience, is to follow precisely the path described above: obtain a working piece of software and then begin to modify the code to see what ensues. Some changes will illuminate what the bits and pieces of the code do; most changes will break it. With the language reference manual in one's lap, one can utilize this technique to evolve a working example into a new (working) example that performs a different function. This strategy allows beginners, for example, to perform complex tasks like running animations with key parameters assigned to sliders, something that a beginner might never even get to run to completion, much less actually perform the desired function. The student can even replace the functions being plotted without understanding how the plots get generated or why the sliders work. Those details can be mastered at some future time, or never, if there is no interest.

This strategy, in large measure, reflects my approach to the material in the text, which leads students through the development of several of the most important discoveries in physics. Few of the exercises ask students to perform a derivation, except in parallel with one that was conducted in the text. The purpose of these derivations is to force students to utilize different notation than that in the text, to see firsthand that there can be different mathematical representations of the same underlying physical ideas. For the most part, the exercises are designed to provide students with a means for visually examining some aspect of the physical system being investigated. Deriving the equations of motion is not the objective; understanding how the various parameters that define those equations affect the trajectory is the goal.

The remaining Chaps. 6–11 may represent more material than can be covered in a second semester. These chapters are largely independent of one another, providing instructors with the option to follow different pathways in the second semester. It is my hope, though, that students' increasing sophistication and competence with the *Mathematica* software will ultimately accelerate their progress and permit them to wade through the text in its entirety.

I recognize that, by limiting the course to such a small number of chapters, some will be critical of important topics that have been left by the wayside. That perspective is understandable but I have made considered choices. My intent is to pique the interest of students who would not have considered a career in physics; perhaps, I will win a few converts. For the remainder of the student population, who are enrolled solely because it is a required course for their major, they will learn some useful mathematical skills and something of the tradecraft of physicists. They will emerge from this course with skills comparable to those obtained in other texts but with an in-depth knowledge of a smaller list of topics, instead of a passing familiarity with a broader list. I believe this to be a worthy goal.

Whether I am successful or not in this enterprise, please understand that all errors in this text are my own. Others have certainly supported me in this effort; foremost of those is my wife Liz, without whose support this text would have never seen the light of day. I am eternally grateful that she has provided unwavering support for my efforts.

Katy, USA   Mark A. Cunningham

Contents

Preface   vii

List of Figures   xv

List of Tables   xix

List of Exercises   xxi

Chapter 1.   Introduction   1
   1.1.   What Is Physics?   2
   1.2.   Space and Time   4
   1.3.   Vocabulary   12

Chapter 2.   On the Motion of Planets   19
   2.1.   Form of the Force Law   20
   2.2.   Inverse Square Law   24
   2.3.   Another Conserved Quantity   31
   2.4.   A Special Frame of Reference   33
   2.5.   Trajectories in More Detail   38
   2.6.   Position as a Function of Time   43
   2.7.   Finite Sizes   48
   2.8.   Unfinished Business   52

Chapter 3.   On the Nature of Matter   55
   3.1.   Hyperbolic Trajectories   56
   3.2.   Asymptotic Behavior   60
   3.3.   Rutherford's Experiments   64
   3.4.   Gauss's Law   66
   3.5.   Nuclear Model of the Atom   74
   3.6.   Finite Nuclear Size   84
   3.7.   Unfinished Business   85

Chapter 4.   On the Nature of Spacetime   87
   4.1.   Statistical Analysis   90
   4.2.   Finite Velocity of Light   96
   4.3.   Wave Equation   101
   4.4.   Lorentz Transform   106
   4.5.   Relativity and Causality   113
   4.6.   Eigenzeit   120
   4.7.   Unfinished Business   128

Chapter 5.   More on the Nature of Matter   131
   5.1.   Vibrations   133
   5.2.   Point Sources   135
   5.3.   Beats   137
   5.4.   Interference   140
   5.5.   Diffraction   144
   5.6.   Spectra   154
   5.7.   Unfinished Business   157

Chapter 6.   Terrestrial Mechanics   159
   6.1.   Motion Near the Earth's Surface   161
   6.2.   Energy Conservation   168
   6.3.   Physics of Baseballs   173
   6.4.   Spin on Baseballs   180
   6.5.   Wind   184
   6.6.   Friction   185

Chapter 7.   Celestial Mechanics   189
   7.1.   Restricted Three-Body Problem   190
   7.2.   Lagrange Points   195
   7.3.   Rocket Equation   197
   7.4.   Launching from the Earth's Surface   203
   7.5.   Gravity Assists   207
   7.6.   Orbital Resonance   212
   7.7.   Chaos   214

Chapter 8.   Constituents of the Atom   217
   8.1.   Lorentz Force   218
   8.2.   Cathode Rays   222
   8.3.   Canal Rays   225
   8.4.   Neutrons   233
   8.5.   The World is not so Simple   239

Chapter 9.   The Classical Electron   251
   9.1.   Currents   251
   9.2.   Magnetic Fields   257
   9.3.   Magnetic Materials   269
   9.4.   Magnetic Field of the Electron   277
   9.5.   Electron Charge   285
   9.6.   Unfinished Business   288

Chapter 10.   Modern Technology   291
   10.1.   Induction   292
   10.2.   Circuits   296
   10.3.   Resonance   305

10.4. Optical Molasses   308
10.5. Telecommunications   310
10.6. Vacuum Tubes   314

Chapter 11. Emergent Phenomena   319
11.1. Probability   320
11.2. Brownian Motion   323
11.3. Statistical Thermodynamics   330
11.4. Applications   339
11.5. Equilibrium   347

Appendix A. Vectors and Matrices   353

Appendix B. Noether's Theorem   363

Index   367

List of Figures

1.1	A timing experiment.	2
1.2	Trajectory of an object.	5
1.3	Coordinate invariance.	9
1.4	Milky Way galaxy (NASA)	11
1.5	Position, velocity and acceleration.	14
1.6	Piecewise constant motion.	18
2.1	Down is relative	19
2.2	Two masses	20
2.3	Planar motion	26
2.4	The eccentricity vector	28
2.5	Eccentricity defines conic sections	30
2.6	Center of mass	33
2.7	The ellipse	39
2.8	Polar coordinates	40
2.9	Kepler's second law	42
2.10	Time along the trajectory	45
2.11	Another view of time along the trajectory	47
2.12	Gravitational field of a rigid sphere	49
2.13	Pythagorean theorem	51
2.14	Rippling shadows (NASA)	53
3.1	The function $1 + e\cos\psi$	58
3.2	Hyperbolic trajectories	59
3.3	Asymptotic behavior	63
3.4	Plum pudding model	65
3.5	Geiger-Marsden experiment	66
3.6	Surface normal	68
3.7	Electric flux	69
3.8	Flux through cubical surface	70
3.9	Gauss's law	71
3.10	Electric field of a sphere	73
3.11	Electric field in the plum pudding model	75
3.12	Electric field in an atom	78
3.13	Scattering parameter	79
3.14	Scattering cross section	80
3.15	Scattering of α particles by gold foil	81
3.16	Scattering of α particles from lead foil	84

4.1   Eclipses of Io   88
4.2   Two sequences of measurements   90
4.3   Normal probability distribution   92
4.4   Neutron lifetime measurements   93
4.5   Predicted and observed Ionian eclipses   98
4.6   Ionian eclipses 1672–1673   100
4.7   Lorentz transform preserves time and space   107
4.8   Lorentz transform produces hyperbolic surfaces   112
4.9   Lorentz transform of time-like vectors   114
4.10   Waves in two dimensions   117
4.11   Doppler effect   118
4.12   The function γ 119
4.13   Relativity at human scale   122
4.14   Electron velocities   128

5.1   Ripples on water surface (Alan Watson-Featherstone)   131
5.2   Displacement on a guitar string   132
5.3   Beats   138
5.4   Interference   140
5.5   Geometry of wave interference   141
5.6   Coherent intensity   142
5.7 N slit interference   144
5.8   Ocean wave diffraction© Google Earth 2012   145
5.9   Huygens wavelets   146
5.10   Lattice vectors   148
5.11   Diffraction geometry   151
5.12   Diffraction from a cubic lattice   153
5.13   Diffraction from a cubic lattice redux   153
5.14   Atomic spectra   155

6.1   Free-body diagram   160
6.2   Local coordinate systems   161
6.3   Aerodynamic forces   164
6.4   Velocity *versus* time   166
6.5   Resistive force   167
6.6   Potential energy is path independent   168
6.7   Energy loss due to resistive force   171
6.8   Work-energy conservation   172
6.9   Trajectory of a baseball   174
6.10   Magnus force   179
6.11   Baseball spin modifies the trajectory   181
6.12   Frictional force   186
6.13   Sliding blocks   187

7.1	Potential energy surface	190
7.2	Two-center potential energy surface	191
7.3	Three-body motion	192
7.4	Unstable three-body motion	193
7.5	Lagrangian points	196
7.6	Rocket equation	199
7.7	Hohmann transfer orbit	201
7.8	Transfer orbit	202
7.9	Surface launch	205
7.10	Circularizing orbits	206
7.11	Gravity assist	209
7.12	Gravity assist redux	210
7.13	Orbital resonance	213
7.14	Chaotic motion	215
8.1	Lorentz force	220
8.2	Helical trajectory	221
8.3	Magnetic deflection	222
8.4	Cathode rays	223
8.5	Thomson'.s experiment	224
8.6	Canal rays	226
8.7	Mass spectrometer	226
8.8	Canal rays redux	229
8.9	Analyzing canal ray data	230
8.10	Aston's apparatus	232
8.11	Mass spectrum	232
8.12	Scattering	235
8.13	Chadwick's apparatus	238
8.14	β-decay of ^{210}Bi	240
8.15	Positron-electron pair production (LBL)	242
8.16	"Elementary" particles	243
8.17	Strange matter (LBL)	245
8.18	Charmed matter	246
8.19	Neutrino event (ANL)	249
9.1	Magnetic field of a wire	252
9.2	Charge distribution	254
9.3	Drude model	255
9.4	Biot-Savart equation	257
9.5	Magnetic field of a current loop	260
9.6	Magnetic field as a function of radius	265
9.7	Magnetic field lines of a current loop	266
9.8	Ampère's law	268

9.9   Equivalence of magnetic fields   270
9.10   Spinning charge creates a magnetic field   271
9.11   Magnetic field lines of a rotating, charged sphere   276
9.12   Precession of magnetic moment   280
9.13   Stern-Gerlach apparatus   281
9.14   Stern-Gerlach experimental results. (Springer
       Science+Business Media)   284
9.15   Change in electric charge on an oil drop   287

10.1   Eastern Mediterranean at night   291
10.2   Faraday transformer   293
10.3   Magnetic induction   295
10.4   Electrical circuit   296
10.5   A more complex circuit   297
10.6   Time-varying sources   300
10.7   Circuit phase   301
10.8   Leyden jar   302
10.9   Capacitance   303
10.10 RC circuit   304
10.11 LRC circuit   305
10.12 LRC circuit resonance.   306
10.13 Resonance avoidance   308
10.14 Radio receiver   311
10.15 Transducer   311
10.16 AM modulation   311
10.17 Vacuum tube diode   315
10.18 Vacuum tube triode   316

11.1   Random walk   323
11.2   Brownian motion   325
11.3   Diffusion   327
11.4   Square root of time behavior   329
11.5   Temperature is not always linear   332
11.6 N particles in a box   335
11.7   Effusion   337
11.8   Maxwell-Boltzmann distribution   338
11.9   Specific heat of diamond   345
11.10 Specific heat of aluminum   345
11.11 Chemical reaction   348

A.1   Surface normal   356
A.2   Cylindrical coordinate system   357
A.3   Spherical coordinate system   359

List of Tables

2.1	Solar system orbital data	43
3.1	Electron positions	75
3.2	α scattering from silver	82
4.1	Rømer's observations of eclipses of Io	89
4.2	Experimental data	92
4.3	Neutron lifetimes τ from the Particle Data Group	94
4.4	More experimental data	100
7.1	Orbital parameters	195
8.1	Particle lifetimes	249
9.1	Millikan's data on drop #6	286
11.1	Perrin's displacement data for 0.367 μm-radius particles	328
11.2	Velocity distribution of 944 K thallium atoms	338
11.3	Molar specific heats of diatomic molecules	342
11.4	Molar specific heats of metals. Units are J/mol-K	343
11.5	Specific heat for copper in units of J/mol·K	346

List of Exercises

Exercise 1.1	3	Exercise 2.26	41
Exercise 1.2	4	Exercise 2.27	42
Exercise 1.3	6	Exercise 2.28	42
Exercise 1.4	6	Exercise 2.29	44
Exercise 1.5	6	Exercise 2.30	44
Exercise 1.6	7	Exercise 2.31	45
Exercise 1.7	9	Exercise 2.32	46
Exercise 1.8	12	Exercise 2.33	46
Exercise 1.9	15	Exercise 2.34	47
Exercise 1.10	15	Exercise 2.35	47
Exercise 1.11	16	Exercise 2.36	47
Exercise 1.12	17	Exercise 2.37	50
Exercise 1.13	18	Exercise 2.38	50
		Exercise 2.39	52
Exercise 2.1	22	Exercise 2.40	54
Exercise 2.2	23		
Exercise 2.3	25	Exercise 3.1	57
Exercise 2.4	27	Exercise 3.2	57
Exercise 2.5	27	Exercise 3.3	60
Exercise 2.6	27	Exercise 3.4	60
Exercise 2.7	27	Exercise 3.5	61
Exercise 2.8	28	Exercise 3.6	62
Exercise 2.9	28	Exercise 3.7	62
Exercise 2.10	29	Exercise 3.8	64
Exercise 2.11	30	Exercise 3.9	64
Exercise 2.12	30	Exercise 3.10	64
Exercise 2.13	30	Exercise 3.11	69
Exercise 2.14	32	Exercise 3.12	69
Exercise 2.15	33	Exercise 3.13	70
Exercise 2.16	33	Exercise 3.14	70
Exercise 2.17	35	Exercise 3.15	72
Exercise 2.18	35	Exercise 3.16	74
Exercise 2.19	36	Exercise 3.17	74
Exercise 2.20	36	Exercise 3.18	76
Exercise 2.21	38	Exercise 3.19	77
Exercise 2.22	39	Exercise 3.20	78
Exercise 2.23	39	Exercise 3.21	79
Exercise 2.24	39	Exercise 3.22	82
Exercise 2.25	40	Exercise 3.23	82

Exercise 3.24	83		Exercise 4.38	128
Exercise 3.25	84		Exercise 4.39	130
Exercise 3.26	85		Exercise 4.40	130
Exercise 3.27	86			
Exercise 3.28	86		Exercise 5.1	134
			Exercise 5.2	134
			Exercise 5.3	136
Exercise 4.1	91		Exercise 5.4	136
Exercise 4.2	93		Exercise 5.5	137
Exercise 4.3	94		Exercise 5.6	137
Exercise 4.4	97		Exercise 5.7	138
Exercise 4.5	99		Exercise 5.8	139
Exercise 4.6	99		Exercise 5.9	140
Exercise 4.7	99		Exercise 5.10	141
Exercise 4.8	100		Exercise 5.11	141
Exercise 4.9	102		Exercise 5.12	141
Exercise 4.10	103		Exercise 5.13	142
Exercise 4.11	103		Exercise 5.14	143
Exercise 4.12	103		Exercise 5.15	143
Exercise 4.13	104		Exercise 5.16	144
Exercise 4.14	104		Exercise 5.17	144
Exercise 4.15	104		Exercise 5.18	146
Exercise 4.16	106		Exercise 5.19	146
Exercise 4.17	106		Exercise 5.20	149
Exercise 4.18	106		Exercise 5.21	149
Exercise 4.19	107		Exercise 5.22	150
Exercise 4.20	110		Exercise 5.23	152
Exercise 4.21	110		Exercise 5.24	154
Exercise 4.22	110		Exercise 5.25	154
Exercise 4.23	111		Exercise 5.26	155
Exercise 4.24	113		Exercise 5.27	156
Exercise 4.25	113		Exercise 5.28	158
Exercise 4.26	114			
Exercise 4.27	115		Exercise 6.1	162
Exercise 4.28	117		Exercise 6.2	163
Exercise 4.29	118		Exercise 6.3	166
Exercise 4.30	122		Exercise 6.4	166
Exercise 4.31	123		Exercise 6.5	167
Exercise 4.32	123		Exercise 6.6	167
Exercise 4.33	124		Exercise 6.7	167
Exercise 4.34	124		Exercise 6.8	167
Exercise 4.35	125		Exercise 6.9	169
Exercise 4.36	126		Exercise 6.10	170
Exercise 4.37	126		Exercise 6.11	170

Exercise 6.12	173		Exercise 7.19	202
Exercise 6.13	173		Exercise 7.20	204
Exercise 6.14	173		Exercise 7.21	205
Exercise 6.15	174		Exercise 7.22	206
Exercise 6.16	175		Exercise 7.23	207
Exercise 6.17	176		Exercise 7.24	207
Exercise 6.18	177		Exercise 7.25	209
Exercise 6.19	177		Exercise 7.26	210
Exercise 6.20	178		Exercise 7.27	211
Exercise 6.21	178		Exercise 7.28	213
Exercise 6.22	182		Exercise 7.29	214
Exercise 6.23	182		Exercise 7.30	214
Exercise 6.24	182		Exercise 7.31	215
Exercise 6.25	183		Exercise 7.32	215
Exercise 6.26	183			
Exercise 6.27	184		Exercise 8.1	219
Exercise 6.28	184		Exercise 8.2	219
Exercise 6.29	184		Exercise 8.3	220
Exercise 6.30	185		Exercise 8.4	221
Exercise 6.31	185		Exercise 8.5	222
Exercise 6.32	186		Exercise 8.6	224
Exercise 6.33	187		Exercise 8.7	225
Exercise 6.34	188		Exercise 8.8	225
Exercise 6.35	188		Exercise 8.9	226
			Exercise 8.10	227
Exercise 7.1	191		Exercise 8.11	227
Exercise 7.2	192		Exercise 8.12	228
Exercise 7.3	193		Exercise 8.13	229
Exercise 7.4	194		Exercise 8.14	229
Exercise 7.5	194		Exercise 8.15	229
Exercise 7.6	194		Exercise 8.16	231
Exercise 7.7	195		Exercise 8.17	231
Exercise 7.8	196		Exercise 8.18	232
Exercise 7.9	196		Exercise 8.19	234
Exercise 7.10	197		Exercise 8.20	234
Exercise 7.11	197		Exercise 8.21	237
Exercise 7.12	198		Exercise 8.22	237
Exercise 7.13	199		Exercise 8.23	238
Exercise 7.14	200		Exercise 8.24	239
Exercise 7.15	200		Exercise 8.25	241
Exercise 7.16	201		Exercise 8.26	244
Exercise 7.17	201		Exercise 8.27	245
Exercise 7.18	202		Exercise 8.28	246

Exercise 8.29   247

Exercise 8.30   247

Exercise 8.31   248

Exercise 8.32   248

Exercise 8.33   249

Exercise 8.34   250

Exercise 9.1   256

Exercise 9.2   256

Exercise 9.3   259

Exercise 9.4   259

Exercise 9.5   260

Exercise 9.6   261

Exercise 9.7   262

Exercise 9.8   263

Exercise 9.9   264

Exercise 9.10   264

Exercise 9.11   265

Exercise 9.12   266

Exercise 9.13   266

Exercise 9.14   269

Exercise 9.15   269

Exercise 9.16   271

Exercise 9.17   272

Exercise 9.18   272

Exercise 9.19   273

Exercise 9.20   274

Exercise 9.21   274

Exercise 9.22   275

Exercise 9.23   275

Exercise 9.24   275

Exercise 9.25   276

Exercise 9.26   276

Exercise 9.27   277

Exercise 9.28   278

Exercise 9.29   278

Exercise 9.30   280

Exercise 9.31   280

Exercise 9.32   283

Exercise 9.33   283

Exercise 9.34   284

Exercise 9.35   284

Exercise 9.36   286

Exercise 9.37   287

Exercise 9.38   287

Exercise 9.39   287

Exercise 10.1   294

Exercise 10.2   298

Exercise 10.3   298

Exercise 10.4   298

Exercise 10.5   298

Exercise 10.6   301

Exercise 10.7   301

Exercise 10.8   301

Exercise 10.9   304

Exercise 10.10   304

Exercise 10.11   304

Exercise 10.12   304

Exercise 10.13   306

Exercise 10.14   306

Exercise 10.15   306

Exercise 10.16   307

Exercise 10.17   309

Exercise 10.18   310

Exercise 10.19   312

Exercise 10.20   312

Exercise 10.21   313

Exercise 10.22   313

Exercise 10.23   313

Exercise 10.24   315

Exercise 10.25   317

Exercise 10.26   317

Exercise 11.1   320

Exercise 11.2   321

Exercise 11.3   322

Exercise 11.4   322

Exercise 11.5   323

Exercise 11.6   324

Exercise 11.7   326

Exercise 11.8   326

Exercise 11.9   326

Exercise 11.10   327

Exercise 11.11   328

Exercise 11.12   328

Exercise 11.13   329

Exercise 11.14   334
Exercise 11.15   335
Exercise 11.16   336
Exercise 11.17   337
Exercise 11.18   338
Exercise 11.19   339
Exercise 11.20   339
Exercise 11.21   340
Exercise 11.22   342
Exercise 11.23   342
Exercise 11.24   342
Exercise 11.25   344
Exercise 11.26   344
Exercise 11.27   346
Exercise 11.28   347
Exercise 11.29   347
Exercise 11.30   347
Exercise 11.31   349
Exercise 11.32   350
Exercise 11.33   352

Exercise B.1   364
Exercise B.2   364

I

Introduction

It is surprisingly difficult to provide a concise definition of *physics*. Most authors (physicists) claim that physics is the most fundamental of the sciences; most students will attest that it is the most arcane. In perusing the literature, we can find the word physics applied in many seemingly unrelated circumstances. For example, there are books and journals in which authors speak of the physics of subatomic particles, the physics of musical instruments and the physics of baseball. Actually, all of these cases are perfectly reasonable uses of the word physics, even though it may not be immediately obvious what baseball has to do with pianos or electrons.

As we shall see during the course of this text, physics is not so much a separate scientific discipline, like biology or chemistry, but is instead a *process* by which we systematically conduct experiments and methodically construct mathematical representations of the behavior of physical systems. In this regard, physics is rightfully considered a difficult subject; practitioners must possess a fluency in mathematics that is not generally found in the population at large. In studying physics, most students will find that they utilize every shred of mathematical knowledge they possess and must work diligently to expand their mathematical skills. Moreover, students must learn to think abstractly, which is a skill that is not often emphasized elsewhere.

A further difficulty that we shall encounter stems from the fact that we use English words (for the most part) to define our physics concepts; those words often have other, conflicting definitions. One could argue that physicists should invent their own words to describe their concepts and physicists have made some effort in that direction. The word "quantum," for example, was initially used by the Romans to mean "how much" of a quantity you possess. As few people now converse in Latin, physicists appropriated the word quantum to identify the small, discrete difference between two atomic states. As such, quantum represents an infinitesimal (but distinguishable from zero) quantity.

© Mark A. Cunningham 2015

M.A. Cunningham, *Neoclassical Physics*, Undergraduate Lecture Notes in Physics, DOI 10.1007/978-3-319-10647-2_1

We find now that the word has passed back into common usage, e.g., "This new product represents a quantum leap in dish-washing detergent technology." This common use of the word means a significant or major change—exactly the opposite of the physicists' usage. So, even when physicists invent their own words (or steal them from Latin or ancient Greek to demonstrate their classical educations), we cannot control usage outside our own domain. We shall attempt to be precise in our usage of terminology in this text; students will simply have to learn to recognize key words and their contexts.

1.1. What Is Physics?

As a first step in trying to understand the process of physics, imagine an experiment in which a number of students are provided stopwatches and then are positioned at equal distances along some path. Each student's objective is to record the time that it takes a runner to reach his or her position along the path. To begin the experiment, a starter waves a flag; the runner starts running; the students start their stopwatches and each student stops his or her watch when the runner passes. The students then reassemble and their measured times are recorded in the order in which the students were aligned on the path. We've plotted the results of this experiment in the figure below.

FIGURE 1.1. A timing experiment. Students spaced an equal distance along some path measure the time it takes a runner to reach their position. These values are shown as dots. A linear fit to the data is also shown as the solid line

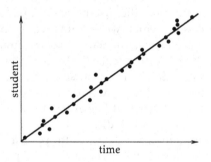

What we observe is that there is a general trend to the data; we have included a linear fit to the data to emphasize that observation. The fit is not perfect but we suspect that not all of the students were equally proficient at timing the runner. Probably replacing the students with a series of electronic measuring devices would reduce the scatter but an essential element of measurement is understanding that all experiments have finite precision and all experiments are afflicted with noise. It is never possible to know something *exactly*.

Our first example demonstrates a few pertinent facts about the process of physics. First, our objective is always to understand the behavior of some physical system and, as we shall see, the definition of a physical system can be quite broad. Second, physics is an *experimental* science; our development of mathematical models is always motivated by experiment. This point may not be always obvious in an introductory course, where there has been significant time to refine our notation and the presentation of our results. Yet, we should not lose sight of the fact that experimental observation drives science. Third, physics is *not* curve fitting. While we may make use of such tools as a means for illuminating the behavior of a system, we will never be content with just the values of fit parameters.

To begin our analysis of the experiment represented by figure 1.1, we know that the equation of a line is $y = mx + b$. If we use the variable t to represent time and the variable s to represent the student, we can recast the general equation of the line into the form $s = mt + b$ that represents our specific experiment. We now have an abstract, mathematical representation of our experiment. This equation describes the motion of the runner over a particular time interval.

Notice, though, that for this equation to make mathematical sense, each of the terms in the equation must have the same dimension.[1] If s represents a distance, then the product mt must also have the dimension of a distance, as must the factor b. Because the product mt has the dimension of a distance, the coefficient m must have the dimension of a distance divided by time. This process of dimensional analysis is an important one that we will revisit periodically throughout the text.

> EXERCISE 1.1.   In the study of object motion, the most common dimensions are mass M, length L and time T. The dimensional equation that corresponds to the equation $s = mt + b$ would be $(L) = (L/T) \cdot (T) + (L)$. Suppose now that we have an equation $s = a_0 + a_1 t + a_2 t^2$. If s has the dimension of length L, and t represents time, what must be the dimensions of the coefficients a_i? What is the corresponding dimensional equation?

Our real objective in physics is to use equations like $s = mt + b$ to predict the position of the runner beyond the domain $t_1 \le t \le t_2$ or to predict the positions of other runners. We would like to understand why the parameters m and b take on particular values. It is essential, though, that we understand what physical concepts the equations represent and what are the limitations of our models. In the experiment depicted in figure 1.1, we wouldn't be surprised if the runner's position continued as a straight

---

[1] Unit systems like *le Système International d'unités* provide precise definitions for dimensions. At this moment, we are not concerned about the units associated with the dimension of length. We shall defer this point until later.

line for a few more seconds but we also wouldn't expect the curve to remain linear indefinitely. Not many runners can maintain a constant pace for long periods of time. We also know, in this particular experiment, that the equation does not allow us to predict the position of the runner prior to the starting time. Presumably, the runner was wandering about in the vicinity of the starting point for a time and then stood motionless at the beginning of the path for a while before the starter waved the flag. That behavior of the runner is not captured by the equation $s = mt + b$. So, an essential part of physics will be to understand the range of validity of the mathematical models that we construct.

EXERCISE 1.2. Many economic models assume compounded growth. That is, if you have a quantity x of something initially, after a year has passed you will have a% more. The quantity at the end of the first year is thus $x \cdot (1 + a)$, where here we represent a as a fraction. After n years, you would therefore have $x \cdot (1 + a)^n$. Suppose that you invest $1000 at 10% interest.

(a) How much money would you have after 10 years?
(b) How much money would you have after 50 years?
(c) (Rhetorical) How likely are things to remain constant for 50 years?

1.2. Space and Time

The value of using mathematics to describe our systems rests upon the ability to make quantitative predictions about the time evolution of whatever system we are studying. Moreover, mathematical descriptions provide us with the means for understanding what parameters control that evolution. For example, the equation $s = mt + b$ permits us to estimate the position of the runner between where we have placed students and, potentially, beyond the timed interval. The parameters m and b characterize the runner's progress in some fashion. If we consider time to be a continuous variable, then we have an equation that represents the runner's position as a continuous function of time: $s = s(t)$. We call such a function a **trajectory**. This is an important, abstract concept that is crucial to our ultimate understanding of the motion of objects.

What should we demand of such a function? First, it seems reasonable to assume that the runner is *somewhere* at each instant in time. Second, we should also impose the condition that the runner is at only *one point* in space at any one time; as appealing an idea as that might be, no one can be in two places at once. Finally, we will make another assumption that from each time increment to the next the runner is somehow close to the previous position. Here, we are specifically thinking of reducing the spacing (and improving the precision) of our timers and taking a mathematical limit. What we are requiring by this last demand on the trajectory is that the runner cannot disappear from one point in space and rematerialize in the next instant at a point elsewhere along the path. These assumptions lead us to the conclusion that the trajectory of an object is a single-valued, continuous function of time. This is a relatively important mathematical point; we'll explore the consequences further as we continue.

At this point, some students may be troubled by our assumptions on the conditions necessary for a function to be a trajectory. The darting, zigzagging flights of bats might seem to provide a counterexample to our requirements that trajectories be smooth, continuous functions. Indeed, human eyes cannot track the apparently frantic motion of bats in flight but with better technology like high-speed cameras, bats' flight patterns can be seen to be smoothly varying. At a sufficiently high time resolution, beyond human capacity, there is nothing to be found in bats' trajectories to indicate that the bats disappear at one point and magically reappear elsewhere. They simply move very rapidly. We will find that all macroscopic objects have trajectories that possess the attributes that we have listed.

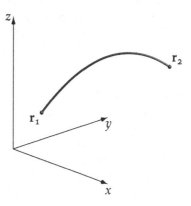

FIGURE 1.2. Trajectory of an object. The dark line represents the position $\mathbf{r}$ in space of an object over the time interval $t_1 \le t \le t_2$. The illustrated coordinate system provides a means for quantifying the trajectory

In order to gain some insights into this concept of a trajectory, consider the curve illustrated in figure 1.2. At some time $t = t_1$, an object is at the position $\mathbf{r}_1$ and at some later time $t = t_2$, the object has moved to position $\mathbf{r}_2$ by taking the path indicated by the dark line. Unfortunately, most real objects do not leave visible tracks like that illustrated in the figure; the trajectory is an abstraction of where the object was located at any particular time. A trajectory is not a physical entity like a piano or a video game controller that can be examined visually and manipulated by hand.

To assign mathematical values to points along the trajectory, or to quantify the trajectory, we now define a coordinate system, as indicated by the three axes labelled x, y and z. In general, the position of the object at the time t requires the specification of three numbers: the values of the coordinates $x(t)$, $y(t)$ and $z(t)$. We can write these values compactly using vector notation. The vector $\mathbf{r}(t)$ is defined as the ordered set of numbers:

$$\mathbf{r}(t) = (x(t), y(t), z(t)).$$

The values of $x(t)$, $y(t)$ and $z(t)$ are called the *components* of the vector $\mathbf{r}(t)$. At the initial time t_1, the position of the object is given by the vector $\mathbf{r}_1 = (x(t_1), y(t_1), z(t_1))$. Similarly, at the final time t_2, we have $\mathbf{r}_2 = (x(t_2), y(t_2), z(t_2))$.

Use of this vector notation will provide us with a compact means of writing the three separate sets of equations that describe the motion of the object along each of the three coordinate axes.

> EXERCISE 1.3. Algebraically, we can define the scalar product of a vector as $a\mathbf{r} - (ax, ay, az)$. The magnitude of the vector $\mathbf{r}_1$ is defined as follows:
>
> $$|\mathbf{r}_1| = r_1 = (x_1^2 + y_1^2 + z_1^2)^{1/2}.$$
>
> Vectors add by components: $\mathbf{r}_1 + \mathbf{r}_2 = (x_1 + x_2, y_1 + y_2, z_1 + z_2)$. The dot product of two vectors is obtained by multiplying their components and summing. The result is a scalar: $\mathbf{r}_1 \cdot \mathbf{r}_2 = x_1 x_2 + y_1 y_2 + z_1 z_2$. We can also define the cross product of two vectors, which results in a vector perpendicular to the initial two vectors:
>
> $$\mathbf{r}_1 \times \mathbf{r}_2 = ((y_1 z_2 - y_2 z_1), (z_1 x_2 - z_2 x_1), (x_1 y_2 - x_2 y_1)).$$

(a) Show that the square magnitude of the vector $\mathbf{r}_1$ can be obtained by taking the dot product of the vector with itself.
The unit vectors in a Cartesian coordinate system can be defined as follows: $\hat{\mathbf{x}} = (1, 0, 0)$ $\hat{\mathbf{y}} = (0, 1, 0)$ $\hat{\mathbf{z}} = (0, 0, 1)$.
(b) What are the results of the following operations: $\hat{\mathbf{x}} \cdot \hat{\mathbf{x}}$ $\hat{\mathbf{x}} \cdot \hat{\mathbf{y}}$ $\hat{\mathbf{x}} \cdot \hat{\mathbf{z}}$?
(c) What are the results of the following operations: $\hat{\mathbf{x}} \times \hat{\mathbf{x}}$ $\hat{\mathbf{x}} \times \hat{\mathbf{y}}$ $\hat{\mathbf{x}} \times \hat{\mathbf{z}}$?

> EXERCISE 1.4. Consider a vector $\mathbf{r}_1 = (x_1, y_1, z_1)$. Using the definitions of the unit vectors defined in the previous exercise, compute the results of the following operations:
>
> $$\hat{\mathbf{x}} \cdot \mathbf{r}_1 \qquad \hat{\mathbf{y}} \cdot \mathbf{r}_1 \qquad \hat{\mathbf{z}} \cdot \mathbf{r}_1$$

> EXERCISE 1.5. Define the vectors $\mathbf{r}_1 = (x_1, y_1, z_1)$ and $\mathbf{r}_2 = (x_2, y_2, z_2)$. Prove that the following relation is true:
>
> $$\mathbf{r}_1 \times \mathbf{r}_2 = -\mathbf{r}_2 \times \mathbf{r}_1.$$

We will often seek to find notation that is concise. There is nothing intrinsically wrong about using a variable name like *position along the* x-*axis*. That notation is much more explicit about what the variable represents but is much more time-consuming to write than x. There is nothing wrong with talking about the *change in position along the* x-*axis per time* but dx/dt is much more concise and $\dot{x}$ is even more concise. We will have to judge from the situation how compact we can make our notation and still understand what it means. So, in choosing to use different notations, we are attempting to simplify how we represent complex ideas. We are also attempting to strike a balance between verbose and succinct and that balance will change with increasing sophistication. An astute choice of notation may also simplify the algebraic manipulations required to solve the problem at hand and reveal more about the mathematical structure of our equations. Learning how to become astute is a lengthy process, one which we shall repeatedly assess during this text.

> EXERCISE 1.6. One of the most astute choices that we can make is to align our coordinate axes with the motion of an object. Suppose that an object moves from $\mathbf{r}_1 = (x_1, y_1, z_1)$ to $\mathbf{r}_2 = (x_2, y_2, z_2)$ along a straight line. This is really a one-dimensional problem. *Define* a new coordinate axis $\hat{\mathbf{x}}'$ by $(\mathbf{r}_2 - \mathbf{r}_1)/|\mathbf{r}_2 - \mathbf{r}_1|$.
>
> (a) What is $\hat{\mathbf{x}}'$ in terms of the components of $\mathbf{r}_1$ and $\mathbf{r}_2$?
>   Points $\mathcal{P}$ along the line connecting $\mathbf{r}_1$ and $\mathbf{r}_2$ are defined by the following expression: $\mathcal{P} = \mathbf{r}_1 + x'\hat{\mathbf{x}}'$.
> (b) If $x' = |\mathbf{r}_2 - \mathbf{r}_1|$, show that this corresponds to the point $\mathbf{r}_2$.
> (c) What is the midpoint of the line connecting $\mathbf{r}_1$ and $\mathbf{r}_2$? What value of x' corresponds to the midpoint?

Now there are a limited number of characters in any alphabet and we shall restrict ourselves to the Latin and Greek alphabets for the most part. Additionally, we will try to use symbols for our concepts that provide some other sort of consistency. For example, a key attribute of an object is its velocity. We will most often utilize the symbol $\mathbf{v}$ to stand for velocity because $\mathbf{v}$ is the first character in the word velocity. We will most often use $\mathbf{r}$ to indicate the position of an object. This is due to historical usage because, obviously, $\mathbf{r}$ is not the first character of the word position. To distinguish two different objects or two different times, we will decorate base variables like $\mathbf{r}$ and $\mathbf{v}$ with subscripts and superscripts. Thus, if you find something in the text like $\mathbf{r}_1$, it is likely that this symbol represents the position of something but use caution when trying to interpret the symbol: you will have to understand the context of its use.

In subsequent chapters, we may use $\mathbf{r}_{\text{sun}}$ and $\mathbf{r}_{\text{earth}}$ to denote the positions of the sun and the earth. Alternatively, we could use $\mathbf{r}_S$ and $\mathbf{r}_E$; these representations are more concise than $\mathbf{r}_{\text{sun}}$ and $\mathbf{r}_{\text{earth}}$ and are easily distinguished from one another. The notational choice will depend on a number of factors: ease of use, distinctness, etc. In preparing this text, the symbol r_{earth} can be defined as a macro, so it doesn't cost the author much to replicate it from line to line; a student having to perform a series of algebraic manipulations may well prefer the shorter r_E. For historical reasons, astronomers will sometimes use $r_\odot$ and $r_\oplus$ to denote the positions of the sun and earth but we will typically avoid straying from Latin and Greek in this text to avoid confusion.[2] Many modern appliances and computer programs have buttons that are decorated with icons rather than English words, making the devices intrinsically marketable around the globe but making life difficult for the user who has not yet learned how to decode the icons. In this text, we shall strive for clarity and consistency of our notation but students should recognize that other authors may adopt somewhat different styles.

The student will need to continually assess what is intended by the mathematical notation, not simply memorize formulas. Notation provides us with the capability to represent complex ideas succinctly but our usage will be context dependent. In the course of this text, we may, for example, define $x_1 = x(t_1)$ because x_1 is quicker to write than $x(t_1)$ and, when we have more complex equations, it may be easier to see the mathematical structure if we simplify our notation. It will therefore be important to understand what the notation represents in the current circumstances. Does x_1 represent the position of an object we've labelled "1" at all times? If so then it is a shorthand for $x_1(t)$, where we have suppressed the time dependence in the representation x_1 but assume that the student will remember that x_1 is a function of time. We may, at times, choose x_1 to represent the position of an object as a specific time $t = t_1$, where we mean $x_1 = x(t_1)$. It is important to remember that this is simply notation—a concise representation of a concept. There is nothing unique in our representations, the variable ξ could also be used to indicate the position of some object and we may choose to use such an unusual symbol if there are a number of other positional variables under consideration like r_1, r_2', etc., and this choice provides some sort of clarity to our notation.

Returning now to the analysis of our trajectory, we should note that the trajectory depicted in figure 1.2 does not depend either on the notation that we use to describe it or the specific coordinate axes that we defined. This latter point can be illustrated by the following example. Consider the trajectory of the basketball launched by the player in figure 1.3. The basketball will follow a trajectory like that shown in figure 1.2. Now suppose that you are sitting in the home section of the bleachers. From your

---

[2]Astronomers use a number of symbols derived from antiquity. Wandering stars (planets) were given the names of the gods, who were represented by symbols like those for Venus and Mars .

FIGURE 1.3. Coordinate invariance. Everyone in the gym will agree on the outcome of the shot: either it goes through the goal or it does not

position, the ball is travelling left to right. If you were asked to quantify that, you might set up a coordinate system with its origin at half court, with the x^H-axis aligned along the length of the floor, the y^H-axis aligned across the court and the z^H-axis extending up from the floor. If, instead, you were sitting in the visitor section on the opposite side of the court, you would describe the ball as travelling right to left. You could well choose a coordinate system that has its origin in some corner of the court, with your x^V- and y^V-axes aligned along the paint lines that define the boundaries of the court.[3]

EXERCISE 1.7. Sketch the basketball court. Now draw the two sets of coordinate systems described in the text. Label one set with a superscript H (for home) and the second with a superscript V (for visitor), as suggested in the text.

Regardless of your seat location, the ball will either go through the goal or it will not. To express this situation mathematically, let us define the position of the ball in the home system to be the vector $\mathbf{r}^H_{\text{ball}}(t)$ and the position of the goal to be the vector $\mathbf{r}^H_{\text{goal}}$. The condition that the ball go through the goal is provided by the following expression[4]:

$$\mathbf{r}^H_{\text{ball}}(t) - \mathbf{r}^H_{\text{goal}} = 0.$$

[3]Notice that we've decorated the variables x^V and y^V with superscript Vs to distinguish them from the other set of coordinates used by the home system.

[4]The rules of basketball also require that the ball enter the goal from above, which imposes an additional constraint but does not affect our current discussion. Finite-sized basketballs and goals relax the equality constraint to an inequality: the difference in positions must be less than some value but this point also does not materially affect the essential point of our argument.

That is, at some point in time, the ball is at the same location of the goal. In the visitor coordinate system, we have a similar requirement:

$$\mathbf{r}^V_{\text{ball}}(t) - \mathbf{r}^V_{\text{goal}} = 0.$$

These two equations are mathematically identical. Moreover, the form of the equations is identical, independent of which coordinate system we choose to represent the flight of the ball.

Consequently, we can see that the outcome of the experiment does not depend on how you choose to describe it mathematically. This might just seem like common sense but the fact that the equations that we use to describe the trajectory are invariant under coordinate transformations has important consequences. The German mathematician Emmy Noether proved in 1915 that, for a large class of physics problems, invariance of the equations of motion under various transformations results in the existence of conserved quantities that correspond to the particular transformations.[5] By *conserved*, we mean that the property is a constant and does not depend upon time. By *invariant*, we mean that the mathematical form of the equation does not change: there are no additional terms that appear in one coordinate system and not the other. In this case of a classical particle (basketball), the fact that the equations describing the motion of the object are independent of coordinate position (translation) and orientation (rotation) means that the quantities of linear and angular momentum (which we shall define shortly) must be conserved. The fact that the trajectory is independent of what we choose to define as the initial time means that the energy of the system must be conserved.

It is difficult for beginning students to appreciate how far-reaching the consequences of these conservation laws are. Indeed, we shall devote a significant portion of this text to understanding the implications. This is, of course, not how the principles of energy and momentum conservation were discovered. Physicists had observed that these quantities seemed to be constant (independent of time) through careful experimentation and formulated the conservation principles as a consequence of their experimental observations. Noether's discovery of a deeper mathematical principle came much later but it is extraordinarily important and has driven much of the development of modern quantum theoretical physics. Now, quantum field theory is a subject that is beyond our current scope of work

---

[5]Noether was working in Göttingen under the direction of the famous mathematicians David Hilbert and Felix Klein, who found themselves in a power struggle with other faculty over the hiring of a woman. Despite the turmoil associated with her appointment, Noether quickly proved what has come to be known to physicists as Noether's theorem, thereby solving a key problem that had stopped Albert Einstein's progress in his pursuit of a general relativistic theory of gravitation.

and further investigations into that aspect of modern physics will have to be deferred until subsequent courses.

FIGURE 1.4. Milky Way galaxy. This negative image, in which stars are represented by *black dots*, represents the best current reconstruction of the structure of our own galaxy. There are two major spiral arms leading from the central bar. Our solar system is located within the *white circle* below the galactic center (Image courtesy of NASA)

For the moment though, consider the consequences if momentum were *not* conserved. In that case, Noether's theorem would dictate that the equations of motion for some object would depend upon the coordinate system chosen. Different observers would disagree on the mathematical description of experiments. This would be an intractable state of affairs. Presumably, there must exist a preferred coordinate system: somewhere in the universe there is a specific coordinate origin. How are we to discover the location and orientation of this coordinate system? As can be seen in figure 1.4, our solar system exists in the far reaches of a modest galaxy amidst the billions of galaxies that exist in the universe. It seems unlikely that we could ever identify this preferred coordinate system.

What if energy were not conserved? As energy conservation is tied to the invariance of our equations to translations in time, then there must exist some special time from which all others must be defined. How can we find that time? Of the thirteen billion or so years that have elapsed since the beginning of the universe, how can we find the special time? Philosophically, the implications of such a dilemma are devastating. It calls into question our basic premise that we can develop theories about the nature of the universe that all observers can test independently.

As this text does not end here, the intrepid student can surmise that physicists have not deemed the problem to be hopeless. We shall make the fundamental assumption that the laws of nature apply equally throughout the universe and that all observers would formulate identical mathematical descriptions of the observed phenomena. If you somehow managed to find yourself standing on the surface of a planet in a galaxy far, far away, we fully expect that whatever physics experiments you choose to conduct will yield results that are consistent with the theories that we have developed here on earth. At present, though, we have no means of transporting ourselves to distant galaxies to prove or disprove that assumption. In lieu of contrary evidence then, we shall embrace the concepts of momentum and energy conservation and exploit them to learn more about our local portion of the universe.

> EXERCISE 1.8. Let us set the origin of a new coordinate system to be the point $O = (x_0, y_0, z_0)$. The vector $\mathbf{r}_1$ to a point $P_1 = (x_1, y_1, z_1)$ from the new coordinate origin is then $\mathbf{r}_1 = (x_1 - x_0, y_1 - y_0, z_1 - z_0)$.

(a) Consider a second point $P_2 = (x_2, y_2, z_2)$. What is the vector $\mathbf{r}_2$?
(b) Show that the vector $\mathbf{r}_2 - \mathbf{r}_1$ does not depend on x_0, y_0 or z_0.

1.3. Vocabulary

Returning now to the trajectory pictured in figure 1.2, what we would like to be able to do as physicists is describe that trajectory mathematically and understand what controls the specific trajectory that we observe. We do so knowing that (i) our knowledge of the actual trajectory is limited by our experimental precision and (ii) our mathematical description will only be approximate. What we are seeking is a definition of the state of the object at any point t in time between some initial t_1 and final t_2 times. Formally, we could obtain such a description from the Taylor series of the vector function $\mathbf{r}(t) = (x(t), y(t), z(t))$ in the vicinity of the time t_1:

$$
(1.1) \qquad \mathbf{r}(t) = \mathbf{r}(t_1) + \sum_{n=1}^{\infty} \left. \frac{d^n \mathbf{r}(t)}{dt^n} \right|_{t=t_1} \frac{(t - t_1)^n}{n!},
$$

provided, of course, that all of the derivatives of the function $\mathbf{r}(t)$ are defined.[6] Not all functions have a convergent Taylor series but, mercifully, we shall find that physics is largely the domain of functions that are suitably well-behaved in a mathematical sense. We've already discussed our assumption that the trajectory is a continuous, single-valued function of time. These requirements provide precisely the sort of mathematical conditions necessary for the derivatives in Equation 1.1 to be defined.

Now because we are seeking an approximate solution to the trajectory $\mathbf{r}(t)$, it should come as no surprise that we intend to truncate the Taylor series shown in Equation 1.1 after a few terms. In fact, we intend to keep only two terms:

$$(1.2) \qquad \mathbf{r}(t) = \mathbf{r}(t_1) + \frac{d\mathbf{r}(t)}{dt}\bigg|_{t=t_1} (t - t_1) + \frac{1}{2}\frac{d^2\mathbf{r}(t)}{dt^2}\bigg|_{t=t_1} (t - t_1)^2.$$

It might seem hugely restrictive to consider only the first two terms from Equation 1.1 but we shall find that many systems are quite adequately described using this simplification. Moreover, given the advent of modern computers, we can extend our range of applicability by making use of the ability of computers to faithfully conduct repetitive calculations. Instead of considering the complete trajectory as one large step from t_1 to t_2, as shown in figure 1.2, we could break it into a sum of smaller steps, from t_1 to t_{1000}, say. We could then assume that on each short segment from t_i to t_{i+1}, we can use the approximation defined by Equation 1.2. We then sum the results over all the steps to recover the complete trajectory. As a practical matter, computers are quite proficient at repetitious calculations and quite complex systems can be studied using the simple approximation used in Equation 1.2.

Because we will use Equation 1.2 so frequently, we will make a few additional definitions to improve our notation.[7] The first time derivative of the position vector is defined to be the **velocity** of the object:

$$(1.3) \qquad \mathbf{v}(t) = \frac{d\mathbf{r}(t)}{dt} = \left(\frac{dx(t)}{dt}, \frac{dy(t)}{dt}, \frac{dz(t)}{dt}\right).$$

---

[6]Here we use the notation that the derivative of the vector is obtained by differentiating each of its components: $d\mathbf{r}/dt = (dx/dt, dy/dt, dz/dt)$.

[7]Names have also been given to some of the higher order derivatives. The *jerk* is the third derivative of position with respect to time. These higher-order terms are not used with much frequency owing to the fact that few real trajectories can be defined by simple functions. If one must resort to numerical evaluation of the system, higher order terms may be included but may play more of a rôle in providing numerical stability than in providing a better approximation of the trajectory.

The second time derivative of the position vector is defined to be the **acceleration** of the object:

$$(1.4) \qquad \mathbf{a}(t) = \frac{d^2\mathbf{r}(t)}{dt^2} = \left(\frac{d^2x(t)}{dt^2}, \frac{d^2y(t)}{dt^2}, \frac{d^2z(t)}{dt^2} \right).$$

A simple consequence of these definitions is that the acceleration is the first time derivative of the velocity:

$$(1.5) \qquad \mathbf{a}(t) = \frac{d\mathbf{v}(t)}{dt}.$$

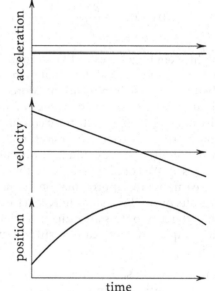

FIGURE 1.5. Position, velocity and acceleration. The velocity is the slope of the position *versus* time curve. The acceleration is the slope of the velocity *versus* time curve

Graphically, the relationships between these quantities are shown (in one dimension) in figure 1.5. Note that all of these quantities: position, velocity and acceleration can be *negative*. A negative position indicates that an object is on the opposite side of the origin from what is defined to be the positive direction. A negative velocity indicates that the object is moving in the direction opposite to what is defined to be the positive direction. A negative acceleration indicates that the velocity is changing in the direction opposite to that defined as positive.

We can now recast Equation 1.2 into a more familiar form:

$$(1.6) \qquad \mathbf{r}(t) = \mathbf{r}(t_1) + \mathbf{v}(t_1)(t - t_1) + \frac{1}{2}\mathbf{a}(t_1)(t - t_1)^2,$$

which is known as a *kinematic* equation.[8] This equation enables us to predict the position **r** of an object at some time t beyond an initial time t_1 if we know the values of (i) the position **r**, (ii) the velocity **v** and (iii) the acceleration **a** at the initial time t_1. We note again that Equation 1.6 could also be used to estimate the position of an object *before* the time t_1. There is nothing mathematically that would preclude the argument $t - t_1$ from being negative but one must understand whether or not that extension is meaningful.

Finally, we should also point out that Equation 1.6 reflects some of the notational difficulties that we face. By the symbol **v**(t_1), we mean that velocity **v** is a function of time and that we want the velocity at the specific time t_1. This is to be multiplied by the time that has elapsed since we started the clocks: $(t-t_1)$. The parentheses here are used as grouping operators. *Mathematica* syntax has specific meanings for the symbols (), [] and {} that we do not follow in the text, as they are not standard mathematical representations. Here we utilize all of these symbols as grouping operators, in an effort to clarify our displayed equations. Students must beware literally translating formulas from the text into *Mathematica* scripts.

> EXERCISE 1.9. Notation is a means for representing complex ideas with a relatively few symbols. The more compact the notation, the more meaning is imbued in each symbol. As a result, compact notation can potentially clarify the mathematical structure at the risk of obscuring the precise meaning of the symbol. *Caveat emptor!*

(a) Write out Equation 1.2 in its component form, expressing the explicit time dependence of the variables and identifying the derivatives.
(b) Rewrite the results of part (a) using the definitions for velocity and acceleration. Use $v_x(t)$ to represent the x-component of the velocity, for example.
(c) Rewrite the results of part (b) suppressing the explicit time dependence. Use a subscript zero to denote the initial time values, e.g., $x(t_1) = x_0$.

> EXERCISE 1.10. Define a function in the *Mathematica* program as follows:

```
F[x_,a_,b_,c_]:= a + b x + c x^2.
```

Define the functions `DF[x_,b_,c_]:= b + 2c x` and `D2F[c_]:= 2c`. Use the `Plot[]` function to plot F and its derivatives over the range

---

[8]The word kinematic is derived from the Greek word κινημα meaning motion. The Greek root also provides us with the English word cinema (moving pictures), Note once again the display of a classical education.

{x,0,1}. Use the `Manipulate[]` function to vary the parameters {a,-1,1}, {b,-2,2} and {c,-1,1}. What happens as you vary the parameters?

The simplest motions of objects, of course, occur when the velocity $\mathbf{v}$ and acceleration $\mathbf{a}$ are independent of time. In particular, if the acceleration is zero, Equation 1.6 reduces to $\mathbf{r}(t) = \mathbf{r}(t_1) + \mathbf{v}(t_1)(t - t_1)$. If we align our coordinate axes[9] such that the direction of $\mathbf{v}(t_1)$ is in the x-direction, then the only non-zero component of $\mathbf{v}(t_1)$ is the x-component: $\mathbf{v}(t_1) = (v_1, 0, 0)$. If we define the position of the object at time t_1 to be $\mathbf{r}(t_1) = (x_1, y_1, z_1)$, we note that only the x-component of $\mathbf{r}$ changes: $x(t) = x_1 + v_1(t - t_1)$. We recognize this expression as the equation of a line. This is just the equation, with a minor change of notation, that we encountered in our initial experiment depicted in figure 1.1. Apparently, our runner was providing an example of an object with a constant velocity and no acceleration.

> EXERCISE 1.11. Use dimensional analysis to identify the like terms in the two expressions: $s = mt + b$ and $x(t) = x_1 + v_1(t - t_1)$. (Hint: Write the dimensional equations.) What is meant by the symbol v_1?

The kinematic equation enables us to describe *how* an object moves in space. Were we to measure motion with more sophisticated tools than students with stopwatches lined up along the sidewalk, a high-speed motion camera perhaps, we could imagine developing a reasonably precise mathematical description of any motion. We might need to break the trajectory into pieces for our approximations to be good enough and we do have a means for accomplishing this using modern computers. In physics, however, we are generally not satisfied with knowing how; we also want to know *why*. We are not going to be content simply with an a posteriori description of motion. We want to be able to predict the motion. It is not enough to know that a lunar eclipse occurred yesterday; we'd like to know when that will happen again.

The seminal advance in our understanding of motion is due to Isaac Newton, whose works we will take up in more detail in the next chapter.[10] Newton's dynamics are based on the notion that an object changes its motion as a result of an external *force*. The force is observed by means of an acceleration of the object, in fact, the force is directly proportional to the

---

[9]Notice we are making an *astute* choice and minimizing our algebraic effort.

[10]Isaac Newton's *Philosphiæ Naturalis Principia Mathematica*, published in 1687 and often referred to as the *Principia*, is one of the most remarkable works in the history of science. In it, Newton not only solves the problem of planetary motion but sets forth the principles that define the modern scientific method.

acceleration. The constant of proportionality we today call *mass*. Mathematically, we can state this relationship as follows:

$$(1.7) \qquad\qquad F = m\mathbf{a}.$$

This simple equation lies at the heart of all that we will cover in this text. Essentially everything can be derived from this rule that is known today as Newton's second law of motion. This equation provides a definition of mass, a fundamental property of matter. We'll find other fundamental properties of matter as we proceed.

Another important quantity that arises in our development of theories of why things move is called the **momentum** and usually is assigned the symbol $\mathbf{p}$[11]:

$$(1.8) \qquad\qquad \mathbf{p} = m\mathbf{v}.$$

We can, from the above definition, immediately note that the force is related to the change of momentum with respect to time:

$$(1.9) \qquad\qquad F(t) = \frac{d\mathbf{p}(t)}{dt},$$

where Equation 1.9 reduces to Equation 1.7, provided that the mass m does not vary with time. This is often the case but we will eventually investigate the behavior of rockets, for which that assumption is not valid. There are, of course, a number of additional vocabulary words we will need but we have enough to get started and will find it most useful to see these concepts applied. It is quite important to understand the meanings of what these words and symbols represent.

> EXERCISE 1.12. Humans are quite proficient at pattern recognition. As a former colleague of mine used to say, humans evolved to recognize tigers. Humans who were not successful in recognizing tigers were, of course, removed from the gene pool. It is important to look at pictures:

(a) Plot the equation $x = 3t + 4$ for the range of $0 \leq t \leq 10$.
(b) Plot the equation $x = 0.01t^2 + 3t + 4$ for the same range as above.
(c) Do the plots differ?
(d) Use the *Mathematica* function Manipulate and change the 0.01 to a parameter a that can take on values in the range of $0 \leq a \leq 1$.
(e) How does changing the value of a affect the trajectory?

---

[11]This notation perhaps has its origins in the Latin *petere*, meaning to rush forward. Newton's phrasing in the *Principia* was quantity of motion. It might seem that **m** would be a better symbol for momentum but there is an obvious notational conflict with the use of m to mean mass.

Exercise 1.13. We can expand the use of Equation 1.6 to more complex behaviors by considering trajectory functions that are *piecewise*
continuous. That is, in each of several different subintervals we will
consider the trajectory to be a continuous function. This will enable
us to approximate the behavior of a car travelling at a constant velocity for a time and then the brakes being applied. Braking does
not happen instantaneously nor is the acceleration constant but, for
purposes of constructing a simple model, we shall assume that both
are true. Such approximations are quite useful in understanding the
gross behavior of objects. If a more detailed description of an object's motion is required, then a more detailed model will have to be
constructed.

Analyze the behavior of the object indicated in figure 1.6. Initially,
the object is moving in a positive direction with a constant velocity.
Sketch the velocity and acceleration in each of the subintervals.

Figure 1.6. Position, velocity and acceleration of an
object. In each subinter-
val, indicated by the vertical dashed lines, the behavior can be explained by
Equation 1.6

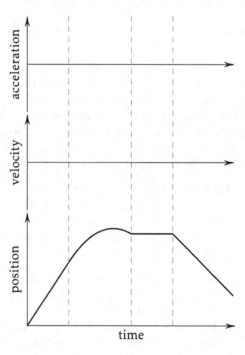

On the Motion of Planets

Most of us have heard some version of the story that results in Isaac Newton being plunked on the head by an apple and discovering gravity.[1] This rather stylized and not particularly accurate rendition of the historical record omits Newton's major *physical insights* into the problem of planetary motion. In some broad sense, we'll follow the reasoning that Isaac Newton utilized in the late 1600s to discover what is now referred to as the law of universal gravitation. In the subsequent 300 years, though, some significant improvements to mathematical notation have been developed; we'll make use of those ideas as well.

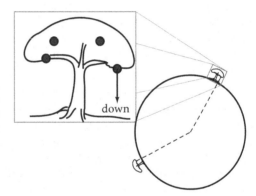

FIGURE 2.1. The direction *down* depends upon your position on the surface of the earth

If we consider an apple falling from a tree, we would observe that it falls *down*. This is, perhaps, not an earth-shattering observation but Newton realized that the direction down depends upon one's location on the surface of the earth. People in Europe or Asia or the Americas would all

---

[1]In October 1665, the Great Plague that beset London spread to the surrounding countryside and forced the closure of Cambridge University. As a result, Newton was required to abandon his formal studies and returned to his family's home in Woolsthorpe Manor, Lincolnshire. During his two years in relative exile, while still in his early twenties, Newton contemplated the problem of planetary motion and constructed a new form of mathematics that we today call calculus.

© Mark A. Cunningham 2015

M.A. Cunningham, *Neoclassical Physics*, Undergraduate Lecture Notes in Physics, DOI 10.1007/978-3-319-10647-2_2

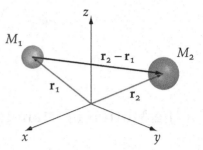

FIGURE 2.2. Two masses M_1 and M_2 are located at positions $\mathbf{r}_1$ and $\mathbf{r}_2$, respectively

describe the same behavior of an apple falling down to the ground. Yet, as depicted in figure 2.1 the direction down is not the same on opposite sides of the earth. To explain the different trajectories, Newton recognized that each apple must fall along a line that extends radially from the center of the earth. Today, we would say that gravity exerts a *central force* on the apple.

In addition to realizing this basic aspect of the gravitational force, Newton looked up at the moon from his home in the English countryside and asked himself just how far does this gravity extend? Could the force acting on the apple extend as far as the moon? Could it extend as far as the sun? If so, then the motions of the planets would be governed by the same force that causes the apple to fall to the ground. This is a remarkable intellectual leap: extending the behavior of local objects to the universe at large.

In trying to understand the nature of the force, Newton also reasoned that the magnitude of the force exerted by the earth on the apple should be the same as the magnitude of the force exerted by the apple on the earth; however, the directions of the forces would be opposite. This principle is deemed Newton's third law of motion. Today, this concept that every action produces an equal and opposite reaction is embedded in our culture and is used to explain everything from political decision making to the outcomes of athletic contests but, in Newton's time, the idea was remarkable and revolutionary and controversial.

2.1. Form of the Force Law

We now want to put these postulates into a mathematical form, to develop an equation that describes our ideas. To begin, let us first define the vector $\mathbf{r}_1$ to be the position of the center of a mass M_1, in some coordinate system, and $\mathbf{r}_2$ to be the position of the center of another mass M_2. For the moment, even though the objects depicted in figure 2.2 have finite sizes, we shall regard them as just points.[2] In this generic notation, we can specify the meanings of the symbols M_1 and M_2 as we see fit. For example, M_1 could represent the sun and M_2 the earth or M_1 could represent the earth and M_2 an apple, or vice versa.[3]

The vectors $\mathbf{r}_1$ and $\mathbf{r}_2$ will generally be functions of time: $\mathbf{r}_1 = \mathbf{r}_1(t)$ and $\mathbf{r}_2 = \mathbf{r}_2(t)$ but we'll suppress the explicit time dependence of the position vectors for now. The force will depend upon these quantities and the time, so we can write that the force on the apple (M_2) due to the earth (M_1) will be a function $\mathbf{F} = \mathbf{F}(t, M_1, M_2, \mathbf{r}_1, \mathbf{r}_2)$. Newton's postulate that the gravitational force on the earth due to the apple be opposite can be expressed mathematically as follows:

$$\mathbf{F}(t, M_2, M_1, \mathbf{r}_2, \mathbf{r}_1) = -\mathbf{F}(t, M_1, M_2, \mathbf{r}_1, \mathbf{r}_2).$$

That is, the function $\mathbf{F}$ must change sign when we exchange the meanings of the indices 1 and 2.

Newton's first postulate, that the gravitational force is a central force, means that the force must depend only on the vector $\mathbf{r}_2 - \mathbf{r}_1$ or its magnitude $|\mathbf{r}_2 - \mathbf{r}_1|$. By definition, $\mathbf{r}_2 - \mathbf{r}_1$ defines the line connecting the centers of the two masses. Other combinations of $\mathbf{r}_1$ or $\mathbf{r}_2$ can be excluded from consideration. This has two immediate consequences.

First, we note that the direction of the force is determined by the direction of the vector $\mathbf{r}_2 - \mathbf{r}_1$ and this quantity changes sign if we interchange the definitions of masses 1 and 2. Mathematically, we have the simple relation $\mathbf{r}_2 - \mathbf{r}_1 = -(\mathbf{r}_1 - \mathbf{r}_2)$. The concept that the apple is pulled down by the earth and the earth is pulled up by the apple, is exactly represented mathematically by the fact that interchanging the roles of the masses (flipping the indices) causes the term $\mathbf{r}_2 - \mathbf{r}_1$ to change sign.

Second, the vector $\mathbf{r}_2 - \mathbf{r}_1$ does not depend on the coordinate system used to define the vectors $\mathbf{r}_1$ and $\mathbf{r}_2$. (See Exercise 1.8.) Imagine grabbing the coordinate origin illustrated in figure 2.2 and moving it around on the page. The vectors $\mathbf{r}_1$ and $\mathbf{r}_2$ will certainly change as the origin moves but their difference will not. The same will also be true if you spin the coordinates around some axis. The definitions of the components of $\mathbf{r}_2 - \mathbf{r}_1$ will change but the vector itself will not. We shall make further use of this fact shortly.

---

[2] We now know that the planet Venus has a size and mass that is roughly comparable to the size of the earth. When observed from earth, Venus is often called the evening star; it appears quite point-like in the sky. This is not true for the sun and moon but we shall justify the point assumption eventually.

[3] In some cases we may use more explicit notation like $\mathbf{r}_{\text{apple}}$ and $\mathbf{r}_{\text{earth}}$ for the specific case of an apple and the earth but indices 1 and 2 are faster to type and have more general applicability.

EXERCISE 2.1. Consider the points $P_1 = (3.2, 1.7)$ and $P_2 = (4.6, 3.3)$, that are defined in a particular coordinate system where the origin is at the point $O = (0, 0)$. In the *Mathematica* program, define a new coordinate origin $O' = (xo, yo)$.

(a) Plot the points P_1, P_2 and O', using the `Manipulate` function to vary the ranges $-3 \leq xo \leq 3$ and $-3 \leq yo \leq 3$.
(b) Compute the vectors $\mathbf{r}_1$ from O' to P_1 and $\mathbf{r}_2$ from O' to P_2.
(c) Compute the vector $\mathbf{r}_2 - \mathbf{r}_1$.
(d) What happens to $\mathbf{r}_2 - \mathbf{r}_1$ as the quantities xo and yo change?

We can place some additional constraints on the functional form of the force by recognizing that masses and positions have different dimensions. As a result, it would make no sense to have a term like $M_1 + \mathbf{r}_1$; one cannot add terms that do not have the same dimensions. Thus the magnitude of the force will have to separate into two parts, two functions $\mathcal{M}$ and $\mathcal{R}$ that depend on the masses and positions independently: $|\mathbf{F}| = \mathcal{M}(M_1, M_2)\mathcal{R}(t, \mathbf{r}_1, \mathbf{r}_2)$, where we have assumed that the masses do not depend on time.

So, the direction of the force is provided by the vector $\mathbf{r}_2 - \mathbf{r}_1$, which we can normalize to have unit magnitude by dividing the vector by its magnitude. The gravitational force then takes the following form:

$$(2.1) \qquad \mathbf{F}(t, M_1, M_2, \mathbf{r}_1, \mathbf{r}_2) = \frac{\mathbf{r}_2 - \mathbf{r}_1}{|\mathbf{r}_2 - \mathbf{r}_1|} \mathcal{M}(M_1, M_2)\mathcal{R}(t, |\mathbf{r}_2 - \mathbf{r}_1|).$$

The direction of the force is provided by the first term, the mass dependence by the function $\mathcal{M}$ and the positional dependence $\mathcal{R}$ we see depends only on the magnitude $|\mathbf{r}_2 - \mathbf{r}_1|$.

Let's turn now to the issue of mass dependence. The function $\mathcal{M}$ must be a symmetric function of the masses.[4] Consequently, the mass dependence of the force must be either the sum $\mathcal{M}(M_1, M_2) = M_1 + M_2$ or product $\mathcal{M}(M_1, M_2) = M_1 M_2$ of the masses.[5] To resolve which is the correct form, let's recall that a force F_2 acting on an object of mass M_2 produces an acceleration a_2 such that $F_2 = M_2 a_2$. Consider now Galileo's experiment in which two separate masses M_2 and M_3 were dropped from a tall tower. Galileo observed that both fell to the ground in equal times. From this

---

[4]The direction component of the force changes sign if we exchange the indices. The position component $\mathcal{R}$ cannot change sign because it depends only on the magnitude $|\mathbf{r}_2 - \mathbf{r}_1|$. Thus, the mass component $\mathcal{M}$ cannot change sign.

[5]As we shall see, the law of universal gravitation can be considered to provide a *definition* of mass. Other mass dependencies like M^α, could simply be redefined to a new effective mass without the power-law dependence: $M_{\text{eff}} \equiv M^\alpha$.

result, we can infer that both masses must have the same gravitational acceleration.[6]

EXERCISE 2.2. Use Equation 1.6 for motion in a single direction to prove the assertion that equal times of travel implies equal acceleration. Assume that two objects have initial velocities of zero and each travels a total distance d in a time t.

Suppose that the gravitational force depends on the sum of the masses. We would have then the following relations:

$$a_2 = \frac{F_2}{M_2} = \frac{M_1 + M_2}{M_2} \mathcal{R}(t, |\mathbf{r}_2 - \mathbf{r}_1|),$$

$$a_3 = \frac{F_3}{M_3} = \frac{M_1 + M_3}{M_3} \mathcal{R}(t, |\mathbf{r}_3 - \mathbf{r}_1|).$$

Because the earth's mass is so much larger than the mass of any objects that Galileo could have manipulated, it is reasonable to estimate that $M_1 + M_2 \approx M_1$. Additionally, because Galileo dropped the masses from the same height above ground, the magnitudes $|\mathbf{r}_2 - \mathbf{r}_1|$ and $|\mathbf{r}_3 - \mathbf{r}_1|$ are essentially equal. Consequently, the position-dependent part of the forces must be equal. Thus, we find that $a_2/a_3 = M_3/M_2$. The accelerations for different masses should be different, in conflict with Galileo's observations.

If we instead assume that the gravitational force depends on the product of the masses, then we would find the following relations:

$$a_2 = \frac{F_2}{M_2} = \frac{M_1 M_2}{M_2} \mathcal{R}(t, |\mathbf{r}_2 - \mathbf{r}_1|),$$

$$a_3 = \frac{F_3}{M_3} = \frac{M_1 M_3}{M_3} \mathcal{R}(t, |\mathbf{r}_3 - \mathbf{r}_1|).$$

Taking the ratio of accelerations here, we do indeed find that $a_2 = a_3$. So we can conclude that $\mathcal{M}(M_1, M_2) = M_1 M_2$.

We have now established that the gravitational force can be written as follows:

$$(2.2) \qquad \mathbf{F} = \frac{\mathbf{r}_2 - \mathbf{r}_1}{|\mathbf{r}_2 - \mathbf{r}_1|} M_1 M_2 \mathcal{R}(t, |\mathbf{r}_2 - \mathbf{r}_1|).$$

To define the positional dependence $\mathcal{R}$, Newton ultimately relied on an additional observation: the planets occupy elliptical orbits around the

---

[6]In the case of uniform acceleration a where the mass begins with no velocity, the mass travels a distance d in a time t given by $d = at^2/2$. Thus, the acceleration would be $a = 2d/t^2$. Consequently, equal times of travel imply equal accelerations.

sun, with the sun at one focus of the ellipse.[7] After extensive calculations, Newton determined that the force must scale like the inverse square of the magnitude $|\mathbf{r}_2 - \mathbf{r}_1|$:

$$(2.3) \qquad \mathcal{R}\left(t, |\mathbf{r}_2 - \mathbf{r}_1|\right) = -\frac{G}{|\mathbf{r}_2(t) - \mathbf{r}_1(t)|^2},$$

where G is a constant of proportionality, known as the universal gravitational constant.[8] This equation and Equation 2.2 together describe the behavior of planets orbiting the sun and apples falling to the ground.

2.2. Inverse Square Law

Let's first prove that this choice for the gravitational force does, indeed, give rise to elliptical orbits for planets. We will use a notation that is explicit but not concise. There are associated exercises that repeat the calculations using a more concise notation. Actually *doing* the exercises should help to illuminate the mathematical structure; the practice is henceforth encouraged.

The force on the mass M_2 due to the mass M_1 gives rise to an acceleration $\mathbf{a}_2 = d^2\mathbf{r}_2/dt^2$ that we can write as follows:

$$(2.4) \qquad \frac{d^2\mathbf{r}_2(t)}{dt^2} = -G\frac{M_1}{|\mathbf{r}_2(t) - \mathbf{r}_1(t)|^3}\left[\mathbf{r}_2(t) - \mathbf{r}_1(t)\right].$$

where we have reorganized the results of Equations 2.2 and 2.3 and divided by the mass M_2. Similarly, the acceleration of the mass M_1 due to the mass M_2 can be written as follows:

$$(2.5) \qquad \frac{d^2\mathbf{r}_1(t)}{dt^2} = -G\frac{M_2}{|\mathbf{r}_1(t) - \mathbf{r}_2(t)|^3}\left[\mathbf{r}_1(t) - \mathbf{r}_2(t)\right].$$

Subtracting the second equation from the first, we can write

$$\frac{d^2\left[\mathbf{r}_2(t) - \mathbf{r}_1(t)\right]}{dt^2} = -G\frac{M_1 + M_2}{|\mathbf{r}_2(t) - \mathbf{r}_1(t)|^3}\left[\mathbf{r}_2(t) - \mathbf{r}_1(t)\right],$$

which represents a second-order differential equation for the time evolution of the vector $\mathbf{r}_2(t) - \mathbf{r}_1(t)$. We have, at this point, satisfied one of our

---

[7]Kepler's *Astronomia nova* was published in 1609. Kepler managed to determine, through painstaking analysis of decades of detailed observations conducted by Tycho Brahe, that the orbit of Mars was an ellipse, with the sun at one focus.

[8]Note that the sign has been chosen to represent the fact that the gravitational interaction is an *attractive* one.

initial goals: we have developed a mathematical description of the behavior of this system, what physicists would call the equations of motion.[9]

We have not, however, completed our task because we also want to understand what sort of solutions exist for those equations of motion and how various types of solutions depend on the parameters that define the system. Now, solving differential equations is somewhat beyond the mathematical capabilities of most students at this juncture but we will now proceed to construct solutions by following a pathway that has been pioneered by our mathematical predecessors. We begin by recasting the three separate equations for the components of $\mathbf{r}_2(t) - \mathbf{r}_1(t)$ into a set of coupled first-order equations as follows:

$$\frac{d\,[\mathbf{r}_2(t) - \mathbf{r}_1(t)]}{dt} = \mathbf{v}_2(t) - \mathbf{v}_1(t)$$

(2.6)
$$\frac{d\,[\mathbf{v}_2(t) - \mathbf{v}_1(t)]}{dt} = -G\frac{M_1 + M_2}{|\mathbf{r}_2(t) - \mathbf{r}_1(t)|^3}\,[\mathbf{r}_2(t) - \mathbf{r}_1(t)].$$

Here, we are using our standard notation: the velocities $\mathbf{v}_1(t)$ and $\mathbf{v}_2(t)$ are the time derivatives of the positions $\mathbf{r}_1(t)$ and $\mathbf{r}_2(t)$, respectively.

EXERCISE 2.3. Let us consider how our results appear using a more concise notation: let $\mathbf{r}_{21} = \mathbf{r}_2(t) - \mathbf{r}_1(t)$ and $r_{21} = |\mathbf{r}_2(t) - \mathbf{r}_1(t)|$. Similarly, we define $\mathbf{v}_{21} = \mathbf{v}_2(t) - \mathbf{v}_1(t)$ and $v_{21} = |\mathbf{v}_2(t) - \mathbf{v}_1(t)|$. Note that this notation suppresses the explicit time dependence of the positions and velocities. Rewrite Equations 2.6 using this concise notation.

For reasons that will become apparent shortly, let us define the vector cross product of the position and velocity vectors:

$$\mathbf{J}(t) = [\mathbf{r}_2(t) - \mathbf{r}_1(t)] \times [\mathbf{v}_2(t) - \mathbf{v}_1(t)].$$

The time derivative of the vector $\mathbf{J}(t)$ is given below:

$$\frac{d}{dt}\mathbf{J}(t) = \frac{d}{dt}\big\{[\mathbf{r}_2(t) - \mathbf{r}_1(t)] \times [\mathbf{v}_2(t) - \mathbf{v}_1(t)]\big\}$$

$$= [\mathbf{v}_2(t) - \mathbf{v}_1(t)] \times [\mathbf{v}_2(t) - \mathbf{v}_1(t)]$$

$$- G\frac{(M_1 + M_2)}{|\mathbf{r}_2(t) - \mathbf{r}_1(t)|^3}\,[\mathbf{r}_2(t) - \mathbf{r}_1(t)] \times [\mathbf{r}_2(t) - \mathbf{r}_1(t)]$$

$$= 0.$$

Here, we have used the results of Equations 2.6 and (twice) the result that the cross product of any vector with itself is zero: $\mathbf{x} \times \mathbf{x} = 0$. Hence, the vector $\mathbf{J}$ must be constant and independent of time; we say that it is

---

[9]In this instance, the equations of motion actually do relate to motion of objects. The phrase "equations of motion" will often be utilized to describe the time evolution of a system, even when there are no physical objects that move.

conserved. This result has an important consequence. The vector $\mathbf{J}$ is perpendicular (by construction) to the vector $\mathbf{r}_2(t) - \mathbf{r}_1(t)$. Therefore, provided that the magnitude J of $\mathbf{J}$ is not zero, we must have that all motion takes place in a plane perpendicular to $\mathbf{J}$, as is depicted in figure 2.3.

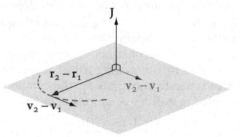

FIGURE 2.3. The (relative) motion of any two masses interacting via the gravitational force is constrained to the plane perpendicular to the vector $\mathbf{J}$

In the case where $J = 0$, the motion will prove to be even simpler. Consider taking the time derivative of the unit vector $[\mathbf{r}_2(t) - \mathbf{r}_1(t)]/|\mathbf{r}_2(t) - \mathbf{r}_1(t)|$. We can use the quotient rule to obtain the following result:

$$\frac{d}{dt}\frac{\mathbf{r}_2(t) - \mathbf{r}_1(t)}{|\mathbf{r}_2(t) - \mathbf{r}_1(t)|} = \left\{ \frac{d[\mathbf{r}_2(t) - \mathbf{r}_1(t)]}{dt}|\mathbf{r}_2(t) - \mathbf{r}_1(t)| \right.$$

$$\left. - [\mathbf{r}_2(t) - \mathbf{r}_1(t)]\frac{d|\mathbf{r}_2(t) - \mathbf{r}_1(t)|}{dt} \right\} \times |\mathbf{r}_2(t) - \mathbf{r}_1(t)|^{-2}.$$

The time derivative of the vector $\mathbf{r}_2(t) - \mathbf{r}_1(t)$ is just the velocity vector $\mathbf{v}_2(t) - \mathbf{v}_1(t)$ and we can show that the following is true:

$$(2.7) \qquad \frac{d|\mathbf{r}_2(t) - \mathbf{r}_1(t)|}{dt} = \frac{[\mathbf{r}_2(t) - \mathbf{r}_1(t)] \cdot [\mathbf{v}_2(t) - \mathbf{v}_1(t)]}{|\mathbf{r}_2(t) - \mathbf{r}_1(t)|}.$$

Putting all of the terms over a common denominator and using the vector identity $\mathbf{x} \cdot \mathbf{x} = x^2$, we can obtain the following result:

$$\frac{d}{dt}\frac{\mathbf{r}_2(t) - \mathbf{r}_1(t)}{|\mathbf{r}_2(t) - \mathbf{r}_1(t)|} = \frac{[\mathbf{r}_2(t) - \mathbf{r}_1(t)] \cdot [\mathbf{r}_2(t) - \mathbf{r}_1(t)]}{|\mathbf{r}_2(t) - \mathbf{r}_1(t)|^3}[\mathbf{v}_2(t) - \mathbf{v}_1(t)]$$

$$(2.8) \qquad - \frac{[\mathbf{v}_2(t) - \mathbf{v}_1(t)] \cdot [\mathbf{r}_2(t) - \mathbf{r}_1(t)]}{|\mathbf{r}_2(t) - \mathbf{r}_1(t)|^3}[\mathbf{r}_2(t) - \mathbf{r}_1(t)].$$

We can simplify this last equation by employing another vector identity: $(\mathbf{a} \times \mathbf{b}) \times \mathbf{c} = (\mathbf{a} \cdot \mathbf{c})\mathbf{b} - (\mathbf{b} \cdot \mathbf{c})\mathbf{a}$. The time derivative of the unit vector can thus be written as follows:

$$\frac{d}{dt}\frac{\mathbf{r}_2(t) - \mathbf{r}_1(t)}{|\mathbf{r}_2(t) - \mathbf{r}_1(t)|} = \frac{\left\{[\mathbf{r}_2(t) - \mathbf{r}_1(t)] \times [\mathbf{v}_2(t) - \mathbf{v}_1(t)]\right\} \times [\mathbf{r}_2(t) - \mathbf{r}_1(t)]}{|\mathbf{r}_2(t) - \mathbf{r}_1(t)|^3}$$

$$(2.9) \qquad = \frac{\mathbf{J} \times [\mathbf{r}_2(t) - \mathbf{r}_1(t)]}{|\mathbf{r}_2(t) - \mathbf{r}_1(t)|^3}.$$

Thus, if $J = 0$, the right hand side of Equation 2.9 vanishes and consequently the unit vector must be a constant vector, independent of time. In this case, motion takes place along a line.

EXERCISE 2.4.   Use the fact that

$$|\mathbf{r}_2(t) - \mathbf{r}_1(t)| = \left\{ [x_2(t) - x_1(t)]^2 + [y_2(t) - y_1(t)]^2 + [z_2(t) - z_1(t)]^2 \right\}^{1/2}$$

and compute the time derivative explicitly. Show that you obtain the result described in Equation 2.7.

EXERCISE 2.5.   Define the vectors $\mathbf{a} = (a_x, a_y, a_z)$, $\mathbf{b} = (b_x, b_y, b_z)$ and $\mathbf{c} = (c_x, c_y, c_z)$. Show that the following relationship holds:

$$(\mathbf{a} \times \mathbf{b}) \times \mathbf{c} = (\mathbf{a} \cdot \mathbf{c})\mathbf{b} - (\mathbf{b} \cdot \mathbf{c})\mathbf{a}.$$

It is important here to note that in the case where motion is along a line that there will be limits to our ability to predict the motion. At some point in time, the masses will collide. This will occur when the vector $\mathbf{r}_2(t) - \mathbf{r}_1(t)$ vanishes. At that time, our expression for the force diverges and beyond that time, we have not yet constructed a theory of behavior. Realistically, we recognize that the outcome of such a collision will depend upon a lot of other factors like the relative sizes of the masses and the collision velocity. A golf ball dropped onto a hard surface like a driveway bounces. If dropped into wet grass, the golf ball does not bounce; an asteroid travelling at high velocity ($\approx 17\,\text{km/s}$) also does not bounce but wreaks significantly more havoc than the golf ball.

EXERCISE 2.6.   Repeat the derivation of Equation 2.9 using the concise notation.

EXERCISE 2.7.   Show that if $J = 0$ and the initial velocity $\mathbf{v}_2(0) - \mathbf{v}_1(0) = 0$ that the masses will collide in finite time. Hint: Multiply both sides of the (scalar) Equation 2.5 by dr_{21}/dt and integrate.

To understand the nature of the motion in the plane, where $J \neq 0$, multiply both sides of Equation 2.9 by $G(M_1 + M_2)$:

$$G(M_1 + M_2) \frac{d}{dt} \frac{\mathbf{r}_2(t) - \mathbf{r}_1(t)}{|\mathbf{r}_2(t) - \mathbf{r}_1(t)|} = J \times \left[G(M_1 + M_2) \frac{\mathbf{r}_2(t) - \mathbf{r}_1(t)}{|\mathbf{r}_2(t) - \mathbf{r}_1(t)|^3} \right]$$

$$= -J \times \frac{d[\mathbf{v}_2(t) - \mathbf{v}_1(t)]}{dt}$$

(2.10)
$$= \frac{d}{dt} \left\{ [\mathbf{v}_2(t) - \mathbf{v}_1(t)] \times J \right\},$$

where we have used Equations 2.6 and the fact that **J** is a constant of the motion to obtain this last result. Both sides of this equation are perfect differentials and we can integrate over time directly to obtain the following result[10]:

$$(2.11) \qquad G(M_1 + M_2)\left[\mathbf{e} + \frac{\mathbf{r}_2(t) - \mathbf{r}_1(t)}{|\mathbf{r}_2(t) - \mathbf{r}_1(t)|}\right] = [\mathbf{v}_2(t) - \mathbf{v}_1(t)] \times \mathbf{J},$$

where **e** is a (vector) constant of integration.

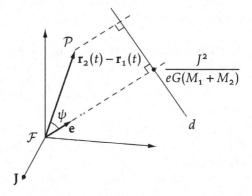

FIGURE 2.4. The vector **e** lies within the plane of motion. In general, there will be some angle ψ between **e** and the position vector $\mathbf{r}_2(t) - \mathbf{r}_1(t)$. The perpendicular to the line extending **e** is known as the directrix d

EXERCISE 2.8.   Repeat the derivation of Equation 2.11 using the concise notation.

To understand the physical meaning of **e**, let's take the dot product of **J** with Equation 2.11:

$$\mathbf{J} \cdot G(M_1 + M_2)\left[\mathbf{e} + \frac{\mathbf{r}_2(t) - \mathbf{r}_1(t)}{|\mathbf{r}_2(t) - \mathbf{r}_1(t)|}\right] = \mathbf{J} \cdot \left\{[\mathbf{v}_2(t) - \mathbf{v}_1(t)] \times \mathbf{J}\right\}$$

$$G(M_1 + M_2)\left[\mathbf{J} \cdot \mathbf{e} + \mathbf{J} \cdot \frac{\mathbf{r}_2(t) - \mathbf{r}_1(t)}{|\mathbf{r}_2(t) - \mathbf{r}_1(t)|}\right] = 0.$$

The right hand side vanishes because **J** is by construction perpendicular to the cross product. We know also that **J** is perpendicular to $\mathbf{r}_2(t) - \mathbf{r}_1(t)$, whereby we have $\mathbf{J} \cdot [\mathbf{r}_2(t) - \mathbf{r}_1(t)] = 0$. Therefore, we must have $\mathbf{J} \cdot \mathbf{e} = 0$ and, consequently, **e** is perpendicular to **J** and lies in the plane of motion. We show a possible value of **e** in figure 2.4.

EXERCISE 2.9.   Define vectors $\mathbf{r}_1 = (x_1, y_1, z_1)$ and $\mathbf{r}_2 = (x_2, y_2, z_2)$. Show explicitly that $\mathbf{r}_1 \cdot (\mathbf{r}_1 \times \mathbf{r}_2) = 0$.

---

[10]The fundamental theorem of calculus provides that $f(x) - f(a) = \int_a^x d\tau \, df(\tau)/d\tau$, when f is defined on the interval $[a, b]$ and $a \le \tau \le b$. The indefinite integrals in Equation 2.10 give rise to constants of integration that can be lumped into the single vector constant **e**.

If we take the dot product of $\mathbf{r}_2(t) - \mathbf{r}_1(t)$ with Equation 2.11, we find

$$[\mathbf{r}_2(t) - \mathbf{r}_1(t)] \cdot \left[\mathbf{e} + \frac{\mathbf{r}_2(t) - \mathbf{r}_1(t)}{|\mathbf{r}_2(t) - \mathbf{r}_1(t)|} \right] = \frac{[\mathbf{r}_2(t) - \mathbf{r}_1(t)] \cdot [\mathbf{v}_2(t) - \mathbf{v}_1(t)] \times \mathbf{J}}{G(M_1 + M_2)}$$

$$[\mathbf{r}_2(t) - \mathbf{r}_1(t)] \cdot \mathbf{e} + |\mathbf{r}_2(t) - \mathbf{r}_1(t)| = \frac{\mathbf{J} \cdot [\mathbf{r}_2(t) - \mathbf{r}_1(t)] \times [\mathbf{v}_2(t) - \mathbf{v}_1(t)]}{G(M_1 + M_2)}$$

where we have employed the vector identity $\mathbf{a} \cdot (\mathbf{b} \times \mathbf{c}) = \mathbf{c} \cdot (\mathbf{a} \times \mathbf{b})$ in this last step. We now can write the following expression that summarizes our results thus far:

$$(2.12) \qquad [\mathbf{r}_2(t) - \mathbf{r}_1(t)] \cdot \mathbf{e} + |\mathbf{r}_2(t) - \mathbf{r}_1(t)| = \frac{J^2}{G(M_1 + M_2)}.$$

EXERCISE 2.10. Define vectors $\mathbf{a} = (a_x, a_y, a_z)$, $\mathbf{b} = (b_x, b_y, b_z)$ and $\mathbf{c} = (c_x, c_y, c_z)$. Show explicitly that $\mathbf{a} \cdot (\mathbf{b} \times \mathbf{c}) = \mathbf{c} \cdot (\mathbf{a} \times \mathbf{b})$.

Suppose now that the integration constant vanishes, that is $\mathbf{e} = 0$. In this case, Equation 2.12 tells us that $|\mathbf{r}_2(t) - \mathbf{r}_1(t)|$ is a constant, with a value of $J^2/G(M_1 + M_2)$; this is the mathematical description of a circular trajectory. If $e \neq 0$, then in general there will be some angle ψ between the vectors $\mathbf{e}$ and $\mathbf{r}_2(t) - \mathbf{r}_1(t)$, as shown in figure 2.4. In this case, the dot product can be written as the magnitudes of the two vectors multiplied by the cosine of the angle between them:

$$[\mathbf{r}_2(t) - \mathbf{r}_1(t)] \cdot \mathbf{e} = e|\mathbf{r}_2(t) - \mathbf{r}_1(t)| \cos \psi.$$

Suppose we extend the vector $\mathbf{e}$ a distance $J^2/eG(M_1 + M_2)$ and construct a line d perpendicular to $\mathbf{e}$, as shown in figure 2.4. If we rearrange Equation 2.12, we can write

$$(2.13) \qquad |\mathbf{r}_2(t) - \mathbf{r}_1(t)| = e\left[\frac{J^2}{eG(M_1 + M_2)} - |\mathbf{r}_2(t) - \mathbf{r}_1(t)| \cos \psi \right].$$

From figure 2.4, we recognize that the term in square brackets on the right hand side of Equation 2.13 is just the distance from $\mathbf{r}_2$ to the line we've constructed.

From analytic geometry, we can recall that the definition of a **conic section** is the locus of points P in the plane of a fixed point $\mathcal{F}$ (called a *focus*) and a fixed line d (called the *directrix*), $\mathcal{F}$ not on d, such that the ratio of the distance from P to $\mathcal{F}$ to the distance from P to d is a constant e (called the *eccentricity*).[11] Equation 2.13 is therefore the equation of a conic section with one focus at the point $\mathbf{r}_1(t)$ and eccentricity e. Furthermore if the

---

[11]This point may be an obscure one but Isaac Newton was a master of geometry. It was not at all obscure to him. Equation 2.13 also justifies our use of the somewhat curious choice of the variable $\mathbf{e}$ to represent the constant of integration in Equation 2.11. The variable e is used historically to represent the eccentricity of conic sections.

value of the eccentricity is less than one ($0 < e < 1$) the trajectory describes an ellipse. If $e = 1$, then the trajectory is a parabola and if $e > 1$ the trajectory is an hyperbola. Trajectories for various values of e are illustrated in figure 2.5.

This result, then, is the proof that the inverse square law proposed by Newton gives rise to elliptical orbits. *Kepler's first law*, that the planets move on ellipses with the sun at one focus, is a consequence of the inverse square force depicted in Equations 2.2 and 2.3.

EXERCISE 2.11.   Repeat the derivation of Equation 2.13 using the concise notation.

FIGURE 2.5. The parameter e controls the type of conic section. For e less than one, one obtains ellipses. For e equal to one, one obtains a parabola. For e larger than one, one obtains hyperbolas. The directrix is shown as the *gray line*. The focus is located at the origin and the angle ψ increases in a counterclockwise manner

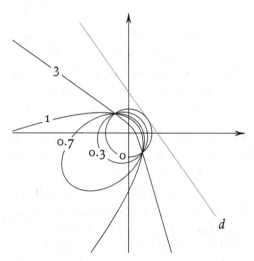

EXERCISE 2.12.   If we rearrange Equation 2.13 to solve for $r_{21} = |\mathbf{r}_2(t) - \mathbf{r}_1(t)|$, we find the following expression:

$$r_{21} = \frac{J^2}{G(M_1 + M_2)(1 + e\cos\psi)}$$

The quantity $J^2/G(M_1 + M_2)$ represents an overall scaling factor. To understand the dependence of r_{21} on ψ, plot the function $F = 1/(1 + e\cos\psi)$ using the *Mathematica* function PolarPlot over the domain $0 \le \psi \le 2\pi$. Use the Manipulate function to vary the eccentricity e over the range $0 \le e \le 4$.

EXERCISE 2.13.   We have demonstrated that an inverse square force law gives rise to elliptical orbits but we have not excluded other possibilities. An inverse cube force law gives rise to spirals

$$r_{21} = \frac{a}{\psi} \qquad \text{or} \qquad r_{21} = e^{a\psi},$$

depending upon the initial conditions. Use the `Manipulate` function to vary the parameter a over the domain $0.1 \le a \le 4$ and plot the distance r_{21} over the domain $0 \le \psi \le 10\pi$.

2.3. Another Conserved Quantity

We have observed that there are two conserved (time-independent) quantities **J** and **e** associated with Newton's universal law of gravitation. It turns out that there is another property of the gravitational force represented by Equations 2.6 that does not depend on time. To identify this quantity, we begin by taking the dot product of the second of the two equations with the vector $\mathbf{v}_2(t) - \mathbf{v}_1(t)$:

$$(2.14) \quad [\mathbf{v}_2(t) - \mathbf{v}_1(t)] \cdot \frac{d[\mathbf{v}_2(t) - \mathbf{v}_1(t)]}{dt}$$

$$= -\frac{G(M_1 + M_2)}{|\mathbf{r}_2(t) - \mathbf{r}_1(t)|^3}[\mathbf{v}_2(t) - \mathbf{v}_1(t)] \cdot [\mathbf{r}_2(t) - \mathbf{r}_1(t)].$$

The left-hand side of this result can be rewritten as follows:

$$[\mathbf{v}_2(t) - \mathbf{v}_1(t)] \cdot \frac{d[\mathbf{v}_2(t) - \mathbf{v}_1(t)]}{dt} = \frac{1}{2}\frac{d}{dt}[\mathbf{v}_2(t) - \mathbf{v}_1(t)] \cdot [\mathbf{v}_2(t) - \mathbf{v}_1(t)]$$

$$= \frac{1}{2}\frac{d}{dt}|\mathbf{v}_2(t) - \mathbf{v}_1(t)|^2.$$

This term is a perfect differential and can be integrated over time directly[12]:

$$(2.15) \quad \frac{1}{2}\int dt \frac{d}{dt}|\mathbf{v}_2(t) - \mathbf{v}_1(t)|^2 = \frac{1}{2}|\mathbf{v}_2(t) - \mathbf{v}_1(t)|^2 + \mathcal{E}_1,$$

where $\mathcal{E}_1$ is a constant of integration.

On the right-hand side of Equation 2.14, the dot product can be rewritten as follows:

$$[\mathbf{v}_2(t) - \mathbf{v}_1(t)] \cdot [\mathbf{r}_2(t) - \mathbf{r}_1(t)] = |\mathbf{r}_2(t) - \mathbf{r}_1(t)|\frac{d}{dt}|\mathbf{r}_2(t) - \mathbf{r}_1(t)|,$$

where we've used the result of Equation 2.7. This result can also be integrated using a somewhat more advanced property of calculus. Consider a function f of a variable x, $f = f(x)$. We can perform a change of variables where f is now considered to be a function of a new variable u, The chain rule provides that the following relation holds:

$$\frac{df(x)}{dx} = \frac{df(u)}{du}\frac{du}{dx}.$$

---

[12]Here we use the notation $\int dx\, f(x)$ to mean the same thing as $\int f(x)dx$. This emphasizes the rôle of integration $\int dx$ as the inverse operator to the differential operator d/dx.

The fundamental theorem of calculus can be rewritten using the chain rule:

$$f(b) - f(a) = \int_a^b dx \, \frac{df(x)}{dx} = \int_a^b dx \, \frac{df(u)}{du} \frac{du}{dx}$$
$$= \int_{u(a)}^{u(b)} du \, \frac{df(u)}{du}.$$

Integrating the right-hand side of Equation 2.14, we find the following result[13]:

$$-G(M_1 + M_2) \int dt \, \frac{1}{|\mathbf{r}_2(t) - \mathbf{r}_1(t)|^2} \frac{d|\mathbf{r}_2(t) - \mathbf{r}_1(t)|}{dt}$$
$$= -G(M_1 + M_2) \int d|\mathbf{r}_2(t) - \mathbf{r}_1(t)| \frac{1}{|\mathbf{r}_2(t) - \mathbf{r}_1(t)|^2}$$
$$= \frac{G(M_1 + M_2)}{|\mathbf{r}_2(t) - \mathbf{r}_1(t)|} + \mathcal{E}_2,$$

where $\mathcal{E}_2$ is another constant of integration. Combining this result with that of Equation 2.15 and multiplying by $M_1 M_2 / (M_1 + M_2)$ we find the following expression[14]:

$$(2.16) \qquad \frac{M_1 M_2}{2(M_1 + M_2)} |\mathbf{v}_2(t) - \mathbf{v}_1(t)|^2 - G \frac{M_1 M_2}{|\mathbf{r}_2(t) - \mathbf{r}_1(t)|} = \mathcal{E},$$

where we have lumped the two (scaled) constants of integration into just a single constant $\mathcal{E}$.

The right-hand side of Equation 2.16 is a constant, so the time derivative must vanish; the quantity $\mathcal{E}$ is therefore independent of time. We can add this to the list of conserved quantities $\mathbf{J}$ and $\mathbf{e}$ that we have already discovered. This quantity will arise again often in our discussions; it defines the (mechanical) **energy** of the system.[15] We recognize that the energy defined in Equation 2.16 has both a velocity-dependent term and a position-dependent term. These are called the *kinetic* and *potential* energies, respectively.

> EXERCISE 2.14. Compute the derivative $d\mathcal{E}/dt$ explicitly and demonstrate that it indeed vanishes.

---

[13]At this point in the semester, students may be unfamiliar with integration techniques. The principle point to be made here is that the right-hand side now is no longer an explicit function of t but of the scalar $|\mathbf{r}_2 - \mathbf{r}_1|$. The result of the integration can be verified by differentiation, which should be familiar.

[14]The need for the multiplicative factor $M_1 M_2 / (M_1 + M_2)$ arises from a dimensional convention that will be discussed in later sections.

[15]The somewhat curious choice of $\mathcal{E}$ for the integration constants is justified by the fact that it is the first letter of the word energy. We use a capital letter to distinguish it from the eccentricity and also to emphasize its importance in our subsequent discussions.

2.4. A Special Frame of Reference

While we have the ability to choose *any* set of coordinates that we like, it turns out to be profitable to not make random choices, Some choices are better than others: reducing the algebra necessary to perform calculations or providing clarity. One such good choice places the origin of the coordinate system at the center of mass of the system. We define the center of mass $\mathbf{r}_{cm}(t)$ as follows:

$$(2.17) \qquad \mathbf{r}_{cm}(t) = \frac{\displaystyle\sum_{i=1}^{N} M_i \mathbf{r}_i(t)}{\displaystyle\sum_{i=1}^{N} M_i},$$

where, in general, there could be N masses. For the case of two masses, this simplifies to $\mathbf{r}_{cm}(t) = (M_1 \mathbf{r}_1(t) + M_2 \mathbf{r}_2(t))/(M_1 + M_2)$, as indicated in figure 2.6.

> EXERCISE 2.15.   If $M_1 = M_2$, show that $\mathbf{r}_{cm}(t)$ is on the line connecting $\mathbf{r}_1(t)$ and $\mathbf{r}_2(t)$. Plot the vectors using a coordinate system in which both lie in the x-y plane and $\mathbf{r}_1(t)$ is aligned with the x-axis. (This will be an astute choice of coordinates.) Use geometry to show that the vector $\mathbf{r}_2(t) + \mathbf{r}_1(t)$ is intersected at its midpoint by the line connecting points $\mathbf{r}_2(t)$ and $\mathbf{r}_1(t)$.

> EXERCISE 2.16.   A more formal proof of collinearity of three points $\mathbf{x}_1, \mathbf{x}_2$ and $\mathbf{x}_3$ is that the cross product vanishes: $(\mathbf{x}_2 - \mathbf{x}_1) \times (\mathbf{x}_3 - \mathbf{x}_1) = 0$. Prove that the center of mass $\mathbf{r}_{cm}(t)$ is collinear with $\mathbf{r}_1(t)$ and $\mathbf{r}_2(t)$.

Now let's use the results of Equations 2.4 and 2.5 to prove an important fact about the center of mass. If we multiply Equation 2.4 by M_2 and add Equation 2.5 multiplied by M_1, we find that

$$M_1 \frac{d^2 \mathbf{r}_1(t)}{dt^2} + M_2 \frac{d^2 \mathbf{r}_2(t)}{dt^2} = 0.$$

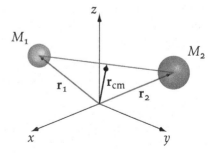

FIGURE 2.6. The center of mass is located on the line that connects the two masses

From this, we can infer the following:

(2.18) $$\frac{d^2}{dt^2}\left[M_1\mathbf{r}_1(t)+M_2\mathbf{r}_2(t)\right]=\frac{d^2\mathbf{r}_{cm}(t)}{dt^2}=0.$$

That is, assuming that the masses are independent of time, the second-order time derivative of the center of mass position vanishes. Consequently, the first-order time derivative of the center of mass position must be some constant, independent of time. We can call this constant something nondescript like $\mathbf{c}$ but we prefer to use something more descriptive: $\mathbf{v}_{cm}$. Mathematically, we can write this as follows:

$$\int dt\,\frac{d}{dt}\left[\frac{d\mathbf{r}_{cm}(t)}{dt}\right]=\frac{d\mathbf{r}_{cm}(t)}{dt}+\mathbf{v}_{cm}.$$

We perform one more integral over time (from t_1 to t_2) and can now write the following expression for the time evolution of the center of mass:

$$\int_{t_1}^{t_2} dt\left[\frac{d\mathbf{r}_{cm}(t)}{dt}+\mathbf{v}_{cm}\right]=\mathbf{r}_{cm}(t_2)-\mathbf{r}_{cm}(t_1)+\mathbf{v}_{cm}(t_2-t_1).$$

Rearranging terms, and substituting the more generic t for t_2 we can now write the following:

(2.19) $$\mathbf{r}_{cm}(t)=\mathbf{x}_{cm}+\mathbf{v}_{cm}(t-t_1),$$

where t_1 is the initial time and $\mathbf{x}_{cm}=\mathbf{r}_{cm}(t_1)$ is the position of the center of mass at time t_1. Thus, the position of the center of mass is a linear function of time.[16] Recall that the definition of linear momentum is the product of mass and velocity: $\mathbf{p}=M\mathbf{v}$. We can reorganize Equation 2.18 as follows:

$$\frac{d^2}{dt^2}\left[M_1\mathbf{r}_1(t)+M_2\mathbf{r}_2(t)\right]=\frac{d}{dt}\left[M_1\frac{d\mathbf{r}_1(t)}{dt}+M_2\frac{d\mathbf{r}_2(t)}{dt}\right]$$

(2.20) $$=\frac{d}{dt}\left[\mathbf{p}_1(t)+\mathbf{p}_2(t)\right]=0.$$

Because this term vanishes, Equation 2.20 is a statement that the total linear momentum of the system does not depend on time. We say that the total linear momentum is *conserved*. Note that this does not imply that the individual momenta $\mathbf{p}_i$ are constants. On the contrary, the $\mathbf{p}_i$ are generally not constant over time but change in such a way that their sum remains constant.

Let us now transform our calculations to the coordinate frame in which the center of mass is at the origin. In this coordinate system, the first

---

[16]There is an unfortunate notational ambiguity in Equation 2.19. The term $\mathbf{v}_{cm}$ is a constant, not a function of the argument $t-t_1$. This will not be the last time we shall have to be conscious of what the notation means.

mass is at a position given by $\mathbf{r}_1(t) - \mathbf{r}_{cm}(t)$ and the second mass is at the position $\mathbf{r}_2(t) - \mathbf{r}_{cm}(t)$. From the definition of the center of mass, we know that $(M_1 + M_2)\mathbf{r}_{cm}(t) = M_1\mathbf{r}_1(t) + M_2\mathbf{r}_2(t)$, which we can rearrange to yield $M_1[\mathbf{r}_1(t) - \mathbf{r}_{cm}(t)] = -M_2[\mathbf{r}_2(t) - \mathbf{r}_{cm}(t)]$.

Because the second time derivative of the center of mass position $\mathbf{r}_{cm}(t)$ vanishes, we can rewrite Equation 2.5 as follows:

$$\begin{aligned}
M_1 \frac{d^2[\mathbf{r}_1(t) - \mathbf{r}_{cm}(t)]}{dt^2} &= -G\frac{M_1 M_2}{|\mathbf{r}_2(t) - \mathbf{r}_1(t)|^3}[\mathbf{r}_2(t) - \mathbf{r}_1(t)] \\
&= -G\frac{M_1 M_2}{|\mathbf{r}_2(t) - \mathbf{r}_1(t)|^3}\big\{[\mathbf{r}_1(t) - \mathbf{r}_{cm}(t)] - [\mathbf{r}_2(t) - \mathbf{r}_{cm}(t)]\big\} \\
&= -G\frac{M_1 M_2}{|\mathbf{r}_2(t) - \mathbf{r}_1(t)|^3} \times \\
&\qquad \Big\{[\mathbf{r}_1(t) - \mathbf{r}_{cm}(t)] + \frac{M_1}{M_2}[\mathbf{r}_1(t) - \mathbf{r}_{cm}(t)]\Big\} \\
&= -G\frac{M_1(M_1 + M_2)}{|\mathbf{r}_2(t) - \mathbf{r}_1(t)|^3}[\mathbf{r}_1(t) - \mathbf{r}_{cm}(t)].
\end{aligned}$$

(2.21)

We can remove the final dependence on $\mathbf{r}_2(t)$ by noting that the following relation holds:

$$|\mathbf{r}_2(t) - \mathbf{r}_1(t)| = |\mathbf{r}_2(t) - \mathbf{r}_{cm}(t)| + |\mathbf{r}_1(t) - \mathbf{r}_{cm}(t)|.$$

We have then that the motion of mass M_1 in the center of mass frame can be written as follows:

(2.22) $$M_1 \frac{d^2[\mathbf{r}_1(t) - \mathbf{r}_{cm}(t)]}{dt^2} = -G\frac{M_1 M_2^3}{(M_1 + M_2)^2} \frac{\mathbf{r}_1(t) - \mathbf{r}_{cm}(t)}{|\mathbf{r}_1(t) - \mathbf{r}_{cm}(t)|^3}.$$

The equation of motion of mass M_1 in the center of mass frame has precisely the same form as our original Equations 2.6. This indicates that the motion of mass M_1 in the center of mass frame behaves as if there were a (suitably scaled) force acting on M_1 from the center of mass. By analogy to our previous derivations, the trajectory of mass M_1 will therefore describe a conic section with one focus at the center of mass location.

EXERCISE 2.17. Prove the following relation:

$$|\mathbf{r}_2(t) - \mathbf{r}_1(t)| = |\mathbf{r}_2(t) - \mathbf{r}_{cm}(t)| + |\mathbf{r}_1(t) - \mathbf{r}_{cm}(t)|.$$

Hint: The three points are collinear. Use this result to prove Equation 2.22.

EXERCISE 2.18. Derive the equivalent to Equation 2.22: the equation of motion for mass M_2 in the center of mass frame.

EXERCISE 2.19. The mass of the earth is approximately 6×10^{24} kg and the mass of the sun is approximately 2×10^{30} kg. The nominal earth-sun distance is approximately 1.5×10^{11} m. Where is the center of mass of the earth-sun system located? How does that compare to the approximate solar radius of $R_{sun} = 7 \times 10^{8}$ m?

The masses of the NASA GOES satellites are about 3250 kg and their geosynchronous orbits are approximately 3.6×10^{7} m above the earth's surface. The earth's radius is approximately 6.4×10^{6} m. Where is the center of mass of the earth-GOES system? How does that compare to the earth's radius?

We can define the *angular momentum* L of a particle by $L = m\mathbf{r} \times \mathbf{v}$. For our two-mass system, the total angular momentum is just given by the following:

$$(2.23) \qquad \mathbf{L} - M_1\mathbf{r}_1(t) \times \mathbf{v}_1(t) + M_2\mathbf{r}_2(t) \times \mathbf{v}_2(t).$$

In the center of mass frame, the total angular momentum consists of the following two terms:

$$\mathbf{L} = M_1[\mathbf{r}_1(t) - \mathbf{r}_{cm}(t)] \times [\mathbf{v}_1(t) - \mathbf{v}_{cm}] + M_2[\mathbf{r}_2(t) - \mathbf{r}_{cm}(t)] \times [\mathbf{v}_2(t) - \mathbf{v}_{cm}].$$

Regrouping terms and recalling the definition of $\mathbf{r}_{cm}(t)$, we can write:

$$\mathbf{L} = \left[M_1\mathbf{r}_1(t) - M_1 \frac{M_1\mathbf{r}_1(t) + M_2\mathbf{r}_2(t)}{M_1 + M_2} \right] \times [\mathbf{v}_1(t) - \mathbf{v}_{cm}]$$

$$+ \left[M_2\mathbf{r}_2(t) - M_2 \frac{M_1\mathbf{r}_1(t) + M_2\mathbf{r}_2(t)}{M_1 + M_2} \right] \times [\mathbf{v}_2(t) - \mathbf{v}_{cm}]$$

$$= \frac{M_1 M_2}{M_1 + M_2} \left\{ [\mathbf{r}_1(t) - \mathbf{r}_2(t)] \times \mathbf{v}_1(t) + [\mathbf{r}_2(t) - \mathbf{r}_1(t)] \times \mathbf{v}_2(t) \right\}$$

$$(2.24) \qquad = \frac{M_1 M_2}{M_1 + M_2} \mathbf{J}.$$

From this result, we can conclude that the total angular momentum L of the system is proportional to the conserved vector J and hence must also be conserved. Alternatively, the statement that J is a constant of the motion and independent of time is equivalent to the statement that angular momentum is conserved.

EXERCISE 2.20. The derivation of Equation 2.24 was not as detailed as others in the text thus far. Go back and repeat the derivation from Equation 2.23 without skipping as many steps. Note, in particular, how did we come by the intermediate results for angular momenta in the center of mass frame:

$$\mathbf{L}_i = M_i (\mathbf{r}_i(t) - \mathbf{r}_{cm}(t)) \times (\mathbf{v}_i(t) - \mathbf{v}_{cm})?$$

Also, what happened to the terms involving $\mathbf{v}_{cm}$ that appear in the intermediate result?

Turning now to the other conserved quantity we have discovered, we define the **kinetic energy** T of a particle to be the velocity-dependent part of the mechanical energy $\mathcal{E}$. For an isolated particle, this is the quantity $T = \frac{1}{2}Mv^2$. In the center of mass frame, we can write that the total kinetic energy T of our gravitational system is the sum of the individual energies:

$$(2.25) \qquad T = \frac{1}{2}M_1|\mathbf{v}_1(t) - \mathbf{v}_{cm}|^2 + \frac{1}{2}M_2|\mathbf{v}_2(t) - \mathbf{v}_{cm}|^2.$$

Using the fact that $\mathbf{v}_{cm} = (M_1\mathbf{v}_1(t) + M_2\mathbf{v}_2(t))/(M_1 + M_2)$, we can show that the following result holds:

$$(2.26) \qquad T = \frac{M_1 M_2}{2(M_1 + M_2)}|\mathbf{v}_2(t) - \mathbf{v}_1(t)|^2.$$

This is just the kinetic energy that we defined in Equation 2.16.

This result is, at first glance, somewhat surprising. The center of mass coordinate system is moving with a velocity $\mathbf{v}_{cm}$ with respect to the original coordinate system used to define the vectors $\mathbf{r}_1(t)$ and $\mathbf{r}_2(t)$. We might have anticipated that a moving coordinate system could change the values of the energy. Instead, we find that the kinetic energy is the same in each coordinate system. To see how this arises, recall that the vector $\mathbf{r}_2(t) - \mathbf{r}_1(t)$ is independent of the coordinate system. The gravitational **potential energy** $\mathcal{U}$ from Equation 2.16 is given by the following:

$$(2.27) \qquad \mathcal{U} = -G\frac{M_1 M_2}{|\mathbf{r}_2(t) - \mathbf{r}_1(t)|}.$$

It depends solely on the magnitude $|\mathbf{r}_2(t) - \mathbf{r}_1(t)|$ that we have already shown is independent of the coordinate system. We know that the total energy $\mathcal{E}$ is conserved, so we should indeed find that the kinetic energy T does not depend on the coordinate system.

We have now managed to demonstrate that Newton's inverse square law makes the following predictions about trajectories:

- Trajectories are conic sections,
- Linear momentum is conserved,
- Angular momentum is conserved,
- The total mechanical energy $\mathcal{E} = T + \mathcal{U}$ is conserved
- The eccentricity vector $\mathbf{e}$ is conserved.

This is remarkable progress.

2.5. Trajectories in More Detail

We have now identified that Newton's inverse square law gives rise to conic section trajectories, this includes specifically the elliptical trajectories observed for planetary motion. What we would like to do now is understand what governs the type of trajectory, elliptical or hyperbolic, etc. that an object would occupy. To illuminate this issue, we can go back to Equation 2.11 and square both sides. We obtain the following result:

$$(2.28) \qquad G^2(M_1 + M_2)^2 \left[e^2 + 1 + 2\mathbf{e} \cdot \frac{\mathbf{r}_2(t) - \mathbf{r}_1(t)}{|\mathbf{r}_2(t) - \mathbf{r}_1(t)|} \right] = |\mathbf{v}_2(t) - \mathbf{v}_1(t)|^2 J^2,$$

where again e is the magnitude of $\mathbf{e}$ and J is the magnitude of $\mathbf{J}$.

> EXERCISE 2.21.   To obtain the above result, we used another vector identity:
> $$(\mathbf{a} \times \mathbf{b}) \cdot (\mathbf{c} \times \mathbf{d}) = (\mathbf{a} \cdot \mathbf{c})(\mathbf{b} \cdot \mathbf{d}) - (\mathbf{a} \cdot \mathbf{d})(\mathbf{b} \cdot \mathbf{c}).$$
> Define the components of vectors $\mathbf{a} = (a_x, a_y, a_z)$, etc., and demonstrate explicitly that the identity holds. Next, if $\mathbf{a} = \mathbf{c} = \mathbf{r}_2(t) - \mathbf{r}_1(t)$ and $\mathbf{b} = \mathbf{d} = \mathbf{J}$, show that we indeed obtain the right hand side of the result illustrated in Equation 2.28.

Now the quantity $\mathbf{e} \cdot [\mathbf{r}_2(t) - \mathbf{r}_1(t)]$ was defined in Equation 2.12 and the quantity $|\mathbf{v}_2(t) - \mathbf{v}_1(t)|^2$ was defined in Equation 2.16. Substituting these results back into Equation 2.28, we obtain the following expression:

$$(e^2 - 1) = \frac{2\mathcal{E}J^2}{G^2 M_1 M_2(M_1 + M_2)},$$

or, solving for the eccentricity,

$$(2.29) \qquad e = \sqrt{1 + \frac{2\mathcal{E}J^2}{G^2 M_1 M_2(M_1 + M_2)}}.$$

The eccentricity e is related to the mechanical energy $\mathcal{E}$, the angular momentum J and the masses. We can see immediately from Equation 2.29 that the type of conic section is governed by the mechanical energy $\mathcal{E}$.[17] To obtain elliptical orbits, where $0 < e < 1$, we must require that the second term under the square root be negative or that $\mathcal{E} < 0$. When $\mathcal{E} = 0$, we obtain parabolic trajectories and when $\mathcal{E} > 0$, we obtain hyperbolic trajectories. So, the type of trajectory depends upon the relative magnitudes of the kinetic (positive) and potential (negative) energies.

Let's focus for now on elliptical trajectories; those are the orbits occupied by planets. We choose a coordinate system in which the $\mathbf{e}$ vector is aligned

---

[17]The masses are positive quantities, as must be the quantities G^2 and J^2.

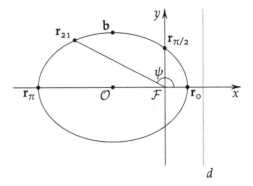

FIGURE 2.7. An ellipse can also be defined by its semi-major axis a and semi-minor axis b. These values are related to the directrix α and eccentricity e

with the x-axis and define the quantity $\alpha = J^2/G(M_1 + M_2)$, then for values of $e < 1$ we obtain ellipses defined by the following equation:

(2.30) $$e r_{21} \cos \psi + r_{21} = \alpha,$$

where we use our compact notation. Points $\mathbf{r}_{21}$ on the ellipse will have coordinates $\mathbf{r}_{21} = (x, y) = \alpha(\cos \psi, \sin \psi)/(1 + e \cos \psi)$. Consider the point where $\psi = 0$. Let's call that point $\mathbf{r}_0$ and we find $e r_0 + r_0 = \alpha$ or $r_0 = \alpha/(1 + e)$. Now consider the point $\mathbf{r}_{\pi/2}$ where $\psi = \pi/2$. Here $\cos \psi = 0$ and, thus, $r_{\pi/2} = \alpha$. This is reassuring because we know that the distance from the focus to the directrix is α/e.

Further, we can define the point $\mathbf{r}_\pi$ as the point where $\psi = \pi$ and note that the points $\mathbf{r}_0$ and $\mathbf{r}_\pi$ define the extent of the ellipse in the x-direction. The geometric center of the ellipse $\mathcal{O}$ can be defined as $\mathcal{O} = (\mathbf{r}_\pi + \mathbf{r}_0)/2$ and the length of the semimajor axis a of the ellipse is defined as $a = |\mathbf{r}_\pi - \mathbf{r}_0|/2$.

EXERCISE 2.22. What are the values of a and $\mathcal{O}$ in terms of α and e?

EXERCISE 2.23. For elliptical orbits, where $\mathcal{E} < 0$, show that the value of α can be written as follows:

$$\alpha = \frac{|1 - e^2| G M_1 M_2}{2|\mathcal{E}|}.$$

The point $\mathbf{b}$ in figure 2.7 marks the maximum extent of the ellipse in the y-direction and determines the minor axis of the ellipse in a coordinate system with $\mathcal{O}$ at the center.

EXERCISE 2.24. What is the value of b in terms of α and e? Hint: the point $\mathbf{b}$ has the same x-coordinate as $\mathcal{O}$.

You might recall that the area of an ellipse is given by the product of the semimajor and semiminor axes: $A = \pi a b$. We can compute the area

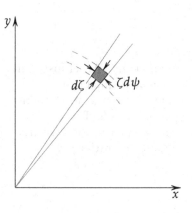

FIGURE 2.8. In polar coordinates, the
differential along the radial direction
is $d\zeta$. For a differential angle $d\psi$,
the length of the differential at a radius
ζ is $\zeta\, d\psi$

using calculus by computing a two-dimensional integral. In rectangular coordinates, we would write

$$A = \int dx \int dy$$

and adjust the limits of the integrals to match the desired area. For a rectangle that extends from $x = 0$ to $x = L$ and from $y = 0$ to $y = w$, we would have the following expression[18]:

$$A = \int_0^L dx \int_0^w dy = Lw,$$

where we recover the well-known result that the area of a rectangle is the product of its length times its width.

EXERCISE 2.25. Consider a right triangle that extends from $x = 0$ to $x = b$ and from $y = 0$ to $y = h$. The x-integral will extend from 0 to b but the y-integral can only extend as far as the line connecting the points $(b,0)$ and $(0,h)$. Show that the equation connecting the points is given by $y = h - hx/b$. The upper limit to the y-integral is then $h - hx/b$. Use this fact to show that the area of the triangle is $A = bh/2$.

For our elliptical trajectories, it will be more convenient to work in polar coordinates, where the path length ds along an arc of length $d\psi$ at a radius ζ is given by $ds = \zeta\, d\psi$, as illustrated in figure 2.8. The area in polar coordinates is then determined from the following equation:

$$A = \int d\zeta\, \zeta \int d\psi = \int d\psi \int_0^{r_{21}} d\zeta\, \zeta$$

---

[18]We have routinely used the integral notation $\int dx\, f(x)$ to mean the same thing as $\int f(x)\, dx$. This practice is especially useful when we encounter multiple integrals because the integration variable is positioned adjacent to the symbol that defines the limits of integration.

where, in the last step, we have used the fact that the radial coordinate that defines the ellipse is a function of ψ and reversed the order of integration. We can perform the radial integration readily: $\int d\zeta\,\zeta = \zeta^2/2$. So, the differential area element as a function of ψ is then $dA = r_{21}^2/2\,d\psi$. This means that the differential change in area for each differential change in time (the time derivative of the area) can be written as follows:

(2.31)
$$\frac{dA}{dt} = \frac{r_{21}^2}{2}\frac{d\psi}{dt}.$$

In our present choice of coordinate system, we note that the vector $\mathbf{r}_2 - \mathbf{r}_1$ has the following form:

(2.32)
$$\mathbf{r}_2(t) - \mathbf{r}_1(t) = (r_{21}\cos\psi, r_{21}\sin\psi, 0).$$

Taking the time derivative, we find that the velocity is given by the following:

(2.33)
$$\mathbf{v}_2(t) - \mathbf{v}_1(t) = \left(\frac{dr_{21}}{dt}\cos\psi - r_{21}\sin\psi\frac{d\psi}{dt}, \frac{dr_{21}}{dt}\sin\psi + r_{21}\cos\psi\frac{d\psi}{dt}, 0\right).$$

The cross product of these two vectors was defined to be the vector $\mathbf{J}$. Evaluating the cross product, we find the following relation:

(2.34)
$$J = r_{21}^2\frac{d\psi}{dt}.$$

Somewhat surprisingly, the terms involving the time derivative of the radial distance cancel. Thus, the magnitude of the angular momentum does not depend on dr_{21}/dt.

EXERCISE 2.26. Compute the vector product $\mathbf{J} = (\mathbf{r}_2 - \mathbf{r}_1) \times (\mathbf{v}_2 - \mathbf{v}_1)$ explicitly. Show that the magnitude J is given by $r_{21}^2\,d\psi/dt$.

Notice that from our results in Equations 2.31 and 2.34 that we must necessarily have

$$\frac{dA}{dt} = J/2.$$

Now J is a constant of the motion, which means that the change in area with respect to time must also be a constant. Integrating, both sides of this last result, we find the following:

$$\int_{t_1}^{t_2} dt\frac{dA}{dt} = \frac{J}{2}\int_{t_1}^{t_2} dt$$

(2.35)
$$A(t_2) - A(t_1) = J(t_2 - t_1)/2.$$

That is, the change in area from some initial time t_1 to a later time t_2 is linearly proportional to the time difference $t_2 - t_1$.

The fact that the area is a linear function of time has two important consequences. First, *Kepler's second law* says that a planet sweeps out equal areas in equal times as it moves along its trajectory. We see that Kepler's second law is equivalent to the statement that angular momentum (J) is constant. Alternatively, Kepler's second law is a manifestation of angular momentum conservation. As we see in figure 2.9, if we measure the time t_{12} that it takes for the planet to move between points $\mathbf{r}_1$ and $\mathbf{r}_2$, the planet will sweep out the shaded area. If we then start our clock ticking when the planet is at position $\mathbf{r}_3$ and stop it after the same time t_{12}, the planet will have moved to the point $\mathbf{r}_4$. The two shaded areas are equal.

FIGURE 2.9. As the planet moves along its trajectory from $\mathbf{r}_1$ to $\mathbf{r}_2$ or from $\mathbf{r}_3$ to $\mathbf{r}_4$, it sweeps out an area denoted by the *shaded areas*. Kepler's second law states that the areas are equal if the times between points are equal

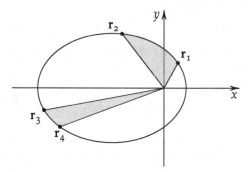

Second, a planet on an elliptical trajectory will eventually return to the same point on its trajectory, no matter where we define that point. The change in area as a result of that motion is just the total area of the ellipse: $A = (\pi)ab$. Using Equation 2.35, the total elapsed time will be $T = t_2 - t_1 = 2\pi ab/J$. We define T to be the **orbital period**; it takes precisely the time T for the planet to complete one orbit and return to its starting point. We see that conservation of angular momentum leads to periodic orbits.

EXERCISE 2.27.   Kepler deduced a third law of planetary motion that says that the square of the period T is proportional to the cube of the semimajor axis a. From the relation $T = 2\pi ab/J$ and the previous result for b, show that the period must be related to a by the following relation:

$$(2.36) \qquad T^2 = \frac{4\pi^2}{G(M_1 + M_2)} a^3.$$

EXERCISE 2.28.   Kepler's third law suggests that the ratio of T^2/a^3 should be a constant for all objects orbiting a common center. Table 2.1 below contains a list of the semimajor axes a, orbital periods T and eccentricities e of the planets in our solar system. The unit used for a is the astronomical unit (AU) that is defined to be the semimajor axis of the earth-sun system. The unit used for orbital period is the earth year (yr). How well does the prediction made by Kepler's third law agree with the experimental observations?

TABLE 2.1. Solar system orbital data

	a (AU)	T (yr)	e
Mercury	0.387	0.241	0.206
Venus	0.723	0.615	0.007
Earth	1.0	1.0	0.017
Mars	1.524	1.881	0.093
Jupiter	5.203	11.86	0.048
Saturn	9.537	29.42	0.054
Uranus	19.19	83.75	0.047
Neptune	30.07	163.7	0.009

2.6. Position as a Function of Time

We have now proven that all three of Kepler's laws are the result of Newton's universal law of gravitation. We know that the planets follow elliptical trajectories and that the motion is periodic but we have not, as yet, specified the location of the planets as a function of time. This will be our next task. Defining position on the orbit as a function of time presents a somewhat challenging problem but one that can yield to modern technology.

We have seen that the two vector Equations 2.6 for the trajectory of a planet can be reduced to a single scalar equation for the angle ψ. We know that, using our compact notation, that $r_{21}^2 \, d\psi/dt = J$ and that $r_{21} = J^2/[G(M_1 + M_2)(1 + e\cos\psi)]$. Hence, we can write the following equation for the time variation of the angle:

$$(1 + e\cos\psi)^{-2} \frac{d\psi}{dt} = \frac{G^2(M_1 + M_2)^2}{J^3}.$$

If we now integrate both sides over time, we find the following result:

$$\int dt (1 + e\cos\psi)^{-2} \frac{d\psi}{dt} = \frac{G^2(M_1 + M_2)^2}{J^3} \int dt$$

$$(2.37) \qquad \tau_e(\psi) \equiv \int d\psi (1 + e\cos\psi)^{-2} = \frac{G^2(M_1 + M_2)^2}{J^3} (t - t_1),$$

where we have defined the function $\tau_e(\psi)$ to be the result of the ψ integration and incorporated the integration constant arising from the right-hand-side integral into an initial time t_1. This can be chosen to be zero without any real consequence to our further discussions.

We find ourselves in a situation that is not quite what we desired but one that will arise with uncomfortable frequency in physics. What we wanted to find was a function $\psi = \psi(t)$ that would complete our description of planetary motion. What we have found instead is a function $\tau_e = \tau_e(\psi)$ that is a function of the angle ψ. Worse, the function τ_e is defined by a relatively complex integral, one that is undoubtedly more complicated than can be solved by students at this point in their academic careers.

So, we find ourselves in something of a mathematical bind: we have a mathematical description of the system under study but not the means to solve the equations. In order to make further progress, we will have to add more mathematical tools to our repertoire.[19] Part of the effort to become proficient in physics will be a continual quest to learn more mathematics. In this modern age of computers, this will include learning numerical methods that might be used to construct solutions. In the present circumstances, it will actually prove possible to find a solution to the integral in terms of simple functions. It will, however, not be possible invert the function $\tau_e(\psi)$ to find the desired function $\psi(t)$, at least not in terms of simple functions.

The $\tau_e(\psi)$ integral can be evaluated in terms of simple functions. We can make use of the *Mathematica* Integrate function to provide the requisite assistance. We will leave the details to subsequent exercises and just state the result here:

$$(2.38) \quad \tau_e(\psi) = \frac{2}{(1-e^2)^{3/2}} \tan^{-1}\left[\left(\frac{1-e}{1+e}\right)^{1/2} \tan(\psi/2)\right] - \frac{e\sin\psi}{(1-e^2)(1+e\cos\psi)}.$$

This expression turns out to be only valid over the interval $0 \le \psi < \pi$, due to the fact that $\tan(\psi/2)$ diverges as $\psi \to \pi$. The function $\tau_e(\psi)$ does have a finite limit:

$$\lim_{\psi \to \pi} \tau_e(\psi) = \frac{\pi}{(1-e^2)^{3/2}}.$$

EXERCISE 2.29. Plot the functions $\tan\psi$ and $\tan^{-1}(x)$. How do these functions behave?

EXERCISE 2.30. Before attempting an integration of an unknown function, it is a good idea to have some understanding of the function being integrated. Plot the function $(1 + 0.5\cos\psi)^{-2}$ over the

---

[19]Albert Einstein's pursuit of a general theory of relativity was hindered by his inability to master the intricacies of non-Euclidean geometry. As he wrote to his friend Arnold Sommerfeld in 1912: "I am now working exclusively on the gravitation problem and believe that I can overcome all difficulties with the help of a mathematician friend of mine here [Marcel Grossmann]. But one thing is certain: never before in my life have I toiled any where near as much, and I have gained enormous respect for mathematics, whose more subtle parts I considered until now, in my ignorance, as pure luxury. Compared with this problem, the original theory of relativity is child's play."

range $0 < \psi < 2\pi$. Convince yourself that the function is symmetric about $\psi = \pi$. Does changing the factor 0.5 to 0.7 alter the symmetry?

EXERCISE 2.31.   The *Mathematica* function Integrate utilizes sophisticated mathematics that is very likely unfamiliar to the typical student, in particular using complex analysis to evaluate integrals.[20] As a result, all undefined coefficients presented to the Integrate function will be interpreted as complex numbers. We can provide information that will aid in the evaluation through the Assumptions option of the Integrate function. For our present circumstances, the parameter e is a real number and, for elliptical trajectories, lies in the range $0 \le e < 1$.

Compute the integral of the function $F = 1/(1 + e\cos\psi)^2$ by using the *Mathematica* Integrate function, providing the appropriate assumptions. Show that the *Mathematica* result can be put into the form shown in Equation 2.38. This will require the identity $i\tan x = \tanh ix$, where i is the square root of negative one.

To extend the range of the function τ to the interval $\pi < \psi \le 2\pi$, we can make use of the fact that the integrand is symmetric about $\psi = \pi$:

$$(1 + e\cos(\psi + \pi))^{-2} = (1 + e\cos(\pi - \psi))^{-2}.$$

To obtain angles in the range $\pi < \psi \le 2\pi$, we can demonstrate that the following relation holds:

(2.39) $\tau_e(\psi + \pi) = 2\tau_e(\pi) - \tau_e(\pi - \psi).$

We illustrate the function τ_e in figure 2.10.

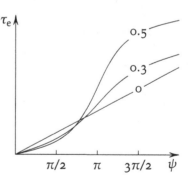

FIGURE 2.10. As the eccentricity e increases from 0 to 0.5, the variation of the time along the pathway τ_e changes significantly

For circular trajectories ($e = 0$), we see that the value of the function τ_e changes linearly as a function of ψ. As a result, the angle ψ will also change linearly as a function of time. Therefore, for circular trajectories,

---

[20]In a very real sense, the mathematics simplifies when we utilize complex numbers, even though the name includes the word complex. Nevertheless, we shall skirt such discussions.

the radius remains constant and so does the magnitude of the velocity. We could anticipate this result from the knowledge that $\mathcal{E} = T + \mathcal{U}$. The position-dependent potential energy does not change during a circular orbit, so the velocity-dependent kinetic energy must also be a constant. Consequently, the magnitude of velocity must remain constant.

> EXERCISE 2.32. If we try to reproduce Equations 2.38 and 2.39 in a *Mathematica* script, we run into a couple of numerical issues. First, for the point $\psi = \pi$, we will find that the function τ_e evaluates to *indeterminate* when the argument is π. This is due to the fact that $\tan \pi/2 = \infty$ and $\tan^{-1} \infty = (n + 1/2)\pi$, where n is any integer. *Mathematica* has no particular way to determine which of the possible roots you want and so reports the *indeterminate* result. There are several ways to avoid this problem but one simple way uses the *Mathematica* function If. Construct a user defined function that computes the value of Equation 2.38 for values where $0 < \psi < \pi$ and $\tau_e = \pi/(1 - e^2)^{3/2}$ when $\psi = \pi$.
>
> The second issue is that we want to extend the range of ψ to $\pi \leq \psi \leq 2\pi$. Construct another function that uses Equation 2.39 to expand the range to $0 \leq \psi \leq 2\pi$. Hint: the *Mathematica* Piecewise function might prove useful. Plot your results and compare to figure 2.10.

For elliptical trajectories, time does not vary in a simple way along the trajectory and there is no simple means of inverting Equation 2.38 to produce the angle ψ as a function of τ_e. As an alternative, we will approximate our desired function $\psi(t)$ by fitting a curve to the function $\tau_e(\psi)$ shown in figure 2.10 and will perform just that activity in the next exercise. This is something of a brute force approach that was not available to early researchers and lacks any particular mathematical elegance. Nonetheless, we should use technology when it is beneficial and, with access to fast computers, we'll find brute force methods to be practical for some problems.

> EXERCISE 2.33. A reasonably simple means for interpolation is to simply compute a large number of points and linearly interpolate between the points. The *Mathematica* function Interpolation uses a more sophisticated technique that employs cubic splines. This method potentially reduces the total number of points required to represent the function and provides continuous derivatives as well, although we won't use that feature at the moment.
>
> Create a list of points (τ_e, ψ) using the *Mathematica* function Table and then fit a spline curve to the points using the Interpolation function. Plot your results and verify that they look like figure 2.10 with the axes reversed.

In figure 2.11, we have plotted a series of points for the ellipse with eccentricity $e = 0.7$ that are equally spaced in time (τ_e). We see that the points near the focus (the origin in this plot) are widely spaced. This means that the planet travels a large distance between each point and therefore has a relatively large velocity. On the opposite end of the trajectory (near $\psi = \pi$) the points are closely spaced. The planet's velocity is correspondingly much smaller at those points.

FIGURE 2.11. Points on the ellipse for $e = 0.7$ are plotted for equal elapsed times. In this case, the period T has been broken into 35 equal intervals

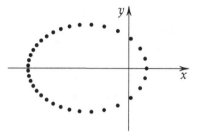

EXERCISE 2.34. It is possible to visualize the orbital motion by using the *Mathematica* function **Animate**. Animate the function $F = 1/(1 + e\cos\psi)$ for various values of the eccentricity. Use the results of the previous exercises to find ψ at constant time intervals. How does the trajectory change as a function of e?

While it is very helpful to visualize the trajectories through animation, we also need to develop static displays of our results for occasions when we do not have access to computers. Figure 2.11 is one such representation in which the position in space is marked by points plotted at equal time intervals. Figure 2.10 is another such representation that portrays another aspect of the solution to the time on orbit problem. We'll find that producing multiple representations of some phenomenon will often bring clarity to our understanding of the mathematics.

To generate further representations of the motion, suppose that we choose a coordinate system like that in figure 2.7, in which motion takes place in the x-y plane. We have then that the position as a function of time is given by Equation 2.32 and the velocity is given by Equation 2.33.

EXERCISE 2.35. Compute and plot the positions $x(t)$ and $y(t)$ for one period of the orbit. Plot the velocities dx/dt and dy/dt and accelerations d^2x/dt^2 and d^2y/dt^2. How do these change as the eccentricity changes?

EXERCISE 2.36. Compute and plot the magnitude $r_{21} = \alpha/(1 + e\cos\psi)$ (scaled by a factor of $1/\alpha$) as a function of time for one period of the orbit. Show that the time derivatives of r_{21} have the following form:

$$\frac{dr_{21}}{dt} = -\frac{eJ\sin\psi}{\alpha} \quad \text{and} \quad \frac{d^2 r_{21}}{dt^2} = -\frac{eJ^2 \cos\psi(1 + e\cos\psi)^2}{\alpha^3}.$$

Plot the (scaled) results as a function of time. How do these change as the eccentricity changes?

2.7. Finite Sizes

We have now completed the initial description of the motion of planets on their orbits around the sun. We have demonstrated that Newton's universal law of gravitation gives rise to elliptical trajectories and have managed to determine that the orbits are periodic and that the position of the planet as a function of time can be determined. All of our work to this point has assumed that the finite sizes of the gravitationally-attracting bodies can be neglected. Unlike other stars that we observe in the night sky, our sun is not a point. Similarly, the size of the moon is not insignificant. So, let us return to the question of does this matter?

As a better approximation to finite-sized objects, let us consider the interaction of homogeneous, rigid spheres. The sun and planets are not particularly homogeneous, deviate in shape from spheres and are not rigid but this approximation illustrates the stepwise approach commonly used in physics to develop models. Rather than trying to solve a hugely complex problem such as the motion of all of the objects in the solar system from the beginning of time, we instead try to break it into a series of simpler, more tractable problems. Our first approximation neglected planet sizes. Now let's see the results of dealing with a somewhat more complex model.

For a rigid, finite-sized object, we can think of chopping it up into small (infinitesimal) pieces. We know how to calculate the force on two particles; that is defined by Equations 2.6. The total gravitational force exerted by one extended object on another would be the sum (integral) over all of the constituent pieces. By constraining the objects to be rigid, we can neglect self-interactions; we need only consider the interactions of one body on another.

Let us approach the problem in a systematic fashion that will prove quite useful in subsequent discussions but may seem somewhat contrived at this point. We shall define the gravitational field $\mathbf{G}_1(\mathbf{r})$ of a particle of mass M_1 located at the point $\mathbf{r}_1$ to be the following:

$$(2.40) \qquad \mathbf{G}_1(\mathbf{r}) = -GM_1 \frac{\mathbf{r} - \mathbf{r}_1}{|\mathbf{r} - \mathbf{r}_1|^3}.$$

The force $\mathbf{F}_{21}$ on a mass M_2 located at the position $\mathbf{r}_2$ is then given by the relation $\mathbf{F}_{21} = M_2 \mathbf{G}_1(\mathbf{r}_2)$. It is a simple exercise to show that this is really just a modest reorganization of our original problem.

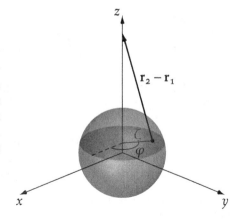

FIGURE 2.12. The gravitational field of a sphere can be constructed from the summation (integral) over a series of thin, circular plates. An infinitesimal point on the plate is illustrated

To generalize the definition of Equation 2.40 to a mass distribution, we can formally replace the summation over a series of individual particles with an integral over a distribution:

$$\mathbf{G}(\mathbf{r}) = -G \sum_i M_i \frac{\mathbf{r} - \mathbf{r}_i}{|\mathbf{r} - \mathbf{r}_i|^3} \rightarrow -G \int dM \frac{\mathbf{r} - \mathbf{r}_1}{|\mathbf{r} - \mathbf{r}_1|^3},$$

where the integral (which generally will be multidimensional) extends over the mass distribution.

Let us now consider the gravitational field of a uniform sphere. To make the problem somewhat more tractable, let us consider forming the sphere from a stack of very thin disks. We shall be astute in our choice of coordinate systems and place the center of the sphere at the origin, and align the point $\mathbf{r}_2$ with the z-axis, as depicted in figure 2.12. Hence, $\mathbf{r}_2 = (0, 0, z_2)$.

An arbitrary point $\mathbf{r}_1$ on a plate located a distance z_1 above the x-y plane will have coordinates $\mathbf{r}_1 = (x_1, y_1, z_1)$. Using polar coordinates on the plate, we have that $x_1 = \zeta \cos \varphi$ and $y_1 = \zeta \sin \varphi$. As a result, we find $\mathbf{r}_2 - \mathbf{r}_1 = (-\zeta \cos \varphi, -\zeta \sin \varphi, z_2 - z_1)$ and $|\mathbf{r}_2 - \mathbf{r}_1| = [\zeta^2 + (z_2 - z_1)^2]^{1/2}$. For an homogeneous sphere, each infinitesimal point will have the same mass δm, so the differential mass element dM is given by $dM = \delta m \, d\zeta \, \zeta \, d\varphi$. The gravitational field due to a plate of radius a is determined by the following expression:

$$\mathbf{G}_{\text{plate}}(\mathbf{r}_2) = -G \, \delta m \int_0^a d\zeta \, \zeta \int_0^{2\pi} d\varphi \frac{(-\zeta \cos \varphi, -\zeta \sin \varphi, z_2 - z_1)}{[\zeta^2 + (z_2 - z_1)^2]^{3/2}}.$$

The integrals over φ can all be computed easily: the x- and y-component terms vanish and the z-component integral yields 2π. We find then that the gravitational field of the plate has only a z-component:

$$\mathbf{G}_{\text{plate}}(\mathbf{r}_2) = -2\pi G\,\delta m(z_2 - z_1)\hat{\mathbf{z}} \int_0^a d\zeta\, \frac{\zeta}{[\zeta^2 + (z_2 - z_1)^2]^{3/2}}.$$

Exercise 2.37. Show that $\int_0^{2\pi} d\varphi\, \sin\varphi = \int_0^{2\pi} d\varphi\, \cos\varphi = 0$. Plot the functions $\sin\varphi$ and $\cos\varphi$ over the range $0 \le \varphi \le 2\pi$. Can you justify the outcomes of the integrations by looking at the plots?

The radial integral can be performed also by noting that the following expression holds when c is a constant:

$$\frac{d}{dx}(x^2 + c^2)^{-1/2} = -\frac{x}{(x^2 + c^2)^{3/2}}.$$

The integrand is thus a perfect differential, so the gravitational field of the plate is then given by the following:

$$\mathbf{G}_{\text{plate}}(\mathbf{r}_2) = 2\pi G\,\delta m\,\hat{\mathbf{z}}\,\frac{z_2 - z_1}{\sqrt{\zeta^2 + (z_2 - z_1)^2}}\bigg|_{\zeta=0}^a$$

$$(2.41) \qquad = 2\pi G\,\delta m\,\hat{\mathbf{z}}\left[\frac{z_2 - z_1}{\sqrt{a^2 + (z_2 - z_1)^2}} - 1\right].$$

Exercise 2.38. The *Mathematica* function **Integrate** that we introduced earlier can be used to perform complex integrations. Integrate the function $F = x/(x^2 + c^2)^{3/2}$ over the interval $[0, a]$. Can you use the result to verify Equation 2.41?

The gravitational field of a sphere of radius R is then obtained by integrating over a series of plates:

$$\mathbf{G}_{\text{sphere}}(\mathbf{r}_2) = -G\,\delta m \int_{-R}^{R} dz_1 \int_0^{\sqrt{R^2 - z_1^2}} d\zeta\,\zeta \int_0^{2\pi} d\varphi\, \frac{(-\zeta\cos\varphi, -\zeta\sin\varphi, z_2 - z_1)}{[r^2 + (z_2 - z_1)^2]^{3/2}}$$

$$(2.42) \qquad = 2\pi G\,\delta m\,\hat{\mathbf{z}} \int_{-R}^{R} dz_1 \left[\frac{z_2 - z_1}{\sqrt{R^2 - z_1^2 + (z_2 - z_1)^2}} - 1\right],$$

where in this last step we utilized the result from Equation 2.41. As we can see from figure 2.13, the variable a that defines the plate radius depends upon z_1. We have that $R^2 = a^2 + z_1^2$. We can perform the z_1 integration if we look at each of the terms individually. First, let us define a new variable $u = \sqrt{R^2 + z_2^2 - 2z_2 z_1}$. Then, we find that the following is true:

$$\frac{du}{dz_1} = -\frac{z_2}{\sqrt{R^2 + z_2^2 - 2z_2 z_1}}.$$

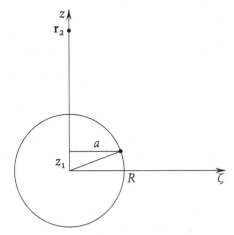

FIGURE 2.13. The gravitational field of a sphere of radius R is obtained from integrating over a series of plates. The radius a of a plate at a height z_1 is defined by the Pythagorean theorem: $R^2 = a^2 + z_1^2$

Hence, the first term in the integrand of Equation 2.42 is a perfect differential:

$$\int_{-R}^{R} dz_1 \frac{z_2}{\sqrt{R^2 + z_2^2 - 2z_2 z_1}} = -\int du$$

$$= -\sqrt{R^2 + z_2^2 - 2z_2 z_1}\,\Big|_{z_1 = -R}^{R}$$

$$= (z_2 + R) - (z_2 - R) = 2R.$$

From the definition of u, the second term in the integrand of Equation 2.42 can be written as follows:

$$\int_{-R}^{R} dz_1 \frac{z_1}{\sqrt{R^2 + z_2^2 - 2z_2 z_1}} = \int du \frac{u^2 - R^2 - z_2^2}{z_2^2} = \frac{1}{z_2^2}\left[\frac{u^3}{3} - (R^2 + z_2^2)u\right]$$

$$= \frac{1}{z_2^2}\left[\frac{[R^2 + z_2^2 - 2z_2 z_1]^{3/2}}{3}\right.$$

$$\left. + (R^2 + z_2^2)\sqrt{R^2 + z_2^2 - 2z_2 z_1}\right]_{z_1 = -R}^{R}$$

$$= \frac{2R^3}{3z_2^2}.$$

The last term in the integral of Equation 2.42 is simple to evaluate and contributes $-2R$ to the total. Summing all of the contributions, we find that the gravitational field of a rigid sphere is given by the following:

$$(2.43) \qquad \mathbf{G}_{\text{sphere}}(\mathbf{r}) = -G\,\delta m\,\hat{\mathbf{z}}\frac{4\pi R^3}{3z_2^2}.$$

We note first that the quantity $4\pi R^3/3$ is the volume V of the sphere and that the mass M_1 of the sphere must be $M_1 = \delta m V$. We also note that, for the coordinates chosen, $\hat{\mathbf{z}} = (\mathbf{r}_2 - \mathbf{r}_1)/|\mathbf{r}_2 - \mathbf{r}_1|$ and that $z_2 = |\mathbf{r}_2 - \mathbf{r}_1|$. Consequently, we can rewrite Equation 2.43 as follows:

$$\mathbf{G}_{\text{sphere}}(\mathbf{r}) = -GM_1\frac{\mathbf{r}_2 - \mathbf{r}_1}{|\mathbf{r}_2 - \mathbf{r}_1|^3}.$$

Surprisingly, this is identical to the gravitational field of a point mass! This result can actually be extended to any spherically-symmetric distribution not just uniform distributions. Thus, we would have to consider, at the least, non-symmetrical mass distributions for the finite size of the masses to affect the resulting trajectory. The theory that we developed using point particles is quite successful despite its simplicity.

EXERCISE 2.39.   Repeat the analysis leading to Equation 2.43 and compute the force on a rigid spherical mass M_2 due to a point mass located at the origin. Show that the force is identical to that of two point masses. Assume that the center of the sphere is located a distance b from the origin, where b is larger than the sphere radius R.

2.8. Unfinished Business

Newton's development of the mathematical tools to describe planetary motion (the calculus) represents a major advance in our ability to provide descriptive, predictive models of the behavior of a complex system. In principle, extension of the results we've obtained thus far to study of the complete solar system should be straightforward. Each pair of bodies interacts via the law of universal gravitation as we have described, so prediction of the trajectories of the bodies amounts to calculating the sum of the forces on each of the planets due to the sun and the other planets:

$$(2.44) \qquad M_i\frac{d^2\mathbf{r}_i}{dt^2} = -GM_i\sum_{j\neq i}\frac{M_j}{|\mathbf{r}_j - \mathbf{r}_i|^3}(\mathbf{r}_j - \mathbf{r}_i).$$

While the statement of the problem may be straightforward, solution of the system of equations represented by Equation 2.44 is extraordinarily challenging. Much to the dismay of Newton and subsequent scientists, there are no closed-form solutions to the general n-body problem except when the number of bodies is two. For three or more masses, the problem must be treated numerically. We shall return to this topic in somewhat more detail in a later chapter.

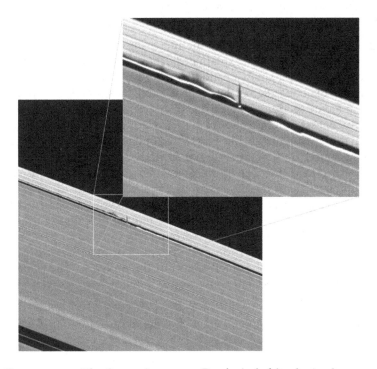

FIGURE 2.14. The Saturnian moon Daphnis (*white dot* in the center of the *inset*) orbits within and is responsible for the Keeler Gap in Saturn's A ring. In this image, taken by the Cassini spacecraft narrow-angle camera from 48° above the ring plane, Daphnis casts a shadow (*dark vertical line*) on adjacent ring material. Ripples caused by the recent passage of Daphnis also cast shadows, revealing the three-dimensional structure of the features (Image courtesy of NASA)

To illustrate the potential complexities, let us first note that the rings of Saturn are a remarkable astronomical phenomenon. The rings were initially observed by the Italian astronomer Galileo Galilei in 1610, who at first thought the rings were large moons of Saturn due to the low resolution of his telescope. Galileo was perplexed when, observing Saturn several years later, the moons seemed to have disappeared. Subsequent observations by the Dutch astronomer Christiaan Huygens about fifty years later led him to propose that the observations of Saturn could be explained if Saturn were encircled by a large, flat disk that was inclined to the plane of the ecliptic.[21] Subsequent observations with increasingly high resolution instruments have found increasing amounts of structure in the ring system, including the 2009 observation of the rings made by the NASA Cassini probe shown in figure 2.14.

The picture demonstrates vividly the increasing levels of sophistication we must achieve if we are to explain everything visible in the picture. The image contains just a small portion of the ring system that spans a radial distance of about 282,000 km yet has a thickness that is generally less than 1 km (some sections like the one displayed in the figure are roughly 30 m in thickness). There are spiral density variations in the ring thickness visible in the Cassini image, like tracks in a phonograph record, We have a clue to the source of these fluctuations from the three-dimensional structures observed near the edge of the ring where the moon Daphnis orbits. Clearly, the passage of the moon through the rings affects the ring material. Yet, the rings are in some sense stable: they have been observed since the 1600s. We know that Newton's laws of motion explain the trajectory of Saturn as it glides through space around the sun and must somehow also explain the behavior of the ring system. For now, we will content ourselves with proving one small aspect of the problem. The remainder will have to await subsequent courses and, in truth, some aspects remain active areas of research.

EXERCISE 2.40.   Use Kepler's third law (Equation 2.36) to prove that the rings cannot be solid disks.

---

[21]Huygens published his *Systema Saturnium* in 1659, in which he explained that the curious observed behavior of the objects first seen by Galileo was consistent with a flat disk not separate moons.

III

On the Nature of Matter

In the previous chapter, we investigated the nature of the gravitational force that confines the planets to their orbits around the sun. We found that the elliptical trajectories of the planets first identified by Kepler could be explained with the universal law of gravitation developed by Newton. Specific trajectories are defined by the angular momentum J and the masses of the gravitating bodies, along with a constant of proportionality G that we take to be universal. Mass was identified as a fundamental property of matter: any two objects that have mass will interact through the gravitational force.

Another fundamental property of matter is the electric charge. Charged matter interacts through the electromagnetic force, which has been the subject of intensive investigation for many years. While the motions of stars in the night sky are indeed fascinating, lightning strikes have captivated humans since the dawn of time. These unpredictable, explosive rearrangements of electric charge between the clouds and ground can have disastrous consequences for anyone in the vicinity, much more so than any distant realignment of bright dots in the sky. As a matter of self-preservation, if not innate scientific curiosity, developing some sort of understanding of the vast forces at work in our atmosphere has been a longstanding task.

A significant, pivotal advance in our understanding of the nature of the electromagnetic force was provided by the French natural philosopher Charles Augustin de Coulomb. Coulomb was trained as a mathematician and spent much of his career as an officer in the *Corps de Genie* working as an engineer designing and building fortifications. In this rôle, he was not content to simply replicate the designs of his predecessors but based his plans on scientific theories of soil mechanics and frictional forces that he developed. Upon Coulomb's return to France in 1781 from

© Mark A. Cunningham 2015
M.A. Cunningham, *Neoclassical Physics*, Undergraduate Lecture
Notes in Physics, DOI 10.1007/978-3-319-10647-2_3

his foreign posting, he became interested in somewhat less practical and more esoteric pursuits about the natural world, especially those relating to electric and magnetic phenomena.[1]

The key to his success in these studies was Coulomb's invention of a torsion balance that was capable of making precise measurements of small forces. He used this balance to demonstrate that the force between two electrically charged objects could be written in the following form:

$$(3.1) \qquad \mathbf{F} = \kappa Q_1 Q_2 \frac{\mathbf{r}_2 - \mathbf{r}_1}{|\mathbf{r}_2 - \mathbf{r}_1|^3},$$

where κ is a constant of proportionality and Q_1 and Q_2 represent the charges of the two objects. Except for a modest shift in notation, we can recognize that Coulomb's law relating the force acting on two charged objects has exactly the same form as that of Newton's law of gravitation! The electromagnetic force between two charges is a central force that depends on the inverse square of the distance between the charges.

This is quite a fortunate development, as it means that all of the work from the previous chapter can be applied to the problem of the motion of two charges, albeit with one caveat. Unlike the gravitational problem where all masses have a positive sign and the overall force is attractive, electric charges can be either positive or negative. This means that the electromagnetic force can be either attractive or repulsive. This behavior is represented mathematically by defining the force in Equation 3.1 to have an overall positive sign. Two charges with opposite signs will result in a negative (attractive) force and two charges with the same sign will result in a positive (repulsive) force.

3.1. Hyperbolic Trajectories

If we return to Equations 2.6, the starting point for our analysis of gravitational motion, we see that we can write the following set of equations for charged particles:

$$(3.2) \qquad \begin{aligned} \frac{d[\mathbf{r}_2(t) - \mathbf{r}_1(t)]}{dt} &= \mathbf{v}_2(t) - \mathbf{v}_1(t) \\ \frac{d[\mathbf{v}_2(t) - \mathbf{v}_1(t)]}{dt} &= \kappa \frac{Q_1 Q_2 (M_1 + M_2)}{M_1 M_2} \frac{\mathbf{r}_2(t) - \mathbf{r}_1(t)}{|\mathbf{r}_2(t) - \mathbf{r}_1(t)|^3}. \end{aligned}$$

---

[1]Coulomb submitted his *Premier Mémoire sur l'Electricité et le Magnétisme* to the French Academy of Sciences in 1785, followed by six more memoirs over the next three years that established a number of basic facts about the nature of the electromagnetic force.

We see explicitly that the overall sign of the second term will depend upon the relative signs of the charges Q_1 and Q_2. Now it may occur to the alert student that, because particles have mass, we should include the gravitational interaction in our description of the motion. As a practical matter, the coefficient of proportionality κ for the electromagnetic force is many orders of magnitude larger than the universal gravitational constant G. Consequently, we can neglect the gravitational interaction for most applications involving charged particles.

EXERCISE 3.1. Show that Equation 3.1 leads to Equations 3.2.

Following the pathway we outlined in the previous chapter, we can show from Equations 3.2 that the angular momentum vector $\mathbf{J}$ is conserved. Consequently, motion is again restricted to the plane perpendicular to $\mathbf{J}$. Additionally, when the magnitude J vanishes, the motion of the charged particles lies along a straight line: Equation 2.9 is equally valid for charged particles. So, if we multiply Equation 2.9 by $\kappa Q_1 Q_2 (M_1 + M_2)/M_1 M_2$, we obtain the following result:

$$\kappa \frac{Q_1 Q_2 (M_1 + M_2)}{M_1 M_2} \frac{d}{dt} \frac{\mathbf{r}_2(t) - \mathbf{r}_1(t)}{|\mathbf{r}_2(t) - \mathbf{r}_1(t)|} = \mathbf{J} \times \left[\kappa \frac{Q_1 Q_2 (M_1 + M_2)}{M_1 M_2} \frac{\mathbf{r}_2(t) - \mathbf{r}_1(t)}{|\mathbf{r}_2(t) - \mathbf{r}_1(t)|^3} \right]$$

$$= \mathbf{J} \times \frac{d[\mathbf{v}_2(t) - \mathbf{v}_1(t)]}{dt}$$

$$(3.3) \qquad\qquad = \frac{d}{dt} \{ \mathbf{J} \times [\mathbf{v}_2(t) - \mathbf{v}_1(t)] \}.$$

Note that there is an overall sign change from Equation 2.10 that is now captured in the relative signs of the charges. Again, we can integrate both sides and recover the result analogous to Equation 2.11:

$$(3.4) \qquad \kappa \frac{Q_1 Q_2 (M_1 + M_2)}{M_1 M_2} \left[\mathbf{e} + \frac{\mathbf{r}_2(t) - \mathbf{r}_1(t)}{|\mathbf{r}_2(t) - \mathbf{r}_1(t)|} \right] = \mathbf{J} \times [\mathbf{v}_2(t) - \mathbf{v}_1(t)].$$

EXERCISE 3.2. Show that, as before, the dot product of $\mathbf{J}$ with Equation 3.4 vanishes and that the vector $\mathbf{e}$ must again lie in the plane of motion.

If we now take the dot product of Equation 3.4 with the vector $\mathbf{r}_2(t) - \mathbf{r}_1(t)$, we obtain the following result:

$$(3.5) \qquad [\mathbf{r}_2(t) - \mathbf{r}_1(t)] \cdot \mathbf{e} + |\mathbf{r}_2(t) - \mathbf{r}_1(t)| = -\frac{J^2 M_1 M_2}{\kappa Q_1 Q_2 (M_1 + M_2)}.$$

Except for the additional sign on the right-hand side, this result has the same form as we obtained in Equation 2.12. Consequently, for charges of opposite sign, we have precisely the same results as we obtained for gravitational motion. Depending upon the value of e, the trajectories are ellipses, parabolas or hyperbolas. So, other than a different overall scale factor, there is nothing different about the behavior of charged particles with opposite signs.

Let us take some time to examine the hyperbolic trajectories that we omitted from our previous discussion on the motion of planets. There are relatively few astronomical objects of interest with hyperbolic trajectories in the solar system but we will soon find applications in the realm of charged particles. Using our compact notation and working in polar coordinates, we can rewrite Equation 3.5 as follows:

$$(3.6) \qquad r_{21}(1 + e\cos\psi) = -\frac{J^2 M_1 M_2}{\kappa Q_1 Q_2 (M_1 + M_2)}.$$

As before, the term r_{21} represents the distance between the two charges and, as such, is a positive number. For oppositely charged particles, the right-hand side of Equation 3.6 is also a positive number. In figure 3.1, we plot the value of the function $1 + e\cos\phi$ for two values of e. We can see that for elliptical trajectories, where $e < 1$ the value of the function is always positive.

FIGURE 3.1. The function $1 + e\cos\psi$ takes on negative values when e is larger than one

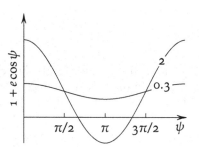

For hyperbolic trajectories, where $e > 1$, the function is negative in the region around $\psi = \pi$ and vanishes at two points where $1 + e\cos\psi = 0$. This means that, upon dividing both sides of Equation 3.6 by $1 + e\cos\psi$, the distance r_{21} would have to be negative in the vicinity of $\psi = \pi$. This is clearly impossible: r_{21} is the polar distance from the origin and has a range $0 \leq r_{21} \leq \infty$. Consequently, the region where $1 + e\cos\psi$ is negative cannot represent a real solution to the equation describing the motion of two charges of opposite sign. We term this an *unphysical solution* to the problem. Students who actually solved Exercise 2.12 undoubtedly found that the *Mathematica* program produced two branches of the hyperbola, as illustrated in figure 3.2, which might have been somewhat confusing at the time. We only plotted the physical branch in figure 2.5 and omitted the unphysical one.[2]

We will encounter unphysical solutions from time to time during the semester. They arise for various reasons having to do with the mathematical representations that we utilize. In the previous chapter, we noted that an object uniformly accelerated from rest will travel a distance d in a time t given by $d = at^2/2$, whereupon $t = \pm(2d/a)^{1/2}$. We generally discount the negative time solution as being unphysical: the object cannot have travelled any distance prior to being released. Deciding which solutions are real and which are unphysical requires physical insight. Again, one cannot simply perform algebra (or calculus) and be oblivious to what the mathematics is intended to represent.

If we now consider the case of two charges of the same sign, we observe that the sign of the right-hand side of Equation 3.6 is negative. The distance r_{21} must be a positive number, so this has two main consequences: First, elliptical and parabolic solutions are not possible: the value of $1 + e\cos\psi$ is always positive for elliptical trajectories and vanishes only at $\psi = \pi$ for parabolic trajectories but is never negative. Second, the only possible solutions are for hyperbolic trajectories in the region where $1 + e\cos\psi$ is negative. Thus, solutions of the equation for a repulsive force occupy what was the forbidden region of ψ for an attractive force.

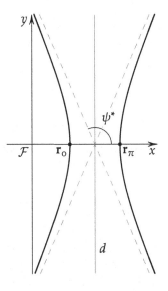

FIGURE 3.2. The branches of hyperbolic trajectories ($e = 2.5$) represent the physical solutions for attractive (*left branch*) and repulsive (*right branch*) forces. The *dashed lines* are the asymptotes for the hyperbolas. The angle ψ^* is defined by $1 + e\cos\psi^* = 0$. The point $\mathbf{r}_0$ represents the distance of closest approach ($\psi = 0$) for an attractive force. The point $\mathbf{r}_\pi$ represents the distance of closest approach ($\psi = \pi$) for a repulsive force

[2]This does not constitute a lie, at least in this author's view. We have ultimately reconciled the matter and apologize for any unnecessary anxiety caused by deferring the discussion of a subtlety to a more appropriate time.

We can develop a deeper understanding of the behavior of the system if we compute the energy of the system. We can follow the logic that led to Equation 2.16 for the gravitational force. We start by taking the dot product of the velocity vector $\mathbf{v}_2(t) - \mathbf{v}_1(t)$ with the second of Equations 3.2:

$$[\mathbf{v}_2(t) - \mathbf{v}_1(t)] \cdot \frac{d[\mathbf{v}_2(t) - \mathbf{v}_1(t)]}{dt} =$$

$$[\mathbf{v}_2(t) - \mathbf{v}_1(t)] \cdot \kappa \frac{Q_1 Q_2 (M_1 + M_2)}{M_1 M_2} \frac{\mathbf{r}_2(t) - \mathbf{r}_1(t)}{|\mathbf{r}_2(t) - \mathbf{r}_1(t)|^3}.$$

If we multiply both sides by $M_1 M_2/(M_1 + M_2)$ and perform some additional algebraic steps, we eventually obtain the following result:

$$(3.7) \qquad \mathcal{E} = \frac{M_1 M_2}{2(M_1 + M_2)} |\mathbf{v}_2(t) - \mathbf{v}_1(t)|^2 + \kappa \frac{Q_1 Q_2}{|\mathbf{r}_2(t) - \mathbf{r}_1(t)|}.$$

Again, the energy has two components: a kinetic component that depends solely on the velocity and a potential component that depends solely on the distance. We can see that the total energy $\mathcal{E}$ (which determines the eccentricity e) can only be positive when both charges have the same sign. Hence, the only possible solutions for a repulsive force are straight line motion when $J = 0$ and hyperbolic motion when $J \neq 0$. When the charges have opposite signs, the potential energy becomes negative, as it was for the gravitational force. In that case, motion can include elliptical and parabolic motion in addition to hyperbolic and straight-line motion.

EXERCISE 3.3.   Fill in the missing steps in the derivation of Equation 3.7.

EXERCISE 3.4.   Show that we can also write Equation 3.6 as follows:

$$(3.8) \qquad r_{21} = \frac{\kappa Q_1 Q_2 (1 - e^2)}{2\mathcal{E}(1 + e \cos \psi)}.$$

Hint: Begin by squaring Equation 3.4 to find the value for $e^2 - 1$.

Show now that $\mathbf{r}_0 = \kappa Q_1 Q_2 (e - 1)/2\mathcal{E}$ and $\mathbf{r}_\pi = \kappa Q_1 Q_2 (e + 1)/2\mathcal{E}$. Use this to show that the semimajor axis of the ellipse is $e\kappa Q_1 Q_2/2\mathcal{E}$.

3.2. Asymptotic Behavior

Hyperbolic motion is quite different from elliptical motion. It is not periodic: the particle is never at the same position twice. In fact, for the vast majority of time, the motion is approximately the same as straight-line motion. To see how this vastly different behavior arises, let us first note that when the function $1 + e \cos \psi$ vanishes, the distance r_{21} defined by Equation 3.5 becomes infinite. We have already mentioned before that we need to be cautious when interpreting equations that are divergent. For example, when the distance between two particles vanishes, the particles will have impacted, the potential energy becomes infinite and we really cannot say anything about what happens to the system after that time. For macroscopic objects like planets, impacts are going to be complicated events, with complicated consequences like massive firestorms, disruption of global weather patterns and extinction of species. Singular points in the equations mark limits to our ability to predict subsequent (or prior) behavior.

> EXERCISE 3.5.   Plot the function $F = 1/(1 + e \cos \psi)$, where $e > 1$, over the range $-\psi^* \leq \psi \leq +\psi^*$, where $\psi^* = \cos^{-1}(-1/e)$. Show that the point $\psi = 0$ corresponds to the distance of closest approach, that is, $r_{21}(\psi = 0)$ is a minimum. Is the function symmetric around the point $\psi = 0$? Can we infer from this plot that the particles were at the same points in space on opposite sides of $\psi = 0$?

The singularity in Equation 3.5 does not give rise to the sort of apocalyptic events characterized by impacts. Particles somewhere on the physical branch of an hyperbolic trajectory will never actually reach infinity in finite time. To verify that this is the case and to examine the so-called *asymptotic behavior* of the particles, let us return to the problem of position on the trajectory as a function of time. When the eccentricity e is larger than one, we can define a function τ_h analogous to the function τ_e we defined in Equation 2.38:

$$(3.9) \qquad \tau_h(\psi) \equiv \int d\psi (1 + e \cos \psi)^{-2} = \frac{\kappa^2 Q_1^2 Q_2^2 (M_1 + M_2)^2}{J^3 M_1^2 M_2^2} (t - t_1),$$

If we go back to Equation 2.37, the function τ_h looks identical to the function τ_e we defined previously. What we will find, though, is a different value for the integral, stemming from the fact that the eccentricity e is larger than one. Even though the integrand looks the same in the two equations, because here $e > 1$ the integrand $(1 + e \cos \psi)^{-2}$ now diverges at the point $\psi^* = \cos^{-1}(-1/e)$.

In evaluating the function τ_h we will again need to be cautious. For an attractive force, where we are interested in the behavior of the trajectory that includes the portion where $\psi = 0$, this function is only valid in the range $-\psi^* < \psi < \psi^*$. In this case, we find that the function τ_h can be written as follows:

(3.10)

$$\tau_h(\psi) = \frac{2}{(e^2 - 1)^{3/2}} \tanh^{-1} \left[\left(\frac{e-1}{e+1} \right)^{1/2} \tan(\psi/2) \right] + \frac{e \sin \psi}{(e^2 - 1)(1 + e \cos \psi)}.$$

As was the case for elliptical functions, we have obtained a function τ_h that is a complicated function of the angle ψ. As was also the case previously, we will not be able to invert the function τ_h in terms of simple functions to recover the angle ψ as a function of time that we desire. Nevertheless, we can resort to our previous strategy and use interpolation to approximate the desired function $\psi(t)$.

> EXERCISE 3.6. Use the *Mathematica* function **Integrate** to perform the integral defined in Equation 3.9. You will have to provide the assumptions that the eccentricity e is larger than one and real. See if you can obtain the result specified in Equation 3.10

> EXERCISE 3.7. Plot the function $\tau_h(\psi)$ over the range $-\psi^* \le \psi \le +\psi^*$, where $\psi^* = \cos^{-1}(-1/e)$. The point $\psi = 0$ corresponds to the distance of closest approach. What is the value of τ_h at that point?

Analysis of hyperbolic trajectories for a repulsive force is somewhat more challenging. These trajectories contain the point $\psi = \pi$ that is the point of closest approach and are restricted to the range $\psi^* \le \psi \le -\psi^*$, in a counterclockwise sense. Here we run into two problems. First, the tangent function is divergent at the point $\psi = \pi$. We have already encountered this problem and recognize that a solution for points where $\psi > \pi$ can be constructed using the symmetry of the function: $\tau_h(\psi - \pi) = \tau_h(\pi - \psi)$. Second, the argument of the inverse hyperbolic tangent in Equation 3.10 moves outside the range of validity: $\tanh^{-1}(x)$ is only defined in the range $-1 \le x \le 1$. For repulsive forces, we need to use the following expression:

$$(3.11)$$
$$\tau_h(\psi) = \frac{2}{(e^2 - 1)^{3/2}} \left\{ -\pi + \tan^{-1}\left[\left(\frac{e-1}{e+1}\right)^{1/2} \tan(\psi/2) \right] \right\} + \frac{e\sin\psi}{(e^2-1)(1+e\cos\psi)}.$$

Here we see that inverse hyperbolic tangent has been replaced by the inverse tangent.[3] We also (in the first term in the curly brackets) subtracted the value of the limit of the function at $\psi = \pi$. This ensures that the value of $\tau_h(\psi)$ vanishes at $\psi = \pi$, which also represents the distance of closest approach for repulsive forces. This choice is somewhat arbitrary but is convenient. Alternative definitions of when to start the clock ticking can always be made by adjusting the value t_1 from Equation 3.9.

We can now proceed to plotting the values of the distance r_{21} and its derivatives as a function of time. As before, we can calculate the value

---

[3]This discussion would be somewhat less opaque if we plunged into the domain of complex numbers. If we permit the angle ψ to be complex, then τ_h can be shown to be a continuous function of the complex ψ. We shall, nonetheless, defer that discussion to subsequent courses.

of $\tau_h(\psi)$ for a number of values of ψ and then use those values to interpolate to find the desired function $\psi(t)$, which is now a function of time. In figure 3.3, we plot the values of r_{21} and its derivatives for an hyperbolic trajectory in which $e = 2$, for both attractive and repulsive forces. We have focussed on the late-time behavior of the system, so the behavior near the origin is somewhat obscured.

What we observe is that the velocities for both attractive and repulsive trajectories limit to the same constant value. This is evident also from the fact that distance *versus* time plots have the same slopes at late times. We observe also that the accelerations become vanishingly small after relatively short times. This occurs when the particles have moved a radial distance that is a few times the distance of closest approach.

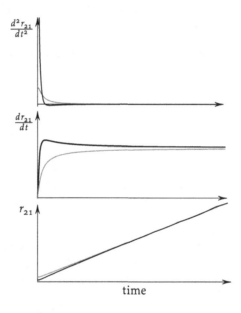

FIGURE 3.3. The asymptotic behavior of both attractive (*black*) and repulsive (*gray*) hyperbolic trajectories are very similar. The origin of the time axis is defined as the time of closest approach

As a result, we can approximate the hyperbolic trajectories at late times by simple straight-line motion in which particle 2 moves off at an angle $\pm\psi^*$ with a constant velocity. This is not strictly true mathematically but real measurements always have finite precision. At some point the difference between the actual and theoretical positions of the particles will drop below the resolution of our instruments. Beyond that point, the motion is effectively linear. As a consequence, the particles can never reach infinite separation in finite time. The distance is simply proportional to time: $r_{21} = vt$, where v is the asymptotic (late time) velocity. To reach infinite separation will require infinite time.

EXERCISE 3.8. Use the *Mathematica* function `Table` to construct a list of (τ_h, ψ) values for the attractive force defined in Equation 3.10. You will want to compute the function over the range $-\psi^* < \psi < \psi^*$, specifically excluding the points $\psi = \pm\psi^*$. Break the range into at least 200 steps in ψ. (τ_h is very non-linear. We need very small increments $\Delta\psi$ to obtain reasonable interpolations as $\psi \to \psi^*$.) Use the `Interpolation` function to obtain the approximation of $\psi(t)$.

Plot r_{21} and its derivatives as functions of time. Can you reproduce the results of figure 3.3? What happens at negative times, i.e., are the curves symmetric around $t = 0$?

The value of dr_{21}/dt is zero at $t = 0$. Does this mean that the velocity is zero?

EXERCISE 3.9. Repeat the previous exercise for the repulsive force defined in Equation 3.11.

EXERCISE 3.10. What is the value of the asymptotic velocity? Calculate the limit of dr_{21}/dt as $\psi \to \psi^*$.

3.3. Rutherford's Experiments

At this point in their academic careers, most students have probably not heard of the New Zealand physicist Ernest Rutherford but his rôle in establishing our modern view of the nature of matter is as significant as Newton's in establishing the universal law of gravitation. In 1895, Rutherford left New Zealand to study physics with J. J. Thomson at Trinity College, Cambridge. Initially, he studied the properties of iron exposed to high-frequency electromagnetic radiation but later focussed his research on on the newly-discovered radioactive properties of uranium. By 1898, Rutherford was able to demonstrate that two types of radiation that he called α- and β-rays emanated from uranium salts.[4] In 1898, Rutherford moved to McGill University in Montreal, where he was able, over the next several years, to demonstrate that α particles were essentially helium atoms and that the uranium atoms were transformed into *different* elements, in a process called **transmutation**.[5] Rutherford was awarded

---

[4]The first two letters in the Greek alphabet are α and β. Uncharged radiation that was subsequently discovered emanating from these radioactive elements was called, not surprisingly, γ-rays.

[5]Rutherford's journey to the scientific backwaters of the colonies was prompted by his desire to marry the fiancé he had left behind in New Zealand. Trinity College denied him a promotion to Fellow and the position at McGill provided a significant boost in salary. Unfortunately, McGill University rules prohibited Rutherford from taking a leave of absence until he had completed a year's service. Rutherford married Mary Newton in 1900.

the Nobel Prize in Chemistry in 1908 "for his investigations into the disintegration of the elements, and the chemistry of radioactive substances." Rutherford was appreciative but nonplussed by the award, as he did not consider himself to be a chemist. (Gabriel Lippmann won the Nobel Prize in Physics in 1908 "for his method of reproducing colours photographically based on the phenomenon of interference.")

In 1907, Rutherford was appointed to a professorship in Manchester University, permitting him to return to a major research institution in what he considered to be the epicenter of scientific thought: Britain. At Manchester, Rutherford and his students conducted a series of experiments that form the backbone of our present discussion. In some sense, these Manchester investigations are more scientifically important than the McGill work Rutherford performed earlier but Rutherford never received a Nobel Prize in Physics for these efforts. Rutherford's objective in this enterprise was to examine the nature of the atom.

His mentor at Trinity College, J. J. Thomson, had earlier shown that β particles were, in fact, electrons and that electrons could be extracted from atomic matter by the application of suitably large electric fields. Thomson was able to measure the mass of the electron and showed that it was a very small fraction of the total mass of the atom. Thomson thereby postulated that the atom could be thought of as a blob of positively charged stuff that was responsible for the lion's share of the mass of the atom in which the much lighter, negatively charged electrons were embedded. This model has come to be called the plum pudding model.

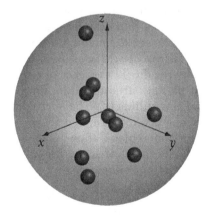

FIGURE 3.4. In the plum pudding model of the atom suggested by Thomson, negatively charged electrons (*dark spheres*) are distributed throughout a heavy, positively charged mass (*gray sphere*). The atom wouldn't necessarily be spherical but is depicted as such

In figure 3.4, we have represented the atom as a sphere and the embedded electrons as smaller spheres randomly distributed throughout the volume. This may not be a particularly accurate representation but it is simple to draw. Rutherford's objective in his Manchester experiments was to see if α particle scattering could shed any light on the distribution of electrons within the atom.

Rutherford and his students constructed a thin beam of α particles by placing a tube containing a few milligrams of radium (that decayed into radon gas) into a lead cask, as illustrated in figure 3.5. A small hole drilled into the side of the cask served to collimate the α particles: any α particles that struck the walls of the cask were absorbed; only those that passed through the hole were able to strike the target. Thin metal foils were used as targets. The α particles were detected by allowing them to impact a thin film of zinc sulfide that emitted a flash of light when struck. Rutherford's students sat in a darkened room and observed the flashes through a microscope that was able to rotate in the horizontal plane, allowing scattering to be measured as a function of the angle θ. The truly remarkable result of Rutherford's experiments was the observation of α particles that bounced from the target foils and were detected on the same side of the foil as the source! Rutherford had, perhaps, anticipated this result because the experiment was designed to permit the microscope to reach angles of $\theta > 90°$. In a lecture describing the experiment, however, he professed astonishment: "It was almost as incredible as if you fired a 15-inch naval shell at a piece of tissue paper and it came right back and hit you".

FIGURE 3.5. The source (S) of radon gas was contained in a small glass tube embedded in a block of lead. A small hole drilled in the block (C) allowed a thin beam of α particles to strike a metal foil (F). A microscope (M) was used to observe the fluorescence of an α particle striking a thin film of zinc sulfide (Z). The microscope could be rotated in an horizontal plane, allowing the measurement of scattering as a function of angle θ

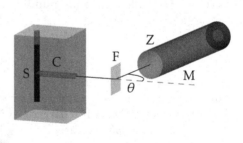

3.4. Gauss's Law

To understand why Rutherford was so astonished at the experimental results, it will first be useful to define the electric field of a charged particle. We define the electric field of a charge Q_1 located at the point $\mathbf{r}_1$ as follows:

(3.12)
$$E_1(r) = \kappa Q_1 \frac{r - r_1}{|r - r_1|^3}.$$

The force F_{21} on a charged particle located at a position r_2 is then given by the following:

(3.13)
$$F_{21} = Q_2 E_1(r_2).$$

These definitions parallel those that we provided previously for the gravitational field of a point mass. Although we haven't, as yet, provided a strong motivation for their introduction, the study of fields will prove to be quite important as we explore phenomena associated with the electromagnetic force. Fields are also quite useful in the description of fluid flows. A common example of a field in this context is the wind, which is a vector velocity field. At each point in space, the velocity of the air is characterized by a direction and a magnitude, which can be quantified abstractly by the mathematical notion of a field.

The German mathematician Carl Friedrich Gauss was interested in the general mathematical properties of fields. Gauss was an extraordinary individual and it can be stated that modern mathematics began with Gauss, whose passion for rigor and precision transformed the field.[6] It is difficult to assess Gauss's motivation: his writing style was polished but terse and unmotivated; Gauss argued that architects do not leave their scaffolding in place once a building is completed, so he felt no compunction to leave traces of his thought processes in his published works.

In the case of the mathematical behavior of fields, Gauss had determined that understanding singular points in the fields was key to understanding their behavior. Away from such singular points, the fields would be generally well-behaved. Mathematically, this amounts to the statement that derivatives of the fields could be constructed and that those derivatives were continuous. For a point charge, we have defined the electric field in Equation 3.12 and we can see that the field is infinite at the source point $r = r_1$. This point is a mathematical singularity in the field. The field and its derivatives are not well-behaved at the source point and, consequently, must be treated specially in any mathematical analysis.

Gauss defined the flux of a field through a surface as the integral of the field over an oriented surface as follows:

(3.14)
$$\Phi = \int dA \cdot E,$$

---

[6]Gauss (Gauß in German) was truly prolific in his work but published only a fraction of his results, in keeping with his personal motto *pauca sed matura* (few but ripe). After his death in 1855, Gauss's unpublished notes were made public and demonstrated, in many cases, that he had proven contemporary mathematical results decades earlier.

where we take the orientation of the surface as the unit vector that is perpendicular to the surface at each point. This is call the *normal* to the surface. Examples of normal vectors are indicated in figure 3.6.

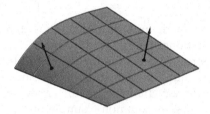

FIGURE 3.6. A larger surface can be broken into smaller pieces as indicated by the *grid lines*. Within each smaller segment, the normal is defined to be the direction perpendicular to the surface, as indicated by the *arrows*

The integral in Equation 3.14 can be interpreted as the limit of the dot product of the field **E** with the unit normals **n** of the surface, multiplied by the differential surface elements $d\mathbf{A} = \mathbf{n}\,dx\,dy$ and then summed over all of the surface elements that make up the complete surface. The dot product selects the component of the electric field that is parallel to the normal vector in each small segment. The field might also have a component that is perpendicular to the normal but that component would be tangential to the surface and cannot be thought of as flowing *through* the surface. Only the normal component penetrates the surface.

To see how to operationally deal with Equation 3.14, let us begin by computing the flux of a point charge located at the origin through a rectangular surface. For simplicity, let us choose a surface in the x-y plane located at some distance z_2 from the origin. A point $\mathbf{r}_2$ on the surface can be defined as $\mathbf{r}_2 = (x_2, y_2, z_2)$. We can allow x_2 to span the range $-a \le x_2 \le a$ and y_2 to span the range $-b \le y_2 \le b$. Now if z_2 is a positive number, we will take the normal **n** to be the unit normal in the z-direction: $\mathbf{n} = \hat{\mathbf{z}}$.

With the source located at the origin $\mathbf{r}_1 = (0, 0, 0)$, we have that $|\mathbf{r}_2 - \mathbf{r}_1| = (x_2^2 + y_2^2 + z_2^2)^{1/2}$. The electric field of a point charge Q_1 at the point $\mathbf{r}_2$ is thereby given by the following expression:

$$\mathbf{E}(\mathbf{r}_2) = \kappa Q_1 \frac{(x_2, y_2, z_2)}{(x_2^2 + y_2^2 + z_2^2)^{3/2}},$$

where we have used the definition of the field from Equation 3.12. Inserting this result into the definition of the flux, we can write the following:

$$\Phi = \kappa Q_1 \int_{-a}^{a} dx_2 \int_{-b}^{b} dy_2 \frac{(x_2, y_2, z_2) \cdot \hat{\mathbf{z}}}{(x_2^2 + y_2^2 + z_2^2)^{3/2}},$$

(3.15)
$$= \kappa Q_1 z_2 \int_{-a}^{a} dx_2 \int_{-b}^{b} dy_2 \frac{1}{(x_2^2 + y_2^2 + z_2^2)^{3/2}},$$

where the result of the dot product is simply the z-component of the vector (x_2, y_2, z_2) and, as z_2 does not depend upon x_2 or y_2, it can be factored out of the integral.

So, we once again find ourselves with a reasonably complicated integral to perform. As before, we can utilize the *Mathematica* program to perform the heavy lifting. We find the following result for the indefinite integral:

$$(3.16) \qquad \int dx_2 \int dy_2 \frac{1}{(x_2^2 + y_2^2 + z_2^2)^{3/2}} = \frac{1}{z_2} \tan^{-1}\left[\frac{x_2 y_2}{z_2 \sqrt{x_2^2 + y_2^2 + z_2^2}}\right].$$

The flux through the surface shown in figure 3.7 can be obtained by inserting the appropriate endpoints for the integrals. For a two-dimensional integral, where the result of the integration is given by some function $f(x, y)$ and the variables have domains $-a \le x \le a$ and $-b \le y \le b$, the value of

FIGURE 3.7. A rectangular surface at the height z is indicated in *gray*. A point $\mathbf{r}_2$ on that surface is defined by the vector $\mathbf{r}_2 - \mathbf{r}_1$. The unit normal $\mathbf{n}$ is the same for all points on the surface

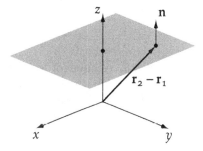

the definite integral is given by $f(a, b) - f(-a, b) - f(a, -b) + f(-a, -b)$. If we note that $\tan^{-1}(-x) = -\tan^{-1}(x)$, then the flux through our surface is given by the following expression:

$$(3.17) \qquad \Phi = 4\kappa Q_1 \tan^{-1}\left[\frac{ab}{z\sqrt{a^2 + b^2 + z^2}}\right].$$

EXERCISE 3.11. Verify the result obtained for the double integral in Equation 3.16 by using the *Mathematica* function Integrate, which can perform multiple integrals. Let $a = 1$ and $b = 1$ and plot the resulting function from $0 \le z \le 5$.

EXERCISE 3.12. Plot the inverse tangent function $\tan^{-1}(x)$ over the domain $-1 \le x \le 1$. What is the relationship between $\tan^{-1}(x)$ and $\tan^{-1}(-x)$?

Now consider what happens if we allow the surface to be at negative z values. The z-component of the vector $\mathbf{r}_2 - \mathbf{r}_1 = (x_2, y_2, -z_2)$ changes sign. If we define the normal $\mathbf{n}$ to be the *outwardly directed* unit vector perpendicular to the surface, then for negative z values, we have $\mathbf{n} = -\hat{z}$ and the picture is essentially identical to that shown in figure 3.7 (with the axis pointing in the $-\hat{z}$ direction). Consequently, the signs in Equation 3.15 cancel and the flux through a surface at $-z$ is identical to the flux through the surface at z.

So, we are now in a position to rediscover Gauss's significant result. For simplicity, consider a cube of side length $2a$ centered on the origin, where a point charge Q_1 is located. The flux through the top z side can be obtained by using $b = a$ in Equation 3.17 and the fact that $\tan^{-1}(1/\sqrt{3}) = \pi/6$. We find that the flux through this surface is just $\Phi = 2\pi\kappa Q_1/3$. Because all of the surfaces of a cube are equivalent, the flux through any of the surfaces will be the same. Consequently, the total flux through the whole cube is six times larger and is given by the following:

$$(3.18) \qquad\qquad \Phi_{\text{total}} = 4\pi\kappa Q_1.$$

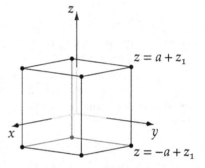

FIGURE 3.8. A cube of side $2a$ is shifted upwards by an amount z_1 from the origin. The two surfaces of constant z of the cube are labelled

Note that this result does not depend upon the size of the cube a. Consequently, the flux through any cube will be that found in Equation 3.18.

EXERCISE 3.13. It is not necessary to restrict our results to cubical surfaces. Suppose that you have a rectangular prism of sides a, b and c. Use the following identity:

$$\frac{\pi}{2} = \tan^{-1}\left[\frac{ab}{c(a^2 + b^2 + c^2)}\right] + \tan^{-1}\left[\frac{bc}{a(b^2 + c^2 + a^2)}\right]$$
$$+ \tan^{-1}\left[\frac{ca}{b(c^2 + a^2 + b^2)}\right]$$

to prove that the flux through the prism is also $\Phi_{\text{total}} = 4\pi\kappa Q_1$.

EXERCISE 3.14. Show that the flux through a surface in the y-z plane is given by the following:

$$\Phi = \kappa Q_1 \tan^{-1}\left[\frac{y_2 z_2}{x_2\sqrt{x_2^2 + y_2^2 + z_2^2}}\right],$$

and that the flux through a surface in the z-x plane is given by the following:

$$\Phi = \kappa Q_1 \tan^{-1}\left[\frac{z_2 x_2}{y_2\sqrt{x_2^2 + y_2^2 + z_2^2}}\right].$$

Use these results to demonstrate that the flux through any side of a cube is the same.

Now suppose that we shift the cube upward a distance z_1, as is depicted in figure 3.8. For the top surface, the flux can be determined to be given by the following:

$$\Phi_{\text{top}} = 4\kappa Q_1 \tan^{-1}\left[\frac{a^2}{(a+z_1)\sqrt{2a^2 + (a+z_1)^2}}\right]$$

$$= 4\kappa Q_1 \tan^{-1}\left[\frac{1}{(1+u)\sqrt{2 + (1+u)^2}}\right].$$

where in the second step we have defined the dimensionless quantity $u = z_1/a$. The parameter u has the value 0 when the box is centered on the charge Q_1 and the value 1 when the box has been shifted by a distance a. For the bottom surface, the flux can be shown to have the following form:

$$\Phi_{\text{bot}} = 4\kappa Q_1 \tan^{-1}\left[\frac{1}{(1-u)\sqrt{2 + (1-u)^2}}\right].$$

The four side surfaces are all equivalent. The flux through any one of them can be shown to have the following form:

$$\Phi_{\text{side}} = 2\kappa Q_1 \left\{\tan^{-1}\left[\frac{1+u}{\sqrt{2 + (1+u)^2}}\right] + \tan^{-1}\left[\frac{1-u}{\sqrt{2 + (1-u)^2}}\right]\right\}.$$

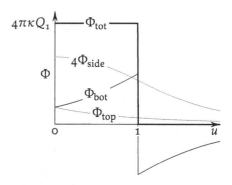

FIGURE 3.9. Gauss's law. The total flux through all six surfaces of the cube is a constant $\Phi_{\text{tot}} = 4\pi\kappa Q_1$ as long as the charge is inside the cube. Otherwise, the flux is zero

We have plotted the flux through each of the surfaces in figure 3.9, along with the total flux Φ_{tot}. Note that the flux through the bottom surface is undefined when $u = 1$, which is the point where the charge is embedded in the surface. We can assign the value o to the flux at that point; this is a somewhat arbitrary decision but the symmetry of the situation makes it a reasonable choice. Surprisingly, the total flux is a constant value $\Phi_{tot} = 4\pi\kappa Q_1$ as long as the charge is inside the box ($u < 1$). When the charge is outside the box ($u > 1$), the total flux vanishes! Gauss's law can therefore be stated as follows:

$$(3.19) \qquad \oint \mathbf{E} \cdot d\mathbf{A} = 4\pi\kappa Q_{\text{enclosed}},$$

where we use the $\oint$ symbol to reflect that the integral extends over a closed surface. This integral vanishes if there is no charge enclosed in the volume enclosed by the surface. The definition of the surface used in Equation 3.19 is quite broad: *any* closed surface will work, not just the cubes and prisms that we have discussed thus far. Spheres or ellipsoids or any random potato-shaped surface would also work. Proof of this assertion is beyond our present mathematical means, so we shall not attempt a proof at this time.

> EXERCISE 3.15.   Use the points defined in figure 3.8 to derive the values of the fluxes through each of the surfaces: Φ_{top}, Φ_{bot} and Φ_{side}. Show that one obtains the values stated in the discussion above. Plot the fluxes from $0 < u < 2$.

One powerful, but limited, application of Gauss's law is to recover the field of a charge or charge distribution when symmetry is present. For example, consider the field of a uniform, charged sphere of radius R and total charge Q_1. Now let us compute the flux through a spherical surface of radius $r > R$. In this case, it behooves us to use spherical coordinates where the differential surface elements are somewhat more complicated than those for Cartesian coordinates. In spherical coordinates, we can write:

$$(3.20) \qquad d\mathbf{A} = \hat{\mathbf{r}}\, r^2 \sin\theta \, d\theta \, d\varphi + \hat{\boldsymbol{\theta}}\, r \sin\theta \, dr \, d\varphi + \hat{\boldsymbol{\varphi}}\, r \, dr \, d\theta.$$

On the surface of a sphere, the outwardly directed normal vector is the radial vector: $\mathbf{n} = \hat{\mathbf{r}}$, as indicated in figure 3.10.

Consider a charge distribution that is only a function of the radius r. By symmetry, the electric field must be constant on the surface at radius r and radially directed: $\mathbf{E}(\mathbf{r}) = E(r)\hat{\mathbf{r}}$.[7] The electric field is not a function of the integration variables θ and φ, so it can be extracted from the integral:

---

[7]This is an example of physical insight. All points on the spherical surface are equivalent. Hence, we conclude that the field cannot be a function of the angular variables.

$$\int \mathbf{E} \cdot d\mathbf{A} = E(r) \int_0^\pi d\theta \int_0^{2\pi} d\varphi \, r^2 \sin\theta$$

(3.21)
$$= 4\pi r^2 E(r).$$

Now, from Gauss's law (Equation 3.19), we also know that the flux is $\Phi = 4\pi\kappa Q_1$. Consequently, we must have that

$$4\pi r^2 E(r) = 4\pi\kappa Q_1$$

$$E(r) = \kappa\frac{Q_1}{r^2}.$$

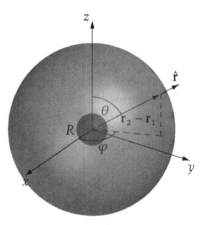

FIGURE 3.10. A charged sphere of radius R is located at the origin. At the radius $|\mathbf{r}_2 - \mathbf{r}_1| = r$, where $r > R$, the electric field must be the same at any point due to the rotational symmetry

The field is exactly that of a point charge, just as we found in the previous chapter.

We can also use Gauss's law to determine the field *inside* the charge distribution. Suppose now that the radius r of the Gaussian surface is less than the radius R of the charge Q_1. We know that the volume of a sphere of radius r is just $V = 4/3\pi r^3$, so the fraction of (uniformly distributed) charge inside the Gaussian surface must be given by

$$Q_{\text{enclosed}} = Q_1\frac{r^3}{R^3}.$$

The field inside the sphere must be given by the following expression:

(3.22)
$$E(r) = \kappa\frac{Q_{\text{enclosed}}}{r^2} = \kappa\frac{Q_1}{R^3}r.$$

That is, the electric field depends linearly on the distance from the center of the charge distribution until the distance reaches the radius of the charge distribution. Thereafter, the electric field falls off as the inverse square of the distance from the center of the charge distribution

EXERCISE 3.16.   Plot the (scaled) electric field ($E(r) = r$ when $r < 1$ and $E(r) = 1/r^2$ when $r > 1$.) from $r = 0$ to $r = 5$. Does the field have a continuous derivative at $r = 1$?

EXERCISE 3.17.   Suppose that the charge density ρ inside the radius R is not constant but instead varies like $\rho = \alpha \cos^2(\pi r/2R)$. If the total amount of charge is Q_1, then integrating over the volume of the sphere should also yield the total charge:

$$Q_1 = \int dV \rho = \int_0^R dr \int_0^\pi d\theta \int_0^{2\pi} d\varphi \, r^2 \sin\theta \, \alpha \cos^2(\pi r/2R),$$

where ρ is the charge density and α is a constant of proportionality. The θ and ϕ integrations can be performed easily and produce a factor of 4π. Perform the r integration and solve for α.

Use Gauss's law to compute the electric field inside the sphere. (Outside, it is the same as a point charge.) You will need to find the total charge inside the radius r. This can be obtained by integrating the charge density from 0 to r. Plot the electric field (assume $R = 1$ and plot from $0 \leq r \leq 5$). How does this compare to a uniform charge distribution? Does the field have a continuous derivative at $r = R$?

3.5. Nuclear Model of the Atom

Rutherford's experiments with α particles cast grave doubt on Thomson's plum pudding model of the atom. If we assume that the negatively charged electrons are uniformly distributed throughout the volume of the atom and that the atom is uniformly filled with positively charged matter, then by Gauss's law, the electric field of the atom will be zero outside the atom and zero inside the atom. Without an electric field, there won't be any scattering of α particles at all! Consequently, the electrons cannot be uniformly distributed throughout the atom if we are to see scattering.

Suppose that we construct an atom that has a total positive charge of $Q_1 = 10$ and radius $R = 1$ and distribute ten negative charges within the volume at the locations indicated in Table 3.1. (This is the distribution illustrated in figure 3.4.) Let's compute the electric field along lines through the $z = 0$ plane for several values of y through the distribution. The results are shown in figure 3.11.

We can observe that the field inside the atom is relatively complex but, if we assume that electrons also have a finite size, the electric field strength is never very large. In the calculations shown in figure 3.11, we assumed that the electron radius was 0.1. The divergent behavior of the inverse

TABLE 3.1. Electron positions

	x_1	y_1	z_1		x_1	y_1	z_1
1	−0.21	−0.53	0.64	6	−0.07	−0.05	−0.11
2	0.43	0.14	−0.59	7	0.53	0.29	0.52
3	−0.09	−0.47	−0.72	8	−0.11	0.61	0.10
4	0.28	−0.03	0.26	9	−0.32	0.10	−0.58
5	0.06	0.19	−0.09	10	−0.39	−0.75	−0.44

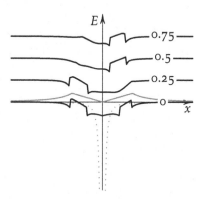

FIGURE 3.11. The relative electric field as a function of x is computed for the distribution shown in figure 3.4. The field of the positively charged background is depicted as the *gray line* and the field of an isolated negative charge is shown as the *dotted gray line*. The fields for different y-values are separated vertically for clarity

square dependence of the electric field on distance is softened dramatically by the linear dependence of the field in the interior of a finite distribution.

How should we now proceed? The field inside the atom is not simple and we have made many assumptions to produce figure 3.11 but let us make a simplifying approximation: let us assume that the field is simply constant throughout the volume. This is clearly not an accurate description of the field inside the atomic volume but we can calculate the result of an α particle traversing a region with a constant field quite simply and this will provide us with a mathematical bound on the size of deflection possible. In any more-realistic calculations, the deflection will prove to be smaller. This is a classic strategy in physics. First, solve a simple problem that will tell you something about the more complex one. The last thing we want to do is to become embroiled in some complex mathematical analysis that turns out to be not particularly useful.[8]

---

[8]This observation is not intended to be sardonic. It is likely that many students find themselves already embroiled in complex mathematical analysis but the process of physics seeks to utilize increasing sophisticated models to represent increasingly accurate descriptions of natural phenomena. We shall try to avoid plunging immediately into the deep end.

An α particle traversing a region of constant electric field $\mathbf{E}$ would feel a force $M_\alpha \mathbf{a} = Q_\alpha \mathbf{E}$. Let us assume that the thickness of the region is given by R and that the α particle has an initial energy $\mathcal{E} = 1/2 M_\alpha v_\alpha^2$. The constant electric field produces a constant acceleration on the particle: $\mathbf{a} = Q_\alpha \mathbf{E}/M_\alpha$. This equation can be integrated over time directly because both sides are independent of the time. Recall that the acceleration $\mathbf{a}$ is the time derivative of velocity $\mathbf{v}$. Hence, the change in velocity arising from the interaction with the constant field is just $\Delta \mathbf{v} = Q_\alpha \mathbf{E} \Delta t/M_\alpha$, where Δt is the amount of time that the particle spends in the field. If the direction of the electric field is transverse to the motion, then $\Delta t = R/v_\alpha$. This generates a component of the velocity transverse to the original velocity and an angle of deflection θ defined by the following expression:

$$(3.23) \qquad \tan \theta = \Delta v/v_\alpha = \frac{Q_\alpha E R}{M_\alpha v_\alpha^2}.$$

EXERCISE 3.18. Consider a particle moving in the z-direction, with velocity $\mathbf{v} = v_z \hat{\mathbf{z}}$. Now at some time $t = 0$, the particle enters a region with a constant acceleration $\mathbf{a} = a_x \hat{\mathbf{x}}$.

(a) What is the velocity $\mathbf{v}$ at some time t_1 later?
(b) What is the change in position of the particle between the initial time $t = 0$ and t_1?
    Suppose now that the acceleration has a z-component as well: $\mathbf{a} = a_x \hat{\mathbf{x}} + a_z \hat{\mathbf{z}}$.
(c) What is the velocity at a time t_2 after the particle acceleration begins?
(d) What is the change in position of the particle between the initial time $t = 0$ and t_2?
(e) What is the time t_2 required then to traverse a distance d in the z-direction?

To compute a value for the angle of deflection, we will need to provide some estimates for the values of the mass, velocity and charge of the α particles, the thickness of the metal foils and the electric field. This was a thorny problem for Rutherford but there is no particular reason that we should not make use of our modern knowledge of these values. At this point, where we are intent on computing a number, it is important for us to choose a consistent system of units for these values. In this text, we shall often utilize the standard system promulgated as Le Système International d'Unités (SI), in which the standard length is the meter (m) and the standard time is the second (s). Nevertheless, we will often have to deal with the fact that different practitioners will often use units that are somehow convenient. For example, were we to ask what is the velocity of a typical garden slug, it is likely that an answer in terms of meters/second will be a very small number, as a garden slug might well take the better part of an hour to travel a single meter. As humans tend to be comfortable with numbers like 1 or 10 or even 100, and less comfortable with fractions or decimals, then we would be likely to choose a system of units in which the slug velocity is a number from 1 to 100.

Nuclear physicists use a system of units in which the energy is expressed in terms of the product of the fundamental charge e and the Volt: the electron-volt (eV).[9] This is convenient because elementary particles all have charges that are multiples of the fundamental charge e and it was common practice to accelerate these particles through known electric potentials V. In these units, the decay of radon to polonium produces α particles with a kinetic energy of $\mathcal{E} = 5.590\,\text{MeV}$. The charge of an α particle is twice that of an electron, $Q_\alpha = +2e$. So, we are left with trying to estimate the thickness of the metal foils R and the electric field in the atom E. Fortunately, our results scale linearly with both R and E. This means that propagating the uncertainty in our estimates into the uncertainty in θ will be straightforward. We shall discuss this in more detail presently.

Gold leaf can be hammered into extraordinarily thin sheets: one ounce of gold can produce a sheet of gold leaf with an area of three hundred square feet. In SI units, this corresponds to a thickness of order $R = 100\,\text{nm}$. Gold atoms can be thought of as having a size of 0.288 nm and gold has a nuclear charge of $Z = 79e$. Let us approximate the field in the atom then as a constant with the value obtained from a point charge at the radius of 0.144 nm with a total charge of $Q = -79e$. For this we shall need the value of the constant κ. In our current set of units, this has the value $\kappa = 1.44 \times 10^{-9}\,\text{V·m}/e$, The electric field E then has the value

$$(3.24) \quad E = \kappa \frac{Q}{r^2} = (1.44 \times 10^{-9}\,\text{V·m}/e) \frac{(-79e)}{(1.44 \times 10^{-10}\,\text{m})^2} = -5.5 \times 10^4\,\text{V/m}.$$

Substituting all of these values back into Equation 3.23, and solving for θ, we find the following result:

$$\theta = \tan^{-1}\left[\frac{Q_\alpha E R}{M_\alpha v_a^2}\right] = \tan^{-1}\left[\frac{(2e)(5.5 \times 10^4\,\text{V/m})(1 \times 10^{-7}\,\text{m})}{2(5.6 \times 10^6\,\text{eV})}\right] \approx 0.3^\circ.$$

That is, we should expect a scattering angle of only a small fraction of a degree!

EXERCISE 3.19. In their experiments with α particles, Rutherford's students Geiger and Marsden utilized thin gold foils. They were able to determine, using precision balances, that some of their foils

---

[9]Here we encounter a notational difficulty. The symbol e is used for both the fundamental charge of the electron and the eccentricity of conic sections. The reader will have to be diligent in order to understand which quantity is meant in each usage of the symbol.

contained 0.26 mg of gold per square centimeter. The density of gold is 19.3 g/cm³. What was the thickness of these films? Is our estimate of $R = 100$ nm utilized above reasonable?

Observations of α particle scattering in excess of 90° cannot be explained by the simple model that we have just described. Yet, the calculations are simple enough that we haven't somehow gone astray performing the algebra. We could try fiddling with some of our estimates: increasing the electric field strength, for example; however, we should recognize that our simplified model of a constant electric field oriented in a single direction orthogonal to the α particle motion undoubtedly *overestimates* the actual deflection. So, how can we explain the experimental observation of large-angle scattering.

EXERCISE 3.20. What value of the field is required to obtain a scattering angle of 30°? The charges of the gold nucleus and the α particle are fixed. What value of r would be required in Equation 3.24 to obtain such a field?

FIGURE 3.12. Rutherford's model of the electric field of an atom. The positive charge is concentrated at the origin and the negative charge is distributed uniformly throughout the atom. Slices through the atom at various y-values are drawn in *black* and labelled as in figure 3.11. For reference, the field of a point (negative) charge is shown in *gray*

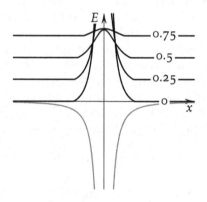

Rutherford recognized that the large-angle scattering of α particles could be explained if the α particles were following hyperbolic trajectories. This was the signature of central force motion and could be achieved if all of the positive charge in the atom were located at a single point at the center of the atom, and the negative charge distributed out to some larger radius (the atomic radius). The electric field seen by incoming α particles would be nearly the same as that of a point (positive) charge, as indicated in figure 3.12. From figure 3.3, we see that the acceleration occurs predominantly when the particle is in close proximity to the charge; this is where the electric field is largest.

As a first approximation, Rutherford postulated that one could simply *ignore* the negative charge due to the electrons. Scattering of α particles at large angles should be explainable by considering the α particle and positively-charged matter that makes up the vast majority of the mass of the atom (what we today call the atomic nucleus) to be point positive charges. Including the effects of the electron cloud around the atom should give rise to relatively small corrections. This is a vastly different model of the atom than the plum pudding model proposed by Thomson. What experimental evidence makes us believe it to be correct?

To explore Rutherford's analysis of α particle scattering, let us rotate figure 3.2 so that the α particle initial direction is aligned with the z axis. All of the α particles in our beam will be moving in the z-direction.

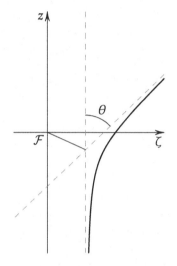

FIGURE 3.13. The parameter ζ is defined as the perpendicular distance between the focus and the incident asymptote. The angle θ that defines the deviation of a particle from a straight line trajectory can be seen to be given by $\theta = 2\psi^* - \pi$, where ψ^* is defined by the equation $1 + e\cos\psi^* = 0$

The asymptote is defined to be a distance ζ from the origin. This α particle will deflect through an angle θ that is given by $\theta = 2\psi^* - \pi$. Recall that the distance from the focus to the directrix is given by $\kappa Q_1 Q_2 e / 2\mathcal{E}$, where here e is the eccentricity. The distance ζ from the focus to the asymptote can be seen from triangle in figure 3.13 to be defined as follows:

$$\frac{2\mathcal{E}\zeta}{\kappa Q_1 Q_2 e} = \sin(\pi - \psi^*) = \cos(\theta/2).$$

The eccentricity e is related to ψ^*: $e = -1/\cos\psi^* = -1/\cos(\pi/2 + \theta/2) = 1/\sin(\theta/2)$. So, finally we can write

(3.25) $$\zeta = \frac{\kappa Q_1 Q_2}{2\mathcal{E}} \cot(\theta/2).$$

EXERCISE 3.21. Redraw figure 3.2 and convince yourself that indeed $\theta = 2\psi^* - \pi$. What is the angle of rotation required to align one of the asymptotes with the vertical axis?

Derive the distance from the focus to the directrix and the perpendicular distance from the focus to the asymptotes.

From the trigonometric addition formulas $\sin(a \pm b) = \sin a \cos b \pm \sin b \cos a$ and $\cos(a \pm b) = \cos a \cos b \mp \sin a \sin b$, convince yourself of the validity of Equation 3.25.

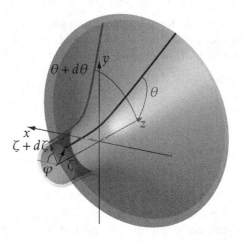

Figure 3.14. ζ is the radial coordinate in the plane normal to the beam direction $\hat{z}$. The angle measured to the trajectory from the x-axis is denoted by φ. The probability of an α particle striking the foil within the range from ζ (*black trajectory*) to $\zeta + d\zeta$ (*gray trajectory*) is proportional to the area of the ring: $2\pi\zeta\, d\zeta$. Scattered particles will emerge in the angular band from θ to $\theta + d\theta$

As we see in Equation 3.25, the angle of deflection is uniquely tied to the parameter ζ, which is known as the *impact parameter*. What we would like to be able to predict is the fraction of the total number of particles that have been scattered at each angle θ from the original beam direction. Physicists term this quantity the cross section, it is usually designated by the symbol σ.

As can be seen from figure 3.14, all particles that approach the nucleus in the thin ring of area $dA = 2\pi\zeta\, d\zeta$ will scatter into the angular range θ to $\theta + d\theta$. At some large distance R_∞ from the scattering source, the scattered particles that emerge in the range θ to $\theta + d\theta$ will occupy an area (recalling Equation 3.20) $dA_{\text{scatt}} = R_\infty^2 \sin\theta\, d\theta\, d\varphi \equiv R_\infty^2\, d\Omega$, where $d\Omega$ is called the solid angle. In this case, where scattering does not depend on the azimuthal angle φ, we have that $d\Omega = 2\pi \sin\theta\, d\theta$. It is customary to omit the factor R_∞^2 and define the differential cross section as follows:

(3.26)
$$\frac{d\sigma}{d\Omega} = \frac{2\pi\zeta}{2\pi\sin\theta} \frac{d\zeta}{d\theta}.$$

From our previous definition of ζ in Equation 3.25, we see that we can write

$$\frac{d\zeta}{d\theta} = \frac{\kappa Q_1 Q_2}{2\mathcal{E}} \frac{d\cot(\theta/2)}{d\theta} = -\frac{\kappa Q_1 Q_2}{4\mathcal{E}} \csc^2(\theta/2).$$

Note that the results involving ζ have trigonometric functions that have $\theta/2$ as arguments, whereas the solid angle is defined in terms of θ. We can use a trigonometric identity to show that $\sin\theta = 2\sin(\theta/2)\cos(\theta/2)$. Substituting back into Equation 3.26, we can write:

$$\frac{d\sigma}{d\Omega} = -\frac{1}{2}\left[\frac{\kappa Q_1 Q_2}{2\mathcal{E}}\right]^2 \frac{\cot(\theta/2)\csc^2(\theta/2)}{2\sin(\theta/2)\cos(\theta/2)}$$

$$(3.27) \qquad = -\left[\frac{\kappa Q_1 Q_2}{4\mathcal{E}}\right]^2 \csc^4(\theta/2) = -\left[\frac{\kappa Q_1 Q_2}{4\mathcal{E}}\right]^2 \frac{1}{\sin^4(\theta/2)}.$$

This is Rutherford's primary result: the angular dependence of the scattering of α particles should scale like the inverse of sine to the fourth power.

FIGURE 3.15. Scattering of α particles by gold foil. The number of scintillations produced by α particles striking the lead sulfide is plotted as a function of microscope angle (*black dots*). The scintillation count was corrected for source decay and geometrical effects. The solid (*gray*) curve is a plot of the function $f(\theta) = 33/\sin^4(\theta/2)$

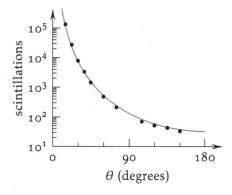

The results shown in figure 3.15 represent the results published in 1913 by Rutherford's students Geiger and Marsden on the scattering of α particles from a gold foil. The trend in the data is clearly well-represented by the $1/\sin^4(\theta/2)$ prediction that is illustrated by the solid line. (An overall normalization is used.) These data clearly support Rutherford's suggestion that the positively-charged matter that constitutes the vast majority of the atomic mass is confined to a small volume in the center of the atom.

Note that we have been somewhat sloppy with our mathematical treatment of Rutherford scattering. We have simply ignored the contributions of the electron cloud to the final scattering angle. We justify this approach based on our observation that scattering from a uniform field does not produce a significant deflection of the α particles unless we choose unrealistic values for the parameters that define the scattering. We recognize that our results do not constitute a mathematical proof of the nuclear structure of the atom but more sophisticated analyses and further experimental observations have provided more evidence that Rutherford's suggestion was, indeed, correct.

EXERCISE 3.22. Geiger and Marsden also reported the results of their experiments with a silver foil target in their 1913 paper; these are listed in Table 3.2. The number of scintillations observed (corrected for source decay and finite detector size) are listed as a function of angle. Plot the data and a function $f(\theta) = a/\sin^4(\theta/2)$. Determine the constant a that best fits the data.

TABLE 3.2. α scattering from silver

θ (deg)	Scintillations	θ (deg)	Scintillations
15	105400	75	136
22.5	20300	105	47.3
30	5260	120	33.0
37.5	1760	135	27.4
45	989	150	22.2
60	320		

While it is impressive to have Rutherford's prediction borne out by the experimental results, we should take some time to understand how the experiment was conducted and just how it relates to the theoretical analysis that we've conducted. The experiment that we should like to perform would consist of sending α particles toward the target foil and controlling the impact parameter ζ. Because we have determined the relationship between ζ and the scattering angle θ in Equation 3.25, we could then determine if the experimental results agree with our model. Such fine control of the beam position is not possible. Indeed, the α particle beam utilized by Geiger and Marsden had a width that spanned thousands of nuclear centers. One might fear that this will somehow vastly complicate matters but all is not lost.

EXERCISE 3.23. The apparatus used by Rutherford and his students (depicted in figure 3.5) had a collimator that permitted only a narrow beam of α particles to exit the source and impact the target foil. If we assume that the beam has a circular cross section with a width of 1 mm and that gold atoms form a rectangular lattice with a 0.288 nm spacing, how many gold atoms fall within the beam footprint in each atomic layer of the target foil?

Suppose that an α particle is (randomly) located within the beam. It is travelling in the z-direction and has some coordinates (x_α, y_α) in the plane transverse to the propagation direction. If for simplicity we assume that the nuclear centers form a rectangular lattice, then the nuclear centers will be found at locations (ia, jb) where i and j are integers and a and b are the distances separating nuclei in the x- and y-directions. The distance of closest approach of the α particle will always be less than $\zeta_{max} = [(a^2 + b^2)/2]^{1/2}$. Consider what happens if we send many α particles through the foil. Each α particle will pass by some nucleus with an impact parameter of ζ. With enough α particles, the complete range of ζ will be covered. Hence, even though we cannot directly control the impact parameter ζ, with enough samples, we can achieve the same result.

EXERCISE 3.24.   To see how this works, consider a simple model in which atoms occupy sites on a square lattice. Let us also, for simplicity, assume that a beam of α particles has a constant intensity and a radius of 100 times the lattice spacing. We can further assume that one α particle in the beam can be located by a point (x, y) that is randomly located within the beam area. We can achieve this by using the *Mathematica* function RandomReal. To obtain a point in the circular beam area, we can obtain the distance of the point from the origin as ζ = RandomReal[100] and the angle from the x-axis as φ = RandomReal[2Pi]. This point will have Cartesian coordinates $(x = \zeta \cos \varphi, y = \zeta \sin \varphi)$.

Produce a list of 1000 points and plot the distribution with the List-Plot function. Are they uniformly distributed?

Now use the Round function to find the lattice point closest to where the α particle is located and calculate the distance

$$d = [(x - \text{Round}[x])^2 + (y - \text{Round}[y])^2]^{1/2}$$

from the α particle to the lattice site. To see how these α particles are distributed around the lattice sites, we can use the BinCounts function to bin the data into 50 bins over the accessible range of $d = 0$ to $d = \sqrt{2}/2$.

Plot the distribution of the 1000 α particles.

Repeat the exercise for 10000 and 100000 α particles. In the limit of large numbers of α particles, do you observe that the particles are uniformly distributed over the range?

This result, while not intuitive, is an example of the large number limit for probabilities. We shall encounter further examples subsequently. For our present purposes, we can state that the physical distribution of a random process becomes the probability distribution when the number of trials is large. As a result, the uniform distribution of α particles within the beam translates into a uniform sampling in ζ. This is a key element in the success of casinos. While the result of any individual coin toss or dice throw is random and, therefore, not knowable, the distribution of millions of dice throws is completely known. Consequently, casinos always make money and gamblers, in total, always lose.

EXERCISE 3.25. Use the `RandomInteger[{1,6},N]` function to simulate rolling a die N times. How many times in $N = 100$ trials do you obtain the number 3? How does that compare to the probability of $N/6$? How many times in $N = 10000$ trials and $N = 100000$ trials do you obtain the number 3?

3.6. Finite Nuclear Size

With the advent of modern particle accelerators, physicists could conduct systematic investigations unavailable to Rutherford and his students. In particular, it was possible to continuously vary the α particle energy, as is depicted in figure 3.16. Rutherford's prediction here, from Equation 3.27, is that the differential cross section should be proportional to the inverse square of the α particle energy, as is seen by the gray curve in the figure. At about 30 MeV, there is a systematic departure from the Rutherford formula. Today, we interpret this as due to nuclear interactions between the α particle and the lead nucleus.

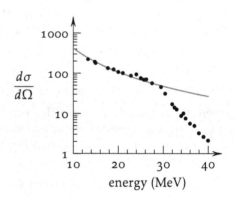

FIGURE 3.16. The differential cross section for α particles scattered from a lead target was measured at a nominal 60° as a function of α particle energy (*black dots*). The Rutherford formula suggests that the cross section should scale like $\mathcal{E}^{-2}$ (*gray curve*). There is a widening discrepancy after 30 MeV

We know from our previous results that the distance of closest approach $\mathbf{r}_\pi$ for the α particle is given by $\mathbf{r}_\pi = \kappa Q_1 Q_2 (e+1)/2\mathcal{E}$. We also know that $e = 1/\sin(\theta/2)$, so for a scattering angle of $\theta = 60°$ we can write:

$$\mathbf{r}_\pi = \kappa Q_1 Q_2 \frac{1+\sin(\theta/2)}{2\mathcal{E}\sin(\theta/2)}.$$

$$= (1.44 \times 10^{-9} \text{ V·m/e})(2e)(82e)\frac{(1+1/2)}{2(30 \times 10^6 \text{ eV})(1/2)}$$

$$= 1.2 \times 10^{-14} \text{ m} = 12 \text{ fm}.$$

So, when the distance between the centers of the α particle and the lead nucleus is about 12 fm, we see deviations from the Rutherford predictions. We can interpret this as an approximation to the size of the lead nucleus.[10] If this is the case, then the nuclear size is vastly smaller than the typical inter-nuclear distance. For gold, we suggested earlier that the nominal atomic size was about 0.288 nm. (We shall revisit this assumption in a later chapter.) Using the value of 12 fm for the size of a gold nucleus, this means that the nucleus is 24,000 times smaller than the atom. This makes physical representations of the atom quite difficult to produce without exaggeration. If we were to draw a nucleus on the blackboard as a 1 cm blob, the nearest neighboring gold nucleus would have to be 240 m distant to faithfully represent the relative distances. This is significantly larger than most classrooms and few faculty are willing to run 240 m down the hallway to prove a point. As a result, when the 1 cm blob is drawn within a 1 m circle, the relative scales have been distorted by a factor of over a hundred.

3.7. Unfinished Business

Rutherford's astonishing discovery of the nuclear structure of the atom gives rise to a significant problem for physicists, one that we will sidestep in this text. That is, how can the nucleus be so small? Atoms are electrically neutral, so the same amount of charge is stored in the nucleus within a volume that is approximately 10^{14} times smaller than the volume of the atom. Our current notions that the nucleus is composed of positively-charged protons and neutral neutrons means that there must be a very strong force holding the charged nuclear constituents together. They certainly feel a strong repulsion due to the Coulomb force. Rutherford's experiments with α particles produces the inevitable conclusion that there must be another force in nature beyond gravity and the electromagnetic force. (This is an obvious application of Newton's fundamental idea that things move for a reason or, in this case, don't fly apart.)

> EXERCISE 3.26. If we make the somewhat dubious assumption that a gold nucleus can be thought of as a proton ($Q_1 = e$) rolling around a platinum nucleus ($Q_1 = 78e$) at a distance of about 8 fm, what is the electromagnetic potential energy of this system. Assume that there is no kinetic energy. How large is this energy compared to the few MeV observed for α particles that result from nuclear decay?

---

[10]A more careful analysis using electron-nuclear scattering suggests that the nucleus is somewhat smaller than we have calculated here. The generally-accepted value is $R \approx r_0 A^{1/3}$, where r_0 is a constant in the range of 1.2–1.5 fm and A is the atomic number. For lead, this leads to a charge radius of about 8 fm.

The atom then is apparently constructed from a small, positively-charged nucleus and is surrounded by negatively-charged electrons. If you have made the observation that the attractive Coulomb force is identical to Newton's gravitational force and concluded that atoms must somehow look like little solar systems with the electrons occupying elliptical orbits around the nucleus, you are to be congratulated. You are now thinking like a physicist. Rutherford's discovery of the nuclear structure of the atom encouraged physicists to embark on precisely this course. All such attempts to construct a classical representation of the atom failed to explain experimental observations. The simplest atom, hydrogen, consists of a single proton and a single electron. Studies of the electromagnetic spectrum of hydrogen identified a series of different, discrete energy states. Apparently, the value of the energy $\mathcal{E}$ can only have specific values but there is nothing in the Coulomb force law that prevents the energy from taking on any value. The resolution of this dilemma required the development of a new theory of microscopic matter known as quantum mechanics. This subject lies well beyond the scope of our present investigations.

EXERCISE 3.27. If we make the assumption that a hydrogen atom consists of a proton ($Q_1 = e$) and an electron($Q_2 = -e$) with masses $m_p = 938\,\text{MeV/c}^2$ and $m_e = 0.5\,\text{MeV/c}^2$, respectively, and that the electron orbits at a radius of 0.1 nm, what is the potential energy of the system? How does this energy compare to nuclear energies?

EXERCISE 3.28. A fundamental problem with the planetary model is that accelerating electrons radiate. The Irish physicist Joseph Larmor showed that the power P radiated by an electron in a circular orbit is given by $P = \mu_0 e^2 a^2/6\pi c$, where a is the magnitude of the acceleration and c is the velocity of light. This radiation subtracts energy from the kinetic and potential energy of the electron, leading to the electron spiralling into the nucleus in a time given approximately by the following:

$$\Delta t = c^3 r_0^3 \left[\frac{m_e}{2\kappa e}\right]^2,$$

where m_e is the mass of the electron and r_0 is the initial orbital radius. How long would a hydrogen atom last, if this equation were true and $r_0 = 0.1$ nm?

IV

On the Nature of Spacetime

One of the most vexing problems for early explorers was their inability to determine longitude. Latitude was readily measurable by astronomical observations of reasonable accuracy but longitude required the measurement of the time difference between the prime meridian in Greenwich or Paris and one's current location.[1] Early clocks were cumbersome mechanical devices that were notably unreliable when placed aboard ships at sea or wagons bouncing across rutted roads. Accurate time-keeping was simply not achievable under the rugged conditions that also vexed early explorers.

In 1610, the Italian mathematician and astronomer Galileo Galilei fashioned a telescope and discovered four moons circling the planet Jupiter. About the same time, the German mathematician and astronomer Simon Marius also began observing the moons of Jupiter and named them Io, Europa, Ganymede and Callisto after companions of the Greek god Zeus, whom the Romans called Jupiter.[2] Today, we refer to these bodies as the Galilean moons of Jupiter due in large part to the fact that Marius waited several years before publishing his observations. For his part, Galileo considered Marius to be a usurper, claiming credit for Galileo's work without, in Galileo's estimation, having ever observed anything.[3] The issue of scientific priority has always been contentious.

---

[1] The earth's circumference at the equator is about 40,000 km. As there are 24 hours in one day, an hour corresponds to a distance at the equator of about 1666 km and a minute to 28 km.

[2] "Io, Europa, the boy Ganymede, and Callisto greatly pleased lustful Jupiter," noted Marius in his 1614 publication *Mundus Iovialis*. There is small doubt that the sex appeal of Marius's names led to their adoption. Galileo suggested naming the moons Cosimo's stars in his 1610 publication *Sidereus Nuncius*, after Cosimo II de' Medichi, the Grand Duke of Tuscany, who supported Galileo's work.

[3] Galileo's concerns were not unwarranted. Marius's student Baldessar Capra published Galileo's manuscript on operation of the sector (a Galileo invention) under his own name in 1607. When his plagiarism was discovered, Capra was expelled from the University of Padua. Marius, in the meantime, had returned (fled?) to Germany.

© Mark A. Cunningham 2015   87
M.A. Cunningham, *Neoclassical Physics*, Undergraduate Lecture
Notes in Physics, DOI 10.1007/978-3-319-10647-2_4

Galileo immediately noted two fundamental consequences of his observations. First, the earth cannot be the center of the Universe if objects are orbiting a distant object.[4] Second, the moons of Jupiter appear to provide a natural system of timekeeping. The moons repeatedly retraced their paths around the planet. So, Galileo reasoned, it should be reasonably straightforward to use observation of the Galilean moons in the service of terrestrial navigation. It turned out, of course, not to be at all straightforward to use timings of the Jovian moons in navigation, despite the diligence and persistence of many observers. Researchers continued to observe the moons of Jupiter throughout the seventeenth century to build and improve ephemeris[5] tables that detailed the daily positions of astronomical objects for navigators.

In 1672, the young Danish astronomer Ole Rømer moved to Paris to study with the noted French/Italian astronomer Giovanni Domenico Cassini. Cassini's observations of the Jovian moons during the years 1666–1668 demonstrated a number of discrepancies in the measurements that Cassini did not understand but initially postulated might be due to a finite propagation velocity for light. Over the next several years, Cassini backed away from this explanation while Rømer embraced it and provided definitive proof in a presentation to the French Academy of Sciences in 1676.[6]

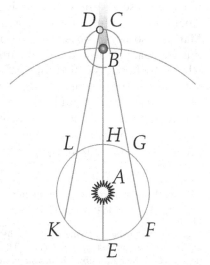

FIGURE 4.1. Rømer measured the times at which the Jovian moon Io either disappeared into the shadow of the planet (C) or emerged from the shadow (D). The earth orbits on the inner trajectory (*gray circle*)

---

[4]The long-held geocentric model of Ptolemy was thereby refuted by direct observation of contradictory behavior. Nevertheless, Galileo was ultimately forced to recant his observations in 1633 by the Inquisition and spent his final years under house arrest. Galileo was pardoned in 1992 by Pope John Paul II.

[5]From the Greek words ἐφήμεροζ for "daily" or ἐφήμεριζ for "diary."

[6]Rømer's presentation was recorded by an anonymous observer.

Rømer's conclusions were based on his observations of eclipses of Jupiter's moon Io. One problem, of course, in making observations of objects in the solar system is that there is no absolute coordinate system in place that enables observers to provide (x, y, z) values. A schematic of how Rømer solved this observational problem is illustrated in figure 4.1. Jupiter (B)

TABLE 4.1. Rømer's observations of eclipses of Io from Paris. The Loc. Column refers to the locations C and D identified in figure 4.1

Date	Time	Loc.	Date	Time	Loc.	Date	Time	Loc.
	1672			1673			1674	
Jan 4	00:42:36	C	Feb 5	05:31:10	C	May 23	20:47:48	D
Jan 11	02:32:14	C	Feb 7	00:00:00	C	May 30	22:41:12	D
Jan 12	20:59:22	C	Feb 14	01:53:20	C	Jun 22	22:49:45	D
Feb 11	22:57:06	C	Feb 28	05:40:10	C	Jul 31	21:19:02	D
Feb 20	19:20:26	C	Mar 2	00:09:01	C		1675	
Mar 7	19:58:25	D	Mar 16	04:00:48	C	Jul 20	20:22:42	D
Mar 14	21:52:30	D	Mar 17	22:28:16	C	Jul 27	22:17:31	D
Mar 23	18:18:14	D	Mar 25	00:24:30	C	Oct 29	18:03:22	D
Mar 29	01:45:30	D	Apr 18	21:22:00	D		1676	
Mar 30	20:14:46	D	Apr 25	23:28:05	D	May 13	02:49:42	C
Apr 6	22:11:22	D	May 11	21:17:39	D	Jun 13	22:56:11	C
Apr 14	00:08:08	D	May 18	23:32:44	D	Aug 7	21:49:50	D
Apr 22	20:34:28	D	Aug 4	20:30:41	D	Aug 14	23:45:55	D
Apr 29	22:30:06	D	Dec 17	18:39:14	C	Aug 23	20:11:13	D
Nov 29	17:37:05	C				Nov 9	17:45:35	D

orbits the sun (A) on the trajectory indicated. Io enters the shadow cast by Jupiter at C and emerges at D. These two points provided Rømer with well-defined points in space. Depending upon the observational geometry, Rømer either timed the emergences of Io from the shadow (when earth was located on its orbit between the points L and K) or the disappearances of Io into the shadow (when earth was located between the points F and G). Observations of Io's eclipses near the extreme point E (referred to by astronomers as opposition) on earth's orbit were not possible because Jupiter was obscured from view by the sun. Observations near the point H (referred to by astronomers as conjunction) were hidden from view by Jupiter itself. Unfortunately, much of Rømer's original notes and manuscripts was destroyed in the Copenhagen fire of 1728 but one small note remains intact to this day. In it, Rømer catalogs his observations of the eclipses of Io; some of those data are depicted in Table 4.1.

4.1. Statistical Analysis

Before we continue our discussion of Rømer's observations, let us take some time to review how scientists analyze measurements. First, it is important to recognize that all measurement apparatus have finite precision. For example, a ruler has gradations indicating millimeters; a stopwatch records time to the hundredth of a second. These represent the limits of precision imposed by the devices themselves. In addition to the inherent limitation of the measurement devices, other random measurement errors can occur. As we discussed in the first chapter, a timing experiment involving flag waving and students with stopwatches will have some intrinsic limitations beyond just the stopwatches themselves. Human beings are not precise to the hundredth of a second, so this introduces an uncertainty in the measurement beyond just the precision of the device.

What we want to do is somehow quantify the uncertainty in our results. In doing so, we must be aware of two similar but distinct concepts: *accuracy* and *precision*. In common usage, these two words are treated as synonyms but we shall make the following distinction. Accuracy is defined as the difference between the measured value and the "correct" value. Determining the correct value with which to compare can be problematic in practice because the purpose of most experiments is to establish the value of some unknown quantity. There may well be no obvious definition of "correct" and, as a result, no simple means of determining the accuracy of our measurement. Precision, on the other hand, can be defined in a systematic fashion as the number of significant figures in the measured value. This will incorporate both the intrinsic limitations of the measurement apparatus and measurement errors.

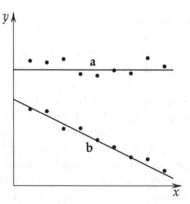

FIGURE 4.2. Two sequences of measurements. In a, the data appear to scatter about a common value. The horizontal line indicates the Şcorrectî value. In b, the data display a trend, indicated by the sloping line.

To quantify our discussion, consider the series of measurements indicated in figure 4.2. In the upper sequence **a**, we observe that the data are scattered about a horizontal line. We can provide a systematic description of the data by computing a series of statistical moments of the data. The first moment is the mean $\bar{y}$:

$$(4.1) \qquad \bar{y} = \frac{1}{N} \sum_{i=1}^{N} y_i,$$

where N is the number of individual measurements y_i. The second moment is defined as the variance $\mathrm{var}(y)$:

$$(4.2) \qquad \mathrm{var}(y) = \frac{1}{N-1} \sum_{i=1}^{N} (y_i - \bar{y})^2.$$

The standard deviation $\sigma(y)$ is defined as the square root of the variance:

$$(4.3) \qquad \sigma(y) = \mathrm{var}(y)^{1/2}.$$

Note that the standard deviation of some quantity has the same dimensionality as the quantity itself. Higher order moments can also be defined. The skew measures the departure of the variance from symmetry:

$$(4.4) \qquad \mathrm{skew}(y) = \frac{1}{N} \sum_{i=1}^{N} \left[\frac{y_i - \bar{y}}{\sigma(y)} \right]^3.$$

The higher-order moments are not used with much frequency in physics applications; however, most statistical arguments assume that the errors are symmetrically distributed about the mean value. Computing the skew can test that assumption.

> EXERCISE 4.1.   Consider the measurements listed in Table 4.2. Compute the means and standard deviations. Plot the data and horizontal lines at the mean value and at the mean plus and minus the standard deviation.
>
> How do your results compare to the "correct" values of $y = 6.4$ and $z = 0.031$?

This series of statistical moments can be interpreted as an estimate of the "correct" value $\bar{y}$ and its uncertainty $\sigma(y)$. It happens that, if the errors in our measurements are random and uncorrelated, they can be described by a normal distribution. The probability of making a measurement y_i is given by the following expression:

$$(4.5) \qquad \mathcal{P}(y_i, \bar{y}, \sigma(y)) = \frac{1}{[2\pi \sigma^2(y)]^{1/2}} e^{-(y_i - \bar{y})^2 / 2\sigma^2(y)}.$$

This function is described as the normal distribution because it is what arises in nearly all measurement systems. If we think back to our original flag-waving and timing experiment, it is a good approximation that each student will act independently (uncorrelated errors.) As a result, each

TABLE 4.2. Experimental data

Sample	y	z	Sample	y	z
1	5.755	0.030967	9	6.104	0.029246
2	6.813	0.032913	10	6.177	0.029562
3	6.183	0.031964	11	6.294	0.032583
4	6.905	0.030649	12	6.679	0.031836
5	6.121	0.032378	13	6.535	0.032742
6	5.890	0.032211	14	5.808	0.033091
7	6.802	0.028817	15	7.001	0.032071
8	5.983	0.032551			

student's determination of when to start timing—at the initial movement of the flag or as the flag crosses the horizontal plane or at the bottom of the flag's sweep—and when to stop timing will generate random errors. In fact, the central limit theorem from statistics proves that, under reasonable circumstances, all distributions limit to the normal distribution when the number of measurements is large. As a result, the Gaussian function used in Equation 4.5 and illustrated in figure 4.3 is used to represent the distribution of errors in the vast majority of cases.

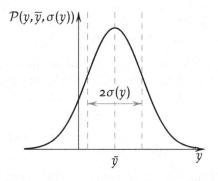

FIGURE 4.3. The normal probability distribution peaks at the mean $\bar{y}$ and is symmetric about the mean. The standard deviation $\sigma(y)$ defines the width of the distribution

The normalization of the probability distribution in Equation 4.5 is chosen so that the total probability is unity. That is, the integral over all possible results is one:

$$\int_{-\infty}^{\infty} dy\, P(y, \bar{y}, \sigma(y)) = 1.$$

The probability then that we can obtain a measurement within a distance y_1 from the mean $\bar{y}$ is given by the following integral:

$$\int_{\bar{y}-y_1}^{\bar{y}+y_1} dy\, P(y, \bar{y}, \sigma(y)) = \mathrm{erf}\left(\frac{\sqrt{2}\,y_1}{2\sigma(y)}\right),$$

where the error function erf(x) is essentially defined by the above integral. *Mathematica* contains a function Erf that will compute its value.

> EXERCISE 4.2. Plot the error function as a function of x. What is the probability of obtaining measurements within two, three and four standard deviations from the mean? What are the numerical values?

We can now turn the problem around and ask the following question: if we measure some mean value $\overline{y}$ and its standard deviation $\sigma(y)$, how significant is that result? Because we are working with statistical inference, there are no absolute answers. We cannot say that subsequent measurements by other experimenters that are several standard deviations removed from our measurement of $\overline{y}$ are totally inconsistent with our results, only that they are not likely to be consistent. If we assume that our measurement errors are normally distributed, then the results of Exercise 4.2 can be interpreted to indicate that only three times out of a thousand measurements will an experimental measurement fall beyond three standard deviations from the mean. Thus, such a measurement is unlikely but not prohibited. As a general rule of thumb, physicists will use three standard deviations as the definition of "significant."

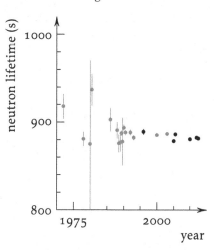

FIGURE 4.4. In 1951, the first "precision" measurement of the neutron lifetime yielded a value of $\tau = 1108 \pm 216$ s. Subsequent experiments have improved the precision of the measurements

An example of measurement uncertainty can be found in the definition of the neutron lifetime.[7] Initial measurements of the neutron lifetime, conducted in the late 1940s and early 1950s, suggested a value of around twenty minutes ($\approx$1100 s). Subsequent experiments have lowered the

---

[7]This is the lifetime of free neutrons. Neutrons in the atomic nucleus appear to be stable in the vast majority of cases. A few radioactive isotopes decay via β particle (electron) emission, in which a neutron in the nucleus does decay.

Table 4.3. Neutron lifetimes τ from the Particle Data Group

Date	τ (s)	$\sigma(\tau)$ (s)	Date	τ (s)	$\sigma(\tau)$ (s)
1951	1108	216	1989	887.6	3.0
1972	918	14	1990	893.6	3.8
1978	881	8	1990	888.4	2.9
1980	875	95	1992	888.4	3.1
1980	937	18	2000	885.4	0.9
1986	903	13	1992	888.4	3.1
1988	891	9	2000	885.4	0.9
1988	876	10	2003	886.8	1.2
1993	882.6	2.7	2010	880.7	1.3
1996	889.2	3.0	2012	882.5	1.4
2005	878.5	0.7	2012	881.6	0.6
2005	886.3	1.2			

value to a consensus (in 2012) of $\tau(n) = 880.0 \pm 0.9$ s. This value is obtained by using the black points illustrated in figure 4.4 (bottom seven entries in Table 4.3). The data, tabulated in Table 4.3, are not all consistent using 3σ as the metric. This gives rise to concerns about potential systematic errors in the recent measurements that affect the accuracy. The error bars in figure 4.4 represent one standard deviation and are comparable to the size of the points for the latest data. As we do not know the exact value for the neutron lifetime, it is not possible to determine which of the recent measurements are in error. All that we can really say is that the most recent measurements are statistically inconsistent.

EXERCISE 4.3.   To determine if two uncertain numbers $a \pm \sigma_a$ and $b \pm \sigma_b$ are comparable, we can define the combined uncertainty as follows:

$$\sigma_{tot} = \sqrt{\sigma_a^2 + \sigma_b^2}.$$

The numbers are significantly different if the quantity $(a - b)/\sigma_{tot}$ is larger than three. Are the two experimental results from 2005 comparable? What about the two results from 2012?

We can hope that further refinements in measurement technology will eventually resolve the discrepancies, or, at least, provide a consistent value for the neutron lifetime. Illustrations like those in figure 4.4 are not terribly uncommon in physics. The Particle Data Group collates experimental information about subnuclear physics and produces regular updates on the currently best available values for various quantities. The group also produces a series of historical graphs depicting the time evolution of the "best available" values. Students may have a mental image that most plots will feature a series of experiments with ever-decreasing standard deviations, with absolutely constant mean values. Such plots do exist but there are many examples where the means are scarcely consistent from one measurement to the next and the mean value varies by a large amount over time. The real world is not simple.

If we now consider the sequence **b** from figure 4.2, we observe that there seems to be a trend in the data. This could, of course, just be an unfortunate coincidence; there may be no real trend, just a larger standard deviation than was observed in sequence **a**. Subsequent measurements could erase the observed trend. In physics though, we are often concerned with measurements where we do indeed expect trends. So, in those cases, how can one assess whether or not the observed trend is statistically significant? Typically, we will try to fit some sort of function to the data. It may be a straight line or some more complicated function that arises from a model that we have constructed. We have proposed that planets travel on elliptical orbits, for example. We could make a number of observations of some planet and then try to determine the parameters **e**, J, etc., that determine the particular orbit.

In general, we can expect the data to represent some function of the independent variables that we group into a vector **x** and some parameters that we group into the vector **a**: $y = y(\mathbf{x}, \mathbf{a})$.[8] What we want to do is fit the model function to the measured data. A reasonably straightforward approach is to minimize some cost function: typically, the square of the difference between the measured and predicted values. This leads to a general approach called least squares. We define a function $\chi^2(\mathbf{x}, \mathbf{a})$ as follows:

$$(4.6) \qquad \chi^2(\mathbf{x}, \mathbf{a}) = \sum_{i=1}^{N} w_i \left[y(\mathbf{x}, \mathbf{a}) - y_i \right]^2 ,$$

where the sum extends over N measured values y_i and the coefficients w_i can be chosen to weight the data in some fashion. Finding the best set of coefficients **a** means finding the set of **a** that minimize Equation 4.6 Without going into the details of that procedure at this time, we can summarize by stating that the result of such a minimization process is a set of estimates of the parameters **a** along with their uncertainties $\sigma(\mathbf{a})$.

To make this idea concrete, let us assume that we have fit the sequence **b** from figure 4.2 with a line: $y = mx + b$. As a result, we would have the values for the slope m and its uncertainty $\sigma(m)$ and the y-intercept b and

---

[8]Note that **x** and **a** are not a vectors in our usual sense: the components that make up the vectors need not have the same dimensionality.

its uncertainty $\sigma(b)$. We can infer that the data describe a straight line if the value of the slope m is more than three standard deviations from zero. In that case, the slope is statistically non-zero. If, instead, the value of the slope was found to be less than three standard deviations from zero, then there would be no statistical support for the assumption that the data depend linearly on x. Notice that our analysis does not prove that the data do (or do not) depend linearly upon x. Subsequent measurements might reverse our inference.

In one relatively recent example of just such a reversal of fortunes, the E288 collaboration at Fermilab published results in 1976 of the purported discovery of a new particle that they designated the Υ. The experimenters were colliding protons accelerated to 400 GeV with a fixed beryllium target and looking for the production of electron-positron pairs.[9] Analysis of the data suggested that a cluster of events found near 6 GeV should be interpreted as evidence for a new state of nuclear matter roughly six times the mass of the proton. The experimenters found that the cluster was more than two standard deviations above the experimental background. Unfortunately, subsequent experiments conducted the next year showed no such cluster of events in the vicinity of 6 GeV but did ultimately find a five standard deviation effect at 9.5 GeV.[10] The mass of the Υ is now understood to be $9.4603(3)\,GeV/c^2$.

4.2. Finite Velocity of Light

We can now return to Rømer's observations of eclipses of Io (Table 4.1). First, we note that Rømer measured the times of the eclipses to the nearest second, suggesting that he believed that he had a precision of seconds in his measurements. There are measurements of Io disappearing into Jupiter's shadow on January 11th and 12th in 1672. These are separated in time by 42 hours, 27 minutes and 8 seconds. This is remarkably close to the modern value of Io's orbital period of 42 hours, 27 minutes and 33.503 seconds. In 1673, Rømer again measured consecutive eclipses in February and March and found elapsed times of 42 hours, 28 minutes and 50 seconds; 42 hours, 28 minutes and 51 seconds; and 42 hours, 27 minutes and 28 seconds, respectively. The average of these four values is 42 hours, 28 minutes and 31 seconds with a standard deviation of 57 seconds.

---

[9]In these units, the rest mass of a proton is $0.938\,GeV/c^2$

[10]Five standard deviations is now the generally accepted standard of significance in the high energy physics field. Leon Lederman, leader of the E288 collaboration, was undoubtedly chagrined to have the $6.0\,GeV/c^2$ non-particle deemed the "Oops-Leon" by uncharitable physicists. His subsequent receipt of the 1988 Nobel Prize in Physics (with Melvin Schwartz and Jack Steinberger) "for the neutrino beam method and discovery of the muon neutrino" undoubtedly softened the blow.

This is a somewhat disappointing result. Cassini led one of the premier astronomical observatories of his day, Rømer was a talented observer and they had access to clocks that measured time to the nearest second. We can infer from the standard deviation of 57 seconds, however, that the measurement errors are not dominated by the precision of the clocks. Nevertheless, even with their one minute precision, Cassini and Rømer would be able to greatly improve the ability of navigators to establish their longitude over what was available at the time.

Of course, the astute student might also reason that the problem with the measurement inaccuracies is simply due to the fact that Io's orbit is not periodic for some reason. It could be that the timing precision is now good enough to reveal that unforeseen aspect of orbital motion. This is a good observation. The fact is that one must understand how the measurements were made and, thereby, decide how they should be interpreted. As it happens in this case, we have a model that predicts periodic motion and are therefore biased towards interpreting the disparate results as "measurement errors." There are many examples in physics, though, where new measurements made with more precise instrumentation has led to the discovery of systematic deviations from established theories and, hence, new physics. In the Io data, it is most likely that the discrepancies are indeed due to uncertainties arising from establishing the precise moment at which Jupiter's shadow hit the Ionian surface or Io first emerged from the Jovian shadow or the long-term accuracy of the clocks. Nonetheless, it is a good practice to ask the question "Is there an alternative explanation for the observed experimental results?"

Closer inspection of Rømer's Io data reveals systematic discrepancies. The Earth and Jupiter were at their closest on March 2, 1672, at the position astronomers call the conjunction (point H in figure 4.1). Rømer was able to observe Io emerging from Io's shadow on March 7 and again on March 14. In that time, Io completed four orbits of Jupiter. If we utilize the 42 hour, 28 minute and 31 second average value for Io's orbital period, we can use this to predict the occurrence of the other Io eclipses. Note that we utilize the difference between predicted and observed in order to observe small differences. If one simply plots the eclipse times, the small differences that we are investigating will be obscured.

> EXERCISE 4.4. Use the observational times from 1672, as listed in Table 4.1. Compute the time differences between the emerging observations (point D) and the observation on March 7. You can use the *Mathematica* function DateDifference. The orbital period will be one quarter of the elapsed time between the March 7 and March 14 observations. Divide the time differences by the orbital period to obtain the number of orbits for each observation. The predicted

FIGURE 4.5. The differ-
ences in time (Δt) between
Rømer's observed Ionian
eclipses and the prediction
based on an Ionian orbital
period of 42:28:31 are plotted as a series of points.
The error bars on the points
represent the standard
deviation of one minute

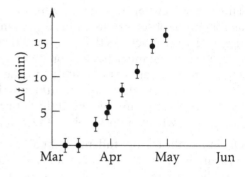

eclipse times can be computed by multiplying the integer number of
orbits (obtained with the Round function) by the orbital period and
using the DatePlus function. Finally, compute the difference Δt between the observed and predicted eclipse times. You should be able
to recover the results in figure 4.5

As indicated in figure 4.5, Rømer's observations indicate that the differences between predicted and observed eclipse times vary systematically
and significantly. From his data, we can estimate that the standard deviation of Rømer's measurements was one minute but the later measurements have differences that are larger than ten minutes, or more than
ten standard deviations. This is statistically highly unlikely. Of course,
a possible explanation is that we have somehow managed to compute the
wrong orbital period. If we used a longer orbital period, we would certainly reduce the later differences.

Consider, though, the first measurement from 1673 that occurs after
conjunction. Using our value for the orbital period, the April 18, 1673
observation occurs after Io has completed an additional 230 orbits of
Jupiter after the March 7, 1672 eclipse. If we have underestimated the
orbital period, then we should see a large difference between the observed and predicted times for this eclipse. Using the orbital period of
42 hours, 28 minutes and 31 seconds that we obtained earlier, we predict
the emergence of Io to occur at 21:18:13, which is a difference of only
3 minutes and 45 seconds from the 21:22:00 time that Rømer observed.
This is very surprising. In the 66 orbits depicted in figure 4.5, there is a
fifteen-minute discrepancy for the April 29, 1672 observation. In another
200 or so orbits, we should have expected that an underestimate of the
orbital period would yield a discrepancy of an hour or more. Yet, Rømer's
observation falls within three minutes of the predicted value. It appears
that we can exclude an underestimate of the orbital period as the source
of the observed variation.

Exercise 4.5.   Use the time difference between the March 7, 1672 and April 18, 1673 (divided by 230 orbits) to define the orbital period. Replot the results from figure 4.5. Is there still a significant variation?

Exercise 4.6.   Use the original 42 hour, 28 minute and 31 second orbital period and all of the observations from 1672 and 1673 to compute Δt. Compute Δt as before for the emergence data. For the immersion data (location C in Table 4.1), use the February 20, 1672 observation as the comparison time.

Rømer recognized that the variation is a minimum near the conjunction (point H in figure 4.1) and increased as the earth moved along its orbit. He predicted that this variation would reach its maximum at opposition, although Jupiter is not visible from Earth at that time. This suggestion is borne out by the results shown in figure 4.6. As the Earth and Jupiter move away from conjunction (March 7 and onward), the observed eclipse times move later and later beyond the predicted times. Given our estimate of the precision of Rømer's observations, these are significant deviations. In the subsequent year, however, the times of emergence are again close to zero initially and then increase as the Earth moves away from conjunction. As the Earth and Jupiter moved toward conjunction (February 20, 1672 and before), the observed eclipse times (moving backwards in time away from the conjunction) were found to be earlier and earlier before the predicted times.

In figure 4.6, we have also plotted the Earth-Jupiter distance d_{EJ} for the years 1672 and 1673, along with the Δt data. The error bars for the eclipses are approximately the size of the data points in this figure and so have been suppressed. The distance data have been scaled and shifted so that the minimum value plots along the x-axis; this does not represent a fit to the data. Nevertheless, the observed time differences Δt match well to the nearly sinusoidal variation of the distance. Now, these distance data were not available to Rømer and Cassini and the observations do not fall perfectly along a sinusoid but Rømer was convinced that these data were compelling: light takes a finite amount of time to reach Earth from Jupiter. Cassini remained unconvinced and the scientific debate continued for some time.[11]

Exercise 4.7.   Use the *Mathematica* function AstronomicalData to obtain the Earth-Jupiter distance for the years that Rømer observed Io. (This will require Internet access.) Plot the (scaled) distance

---

[11]Rømer never published his findings in the journal of the Royal Society, in part due to Cassini's continuing objections to attributing the variations to a finite velocity of light.

FIGURE 4.6. The time differences between predicted and observed eclipses of Io from 1672 and 1673 are plotted as *circles* for emerging *D* and as *triangles* for immersing *C* observations. The Earth-Jupiter distance during this time is illustrated as the *gray curve*. The minimum distance was approximately 6.63×10^{11} m

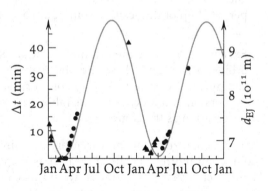

Jan Apr Jul Oct Jan Apr Jul Oct Jan

and Δt and examine the correlation. What happens if you include Rømer's data from the years 1674–1676?

EXERCISE 4.8. Consider the measurements listed in Table 4.4. Use the *Mathematica* function LinearModelFit to fit linear ($f(x) = a + bx$) and quadratic ($f(x) = a + bx + cx^2$) functions. Plot the results. Examine the uncertainties in the fitted parameters. Is the *c* coefficient significant? Can you justify the proposals that the data are described by either a quadratic function or a linear function?

TABLE 4.4. More experimental data

x	y	x	y	x	y	x	y
1	−0.645	6	−6.158	11	−28.422	16	−41.135
2	−10.564	7	−10.979	12	−24.535	17	−31.868
3	1.263	8	−17.010	13	−26.984	18	−46.242
4	−6.536	9	−20.059	14	−29.790	19	−38.129
5	−17.143	10	−23.655	15	−25.605	20	−34.853

In 1704, Isaac Newton published his book *Opticks*, in which he argued for a corpuscular theory of light. That is, Newton imagined that light was composed of small particles and, using this theory, he was able to predict some of the observed behavior of light, such as the refraction (bending) of light beams at interfaces. Given the success of Newton's other scientific theories such as gravitation and his profound scientific reputation, it is not surprising to find that the corpuscular theory of light held sway for a long time. Others believed that a wave theory provided a better description of light, and was capable of explaining phenomena that Newton's corpuscular theory could not, such as diffraction.[12] The exact nature of light was not established until much later.

In 1849, the French physicist Hyppolite Fizeau made the first direct measurement of the velocity of light on earth. A year later, his former colleague Léon Foucault made a more precise measurement of the velocity of light and was able to further demonstrate that the velocity of light was slower in water than in air. This was a key finding in support of a wave theory of light, as the corpuscular theory demanded that light travel more rapidly in water than in air. A complete theory of light as a component of the phenomenon we call *electromagnetism* was provided by James Clerk Maxwell in 1864. We shall investigate more aspects of electromagnetic theory subsequently. For the moment, we shall focus on a key result of Maxwell's theory: the components of the time-varying electric and magnetic fields are solutions of the wave equation.

4.3. Wave Equation

Waves are a manifestation of energy propagating through a system and are key elements in our understanding of many phenomena. The equation that describes waves arises in many areas of physics and was well-studied before Maxwell's application to electromagnetic phenomena. In its simplest form, let us assume that there is some function f that describes a physical property such as density, or pressure, or displacement, of some system. For the moment, let us further assume that f is only a function of one space dimension. The wave equation in one dimension can be written as follows:

$$(4.7) \qquad \frac{\partial^2 f(x,t)}{\partial x^2} - \frac{1}{v_c^2} \frac{\partial^2 f(x,t)}{\partial t^2} = 0,$$

where the symbol ∂ indicates a *partial derivative* and v_c is a characteristic velocity. The partial derivative of a function of multiple variables is obtained by holding the other variables constant and taking the derivative of just the one variable:

$$\frac{\partial f(x,t)}{\partial x} = \lim_{h \to 0} \frac{f(x+h,t) - f(x,t)}{h} \quad \text{and} \quad \frac{\partial f(x,t)}{\partial t} = \lim_{\delta \to 0} \frac{f(x,t+\delta) - f(x,t)}{\delta}.$$

In this simplest description of wave phenomena, waves propagate with a characteristic velocity v_c that depends only upon the physical properties of the material in which the wave is moving and is a constant everywhere in space. A more complex situation arises if we allow the characteristic

---

[12]We shall define these terms in more detail subsequently.

velocity itself to be a function of space and time. For the present, we shall ignore such complications.

Another assumption underlying Equation 4.7 is that the function f represents a relatively small disturbance on some background. Sound waves, for example, travel through air and water (with different characteristic velocities) but the presence of the wave does not drastically or permanently alter the material properties. On the other hand, high-powered lasers will ionize the air through which they are propagating, drastically modifying the material properties of the medium and, thereby, significantly affecting the ability of the beam to continue propagating.[13] In this first exposure to wave phenomena, we shall follow the simpler course and assume that the wave does not significantly modify the medium in which it is propagating. We shall not stretch guitar strings to their breaking point and not investigate situations in which the energy density in the wave is sufficient to ionize the material or break chemical bonds.

Now, as a general rule, partial differential equations are much more difficult to solve than ordinary differential equations. For the wave equation, we can make use of a trick first proposed by the French mathematician Jean-Baptiste le Rond d'Alembert: make a change of variables from x and t to a new set of variables $\zeta = x + v_c t$ and $\xi = x - v_c t$. In this new coordinate system, the wave equation becomes:

$$(4.8) \qquad \frac{\partial^2 f(\zeta, \xi)}{\partial \zeta \, \partial \xi} = 0.$$

We can now write down solutions immediately! *Any* function of ζ or ξ separately is a solution to the wave equation. The most general solution to the wave equation is written as follows:

$$(4.9) \qquad f(x, t) = f_1(x + v_c t) + f_2(x - v_c t),$$

where f_1 and f_2 are any functions.

> EXERCISE 4.9. Changing variables requires repeated application of the chain rule defined as follows for ordinary derivatives:
>
> $$\frac{df}{dx} = \frac{df}{dy}\frac{dy}{dx}.$$
>
> For partial derivatives, where the original variable x can be thought of as a function of both new variables ζ and ξ, we find the chain rule can be written as follows:

---

[13]The US Strategic Defense Initiative, commonly known as Star Wars, investigated using lasers to shoot down incoming nuclear warheads or destroy missiles in their launch phase. In these studies, the simplifying assumption that the propagating wave does not modify or interact with the medium of propagation has to be abandoned.

$$\frac{\partial f}{\partial x} = \frac{\partial f}{\partial \zeta}\frac{\partial \zeta}{\partial x} + \frac{\partial f}{\partial \xi}\frac{\partial \xi}{\partial x}.$$

From d'Alembert's definition of ζ and ξ, we know that $\partial\zeta/\partial x = \partial\xi/\partial x = 1$ and that $\partial\zeta/\partial t = v_c$ and $\partial\xi/\partial t = -v_c$. Use these results to recover d'Alembert's equation 4.8.

EXERCISE 4.10. Show explicitly (by differentiating) that the following functions are solutions to the wave equation:

(a) $\sin(x + v_c t)$ (b) $\cos(x - v_c t)$ (c) $e^{(x-v_c t)}$.

EXERCISE 4.11. Plot the function $\sin(x + v_c t)$ over the domain $0 \le x \le 5$ and use the *Mathematica* Manipulate function to vary a single parameter vct from $0 \le$ vct ≤ 2. How does the function change as time (vct) increases?

Repeat for the function $\sin(x - v_c t)$.

Which of the two functions would be described as the forward propagating solution, if we define "forward" as the positive x-direction? Replace the Manipulate function with the Animate function. Does this help with the definition of forward propagation?

EXERCISE 4.12. Define the function $f(x,t) = \sin[a(x - v_c t)]e^{-b(x-v_c t)^2}$. Plot the function from $-10 \le x \le 10$ and use the *Mathematica* function Manipulate to vary the parameters $1 \le a \le 20$ and $0 \le b \le 10$. Define a single parameter vct that provides the time dependence and use the Animate function to vary this over the domain $-10 \le$ vct ≤ 10. The Gaussian portion of $f(x,t)$ forces the function to have a finite width; such short pulses are found in many wave applications.

How does the behavior of the function change as a and b are varied?

There are many features of the wave equation that we will explore subsequently that will enable us to describe many physical phenomena. For the moment, we shall focus on some of the mathematical properties of Equation 4.7. We have alluded to conservation laws throughout our discussion to this point and have noted that in Newton's law of universal gravitation that there are several quantities that do not change over time. In light of the importance of Noether's theorem to our deeper understanding of physical phenomena, we now want to examine the wave equation for conserved quantities.

Consider transforming to a new coordinate system (x', t'), where the function f that defines our wave is now thought of as a function of these new coordinates: $f = f(x', t')$. Converting to the new coordinate system means making use of the chain rule for partial derivatives that we described above. For the derivative with respect to x, we have the following expression:

$$(4.10) \qquad \frac{\partial f(x,t)}{\partial x} = \frac{\partial f(x',t')}{\partial x'} \frac{\partial x'}{\partial x} + \frac{\partial f(x',t')}{\partial t'} \frac{\partial t'}{\partial x}.$$

If we assume that the transformation between the (x,t) coordinate system and the (x',t') coordinate system is linear, then the second derivative with respect to x can be rewritten as follows:

$$(4.11) \quad \frac{\partial^2 f(x,t)}{\partial x^2} = \frac{\partial^2 f(x',t')}{\partial x'^2} \left(\frac{\partial x'}{\partial x}\right)^2$$
$$+ 2\frac{\partial^2 f(x',t')}{\partial x' \partial t'} \frac{\partial x'}{\partial x} \frac{\partial t'}{\partial x} + \frac{\partial^2 f(x',t')}{\partial t'^2}\left(\frac{\partial t'}{\partial x}\right)^2.$$

EXERCISE 4.13.   Use the chain rule to differentiate Equation 4.10 by x. By *linear*, we mean that $\partial^2 x'/\partial x^2 = 0$. Show that you obtain Equation 4.11.

EXERCISE 4.14.   Derive the analogous expression to Equation 4.11 for the second derivative with respect to time: $\partial^2 f(x,t)/\partial t^2$.

If we consider the wave equation from Equation 4.7, we find that the transformation to the primed coordinate system yields the following result:

$$\frac{\partial^2 f(x,t)}{\partial x^2} - \frac{1}{v_c^2}\frac{\partial^2 f(x,t)}{\partial t^2} = \frac{\partial^2 f(x',t')}{\partial x'^2}\left[\left(\frac{\partial x'}{\partial x}\right)^2 - \frac{1}{v_c^2}\left(\frac{\partial x'}{\partial t}\right)^2\right]$$
$$+ 2\frac{\partial^2 f(x',t')}{\partial x' \partial t'}\left[\frac{\partial x'}{\partial x}\frac{\partial t'}{\partial x} - \frac{1}{v_c^2}\frac{\partial x'}{\partial t}\frac{\partial t'}{\partial t}\right]$$
$$(4.12) \qquad\qquad - \frac{1}{v_c^2}\frac{\partial^2 f(x',t')}{\partial t'^2}\left[\left(\frac{\partial t'}{\partial t}\right)^2 - v_c^2\left(\frac{\partial t'}{\partial x}\right)^2\right].$$

So, for arbitrary choices of coordinate systems, we will not recover the same form of the wave equation. Note that on the right hand side of Equation 4.12, the second term involves a cross term that is not present on the left hand side.

EXERCISE 4.15. Use the results from Equation 4.11 and Exercise 4.14 and show that you obtain Equation 4.12. The derivation requires a fair amount of algebra. Check your results by dimensional analysis. Suppose that f has dimension X. Then $\partial f/\partial t$ has dimension X/T.

Let us choose a general form for the linear transformation that relates the two coordinate systems. Suppose that we choose the following relations[14]:

(4.13) $\qquad x' = a_1 x + a_2 t + a_3 \qquad$ and $\qquad t' = a_4 x + a_5 t + a_6.$

Then, substituting back into Equation 4.12, we obtain the following:

$$\frac{\partial^2 f(x,t)}{\partial x^2} - \frac{1}{v_c^2}\frac{\partial^2 f(x,t)}{\partial t^2} = \frac{\partial^2 f(x',t')}{\partial x'^2}\left[a_1^2 - \frac{a_2^2}{v_c^2}\right]$$

(4.14)
$$+ 2\frac{\partial^2 f(x',t')}{\partial x'\,\partial t'}\left[a_1 a_4 - \frac{a_2 a_5}{v_c^2}\right] + \frac{\partial^2 f(x',t')}{\partial t'^2}\left[a_5^2 - v_c^2 a_4^2\right].$$

If we want to ensure that the form of the wave equation is preserved in the new (primed) coordinate system, then the terms in square brackets on the right hand side of Equation 4.14 must take on particular values:

(4.15) $\qquad a_1^2 - \dfrac{a_2^2}{v_c^2} = 1 \quad$ and $\quad a_1 a_4 - \dfrac{a_2 a_5}{v_c^2} = 0 \quad$ and $\quad a_5^2 - v_c^2 a_4^2 = 1.$

Note that the constants a_3 and a_6 are not constrained and can take on any values. We shall exclude them from further consideration at the moment.[15] We can find solutions to these equations if we first recall that the hyperbolic trigonometric functions satisfy the equation $\cosh^2\zeta - \sinh^2\zeta = 1$.[16] This suggests that we can set $a_1 = \cosh\zeta$, $a_2/v_c = \sinh\zeta$, $a_5 = \cosh\chi$ and $a_4 v_c = \sinh\chi$. These choices will automatically satisfy the first and third of Equations 4.15. We are left with the following result for the cross term, the second of Equations 4.15:

$$\cosh\zeta\,\sinh\chi - \sinh\zeta\,\cosh\chi = 0.$$

We can use the addition formula for the hyperbolic functions to conclude that this leads to $\sinh(\zeta - \chi) = 0$. For real values of ζ and χ, this is true only if $\zeta = \chi$. We can thus write the general transformation that preserves the form of the wave equation as follows:

(4.16) $\qquad v_c t' = v_c t\cosh\zeta + x\sinh\zeta \qquad$ and $\qquad x' = v_c t\sinh\zeta + x\cosh\zeta.$

A number of physicists and mathematicians have separately derived this result but it is known almost universally as the Lorentz transformation.[17]

---

[14]Note that for these equations to make sense, the coefficients must have dimensions.

[15]Setting $a_3 = a_6 = 0$ amounts to making the choice that origins of the coordinate systems coincide at time $t = t' = 0$.

[16]The parameter ζ here is not related to the radial coordinate in a cylindrical coordinate system.

[17]The Dutch physicist Hendrik Antoon Lorentz contributed significantly to our understanding of the mathematical properties of the wave equation. He was awarded the Nobel Prize in Physics (with Pieter Zeeman) in 1902 "in recognition of the extraordinary service they rendered by their researches into the influence of magnetism upon radiation phenomena."

EXERCISE 4.16. The hyperbolic trigonometric functions $\cosh x$ and $\sinh x$ can be defined in terms of the exponential function:

$$\cosh x = \tfrac{1}{2}(e^x + e^{-x}) \qquad \text{and} \qquad \sinh x = \tfrac{1}{2}(e^x - e^{-x}).$$

Use the exponential form to show that $\cosh^2 x - \sinh^2 x = 1$.

Use the exponential forms of the functions to prove the addition formula:

$$\sinh x, \cosh y - \cosh x \sinh y = \sinh(x - y).$$

EXERCISE 4.17. Use the *Mathematica* function `ParametricPlot` to examine the effects of the Lorentz transform (Equation 4.16) on the (x', t') plane. Use the `Manipulate` function to vary the ζ parameter over the domain $0 \le \zeta \le 1$. Changing the characteristic velocity v_c will rescale the time coordinate but not alter any essential behavior, so just use $v_c = 1$.

How do the x' and t' coordinates change as the parameter ζ changes?

4.4. Lorentz Transform

The Lorentz transform is something of a curiosity: it mixes space and time dimensions but it does so in a particular fashion. In order for the transformation to make dimensional sense, we note that it is actually the quantity ct and not time alone that forms the first component of the vector. Let us consider the concept of an *event*. This is actually just a point in space and time: $P_1 = (ct_1, x_1)$, where henceforth we will make use of the common notation in which the magnitude of the velocity of light is denoted by the symbol c. We are often concerned with the relationships between two events, P_1 and P_2. What we can demonstrate is that the Lorentz transformation preserves the quantity $s^2 = (ct)^2 - x^2$.[18] We shall discuss the importance of this requirement soon.

EXERCISE 4.18. Use Equation 4.16 and demonstrate explicitly that the following relation holds:

$$(ct')^2 - (x')^2 = (ct)^2 - x^2.$$

Consider two events P_1 and P_2 that occur at the same point in space x_1 but at two different times t_1 and t_2, where we can assume $t_1 < t_2$. The vector that connects these two spacetime points is given by $P_2 - P_1 = (c(t_2 - t_1), 0)$, which is a point on the positive ct-axis. We would call such a vector a *time-like* vector, as it represents two events separated in time. The effect

---

[18]There is a sign choice to be made here. The Lorentz transform also preserves the quantity $-s^2$. We shall justify our usage soon.

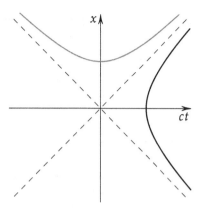

FIGURE 4.7. The Lorentz transform preserves time and space. A time-like quantity (*black curve*) will remain time-like under transformation. Likewise, a space-like quantity (*gray curve*) will remain space-like. Light-like $((ct)^2 - x^2 = 0)$ vectors lie on the *diagonal dashed lines*.

of different Lorentz transformations on the vector $P_2 - P_1$ (different values of ζ in Equation 4.16) will move the vector along a hyperbola in spacetime. This is illustrated by the black curve in figure 4.7. Note that this hyperbola is not the same as the hyperbolic trajectories that we have discussed previously. The Lorentz transform does not define a trajectory. We note that all points along the black curve have the same value of $s^2 = c^2(t_2 - t_1)^2$.

Now consider two events P_3 and P_4 that occur at the same time t_3 but at two different points in space x_3 and x_4, where we can assume $x_3 < x_4$. The vector connecting these two points in spacetime is given by $P_4 - P_3 = (0, x_4 - x_3)$, which is a point on the positive x-axis A Lorentz transform applied to this *space-like* vector will produce a point somewhere along the gray curve illustrated in figure 4.7; all points on the gray curve have the same value of $s^2 = -(x_4 - x_3)^2$.

We note that the time-like vectors have the property that $s^2 > 0$ and that the space-like vectors have the property that $s^2 < 0$ and that the Lorentz transform does not alter that property. The case where $s^2 = 0$ corresponds to the diagonal lines in figure 4.7. Points along these diagonals are called *light-like* vectors.

> EXERCISE 4.19. Use the *Mathematica* ParametricPlot function to plot the Lorentz transforms of the vectors $(ct = 1, x = 0)$ and $(ct = 0, x = 1)$, as a function of the transform parameter ζ. Show that you obtain the curves illustrated in figure 4.7.

Before proceeding further, we should acknowledge that Maxwell's theory of electromagnetism results in electromagnetic fields that are solutions to the three-dimensional wave equation, not the one-dimensional equation that we have studied thus far. So, it is reasonable to ask what happens in two and three dimensions? We again would like to ascertain the types of transformations that leave the equation unchanged. Let's continue our investigations and examine the two-dimensional version of the wave equation.

The wave equation in two spatial dimensions can be written as follows:

(4.17) $$\frac{\partial^2 f(x,y,t)}{\partial x^2} + \frac{\partial^2 f(x,y,t)}{\partial y^2} - \frac{1}{c^2}\frac{\partial^2 f(x,y,t)}{\partial t^2} = 0,$$

where here we are using c to denote the magnitude of the velocity of light. If we again restrict ourselves to linear transformations and exclude constant terms, the general coordinate transformation can be written as the following:

$$ct' = a_1 ct + a_2 x + a_3 y \qquad x' = a_4 ct + a_5 x + a_6 y \qquad y' = a_7 ct + a_8 x + a_9 y$$

We shall forge ahead now and derive the equivalent expression to that we obtained in Equation 4.14. This will provide us with a set of equations for the unknown coefficients $a_1,\ldots,a_9$ but let us try to be somewhat more sophisticated in our analysis.

First of all, let us introduce matrix notation, which will help us see the mathematical structure of the operations more clearly. Let us define vectors $\mathbf{r} = (ct,x,y)$ and $\mathbf{r}' = (ct',x',y')$, and a matrix $\mathbf{A}$. The transformation equations above can be written concisely as $\mathbf{r}' = \mathbf{A}\mathbf{r}$ or, more explicitly as follows:

(4.18) $$\begin{bmatrix} ct' \\ x' \\ y' \end{bmatrix} = \begin{bmatrix} a_1 & a_2 & a_3 \\ a_4 & a_5 & a_6 \\ a_7 & a_8 & a_9 \end{bmatrix} \begin{bmatrix} ct \\ x \\ y \end{bmatrix}.$$

The matrix $\mathbf{A}$ has elements A_{ij}, where the index i runs over the rows and the index j runs over the columns. For example, the element that corresponds to the first row and third column would be $A_{13} = a_3$. The rule for multiplying a matrix by a vector involves multiplying the matrix row elements by the vector column elements and adding. That is, the ith element of the resultant vector is given by the following expression:

$$r_i' = \sum_{j=1}^{3} A_{ij} r_j.$$

So, for the first row, we would find the following:

$$r_1' = ct' = A_{11} r_1 + A_{12} r_2 + A_{13} r_3 = a_1 ct + a_2 x + a_3 y,$$

which is simply the first of the three transformation equations. The matrix notation provides an alternative, more compact representation of the defining equations of the Lorentz transform.

Now to obtain the requirements for transformations that leave the two-dimensional wave equation unchanged, we can recognize that the transformation that we have already derived for the one-dimensional wave equation also leaves the two-dimensional wave equation unchanged. It amounts to the special case where $a_3 = a_6 = a_7 = a_8 = 0$ and $a_9 = 1$ and mixes the ct and x dimensions, leaving the y-dimension unchanged. In matrix form, we can write this as follows:

(4.19)
$$\begin{bmatrix} ct' \\ x' \\ y' \end{bmatrix} = \begin{bmatrix} \cosh \zeta_1 & \sinh \zeta_1 & 0 \\ \sinh \zeta_1 & \cosh \zeta_1 & 0 \\ 0 & 0 & 1 \end{bmatrix} \begin{bmatrix} ct \\ x \\ y \end{bmatrix},$$

where ζ_1 is the parameter defining the Lorentz transform.

We can immediately see that we can construct a transformation in which the ct and y dimensions are mixed, leaving the x dimension unchanged. We can write this as follows:

(4.20)
$$\begin{bmatrix} ct' \\ x' \\ y' \end{bmatrix} = \begin{bmatrix} \cosh \zeta_2 & 0 & \sinh \zeta_2 \\ 0 & 1 & 0 \\ \sinh \zeta_2 & 0 & \cosh \zeta_2 \end{bmatrix} \begin{bmatrix} ct \\ x \\ y \end{bmatrix},$$

where ζ_2 is a parameter that defines the Lorentz transform. Both of these transformations leave the wave equation unchanged and each mixes one of the spatial dimensions with the time dimension.

What remains now is to determine the sort of transformation in which the time dimension is left unchanged while the spatial dimensions are mixed. This amounts to looking for a transformation in which $a_2 = a_3 = a_4 = a_7 = 0$ and $a_1 = 1$. In matrix form, we are looking for a transformation of the following form:

(4.21)
$$\begin{bmatrix} ct' \\ x' \\ y' \end{bmatrix} = \begin{bmatrix} 1 & 0 & 0 \\ 0 & a_5 & a_6 \\ 0 & a_8 & a_9 \end{bmatrix} \begin{bmatrix} ct \\ x \\ y \end{bmatrix},$$

If we ignore the time derivative component of the wave equation for the moment, which will not be affected by the transformation depicted in Equation 4.21, the spatial derivatives must have the following transformation properties:

(4.22)
$$\frac{\partial^2 f(x,y,t)}{\partial x^2} + \frac{\partial^2 f(x,y,t)}{\partial y^2} = \frac{\partial^2 f(x',y',t')}{\partial x'^2}\left[a_5^2 + a_8^2\right]$$
$$+ 2\frac{\partial^2 f(x',y',t')}{\partial x'\partial y'}\left[a_5 a_6 + a_8 a_9\right]$$
$$+ \frac{\partial^2 f(x',y',t')}{\partial y'^2}\left[a_6^2 + a_9^2\right].$$

The wave equation will be left unchanged if the following relations hold:

(4.23) $a_5^2 + a_8^2 = 1$ and $a_5 a_6 + a_8 a_9 = 0$ and $a_6^2 + a_9^2 = 1$.

If we choose $a_5 = \cos\theta$, $a_8 = \sin\theta$, $a_9 = \cos\phi$ and $a_6 = \sin\phi$, then the first and third of Equations 4.23 are satisfied immediately. The addition formula for trigonometric functions allows us to rewrite the second of Equations 4.23 as follows:

$$\cos\theta \sin\phi + \sin\theta \cos\phi = \sin(\theta + \phi) = 0,$$

which is satisfied if $\theta = -\phi$. The transformation that leaves the wave equation unchanged and mixes the spatial dimensions can be written as follows:

(4.24)
$$\begin{bmatrix} ct' \\ x' \\ y' \end{bmatrix} = \begin{bmatrix} 1 & 0 & 0 \\ 0 & \cos\theta & -\sin\theta \\ 0 & \sin\theta & \cos\theta \end{bmatrix} \begin{bmatrix} ct \\ x \\ y \end{bmatrix},$$

This transformation is a rotation of the coordinates!

EXERCISE 4.20.   Fill in the missing steps in the derivation of Equation 4.22. Hint: focus on the spatial derivatives of the two-dimensional wave equation.

EXERCISE 4.21.   Use the *Mathematica* function `ParametricPlot` to examine the effects of the Lorentz transformations (Equation 4.24) on the (x', y') plane. Use the `Manipulate` function to vary the θ parameter over the domain $0 \leq \theta \leq \pi$.

How do the x' and y' coordinates change as the parameter θ changes?

What we shall simply assert without proof is that *any* Lorentz transformation in two space dimensions can be decomposed in terms of the three basic transformations associated with the parameters ζ_1, ζ_2 and θ. The mathematical proof of this assertion is beyond the scope of our present discussion, so we shall proceed apace. The transformations associated with the parameters ζ_1 and ζ_2 are called *boosts* for historical reasons that will become somewhat more apparent subsequently. The transformation associated with the parameter θ is a rotation in the x-y plane.[19] In two space dimensions, the Lorentz transform leaves the quantity $s^2 = (ct)^2 - x^2 - y^2$ unchanged.

EXERCISE 4.22.   Vectors in *Mathematica* are stored as lists and matrices as lists of lists. We can define, for example:

$$r = \{ct, x, y\}$$

---

[19]It is tempting to think of this as a rotation "around" the time ct axis and to think of the boosts as rotations of a sort around the spatial axes.

and

$$B_1[a\_] := \{\{Cosh[a], Sinh[a], 0\}, \{Sinh[a], Cosh[a], 0\}, \{0, 0, 1\}\}.$$

You can use the *Mathematica* function MatrixForm to view the matrix in a more usual format. Matrix-vector multiplication can be performed with the "dot" operator: rp = B1.r, where the result of the operation is another vector rp.

Define matrices for the two boosts B1 and B2 and the rotation R. Define the vector r as shown above and compute the following transforms:

   (a) B1[a].r   (b) B1[a].B2[b].r   (c) B1[a].B2[b].R[c].r

Show in each case that the quantity $(ct')^2 - x'^2 - y'^2$ is unchanged. We haven't explained how to compute the matrix products but you can think of the operations above as a series of matrix-vector products beginning at the right, e.g., first, compute the vector that results from B2[b].r and then compute the vector that results from applying B1[a] to that vector. This works because matrix multiplication is associative: $(\mathbf{AB})\mathbf{C} = \mathbf{A}(\mathbf{BC})$.

So, what we observe for the two-dimensional wave equation is that the Lorentz transform preserves the time-like or space-like nature of vectors. This is illustrated in figure 4.8, where we have depicted the hyperbolic surfaces that are defined by the Lorentz transform. If we consider a time-like vector, say the point where the dark surface intersects the ct-axis, then the result of any Lorentz transform on that vector will be to displace the vector somewhere on the dark surface. All points on the dark surface have the same (positive) value of $(ct)^2 - x^2 - y^2$. Similarly, a space-like vector, say a point where the light surface intersects the positive x-axis, would be transformed somewhere along the light surface. All points on the light surface have the same (negative) value of $(ct)^2 - x^2 - y^2$. As can be seen from the figure, the two surfaces never intersect.

EXERCISE 4.23. Use the ParametricPlot3D function to examine the effect of Lorentz transformations, Equations 4.19, 4.20 and 4.24, on a time-like vector ($ct = 1, x = 0, y = 0$) and a space-like vector ($ct = 0, x = 1, y = 0$). Use the boost parameter ζ for the x-direction and a separate parameter θ for rotations in the x-y plane.

The step now to three space dimensions is straightforward; we have all of the mathematical tools we require. The wave equation in three space dimensions can be written as follows:

$$(4.25) \quad \frac{\partial^2 f(x,y,z,t)}{\partial x^2} + \frac{\partial^2 f(x,y,z,t)}{\partial y^2} + \frac{\partial^2 f(x,y,z,t)}{\partial z^2} - \frac{1}{c^2}\frac{\partial^2 f(x,y,z,t)}{\partial t^2} = 0.$$

FIGURE 4.8. The Lorentz transformation in two space dimensions maps out hyperbolic surfaces. Time-like vectors are mapped into other time-like vectors (*dark surface*). Space-like vectors are mapped into other space-like vectors (*light surface*). The results displayed in figure 4.7 represent a cut along the $y = 0$ plane

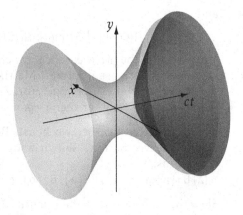

We are again interested in transformations that leave the equation unchanged and these can be constructed now from the basic boost and rotation operations. As we have three space dimensions, we are concerned with transformations of a four-dimensional vector $\mathbf{r} = (ct, x, y, z)$. There will be three boosts $\mathbf{B}$, corresponding to transformations that mix the time and space dimensions and three rotations $\mathbf{R}$ that mix the spatial dimensions. By analogy to our previous efforts, we can write these immediately:

$$\mathbf{B}_1 = \begin{bmatrix} \cosh\zeta_1 & \sinh\zeta_1 & 0 & 0 \\ \sinh\zeta_1 & \cosh\zeta_1 & 0 & 0 \\ 0 & 0 & 1 & 0 \\ 0 & 0 & 0 & 1 \end{bmatrix} \qquad \mathbf{R}_1 = \begin{bmatrix} 1 & 0 & 0 & 0 \\ 0 & \cos\theta_1 & -\sin\theta_1 & 0 \\ 0 & \sin\theta_1 & \cos\theta_1 & 0 \\ 0 & 0 & 0 & 1 \end{bmatrix}$$

$$\mathbf{B}_2 = \begin{bmatrix} \cosh\zeta_2 & 0 & \sinh\zeta_2 & 0 \\ 0 & 1 & 0 & 0 \\ \sinh\zeta_2 & 0 & \cosh\zeta_2 & 0 \\ 0 & 0 & 0 & 1 \end{bmatrix} \qquad \mathbf{R}_2 = \begin{bmatrix} 1 & 0 & 0 & 0 \\ 0 & \cos\theta_2 & 0 & \sin\theta_2 \\ 0 & 0 & 1 & 0 \\ 0 & -\sin\theta_2 & 0 & \cos\theta_2 \end{bmatrix}$$

$$(4.26) \ \mathbf{B}_3 = \begin{bmatrix} \cosh\zeta_3 & 0 & 0 & \sinh\zeta_3 \\ 0 & 1 & 0 & 0 \\ 0 & 0 & 1 & 0 \\ \sinh\zeta_3 & 0 & 0 & \cosh\zeta_3 \end{bmatrix} \qquad \mathbf{R}_3 = \begin{bmatrix} 1 & 0 & 0 & 0 \\ 0 & 1 & 0 & 0 \\ 0 & 0 & \cos\theta_3 & -\sin\theta_3 \\ 0 & 0 & \sin\theta_3 & \cos\theta_3 \end{bmatrix}.$$

The choice of signs of the sine terms in $\mathbf{R}_2$ is predicated on making a rotation in the positive sense around the y-axis.[20]

---

[20]By positive sense, we use a right-hand rule: if your right thumb points along the axis of rotation, your fingers curl in the positive angular direction.

Otherwise, it is reasonably clear that we have simply distributed the boost and rotation operations across an additional spatial dimension. Again we will find that the Lorentz transformations will preserve the quantity $s^2 = (ct)^2 - x^2 - y^2 - z^2$.

> EXERCISE 4.24. Define the vector r= {ct,x,y,z} and the boosts and rotations that are defined by Equations 4.26. Compute the following transformations and show that s^2 is unchanged:
>   (a) R₁[a].R₃[b].r (b) B₁[a].R₂[b].r (c) R₁[a].B₃[b].R₂[c].r

> EXERCISE 4.25. In *Mathematica* define the following points:

$$x_1 = \{1, 0, 0\}, \quad x_2 = \{0, 1, 0\} \quad \text{and} \quad x_3 = \{0, 0, 1\}.$$

> Define the 3 × 3 rotation matrices (bottom right-hand corners of the **R** matrices defined in Equation 4.26).

> Use the *Mathematica* Animate and ListPointPlot3D functions to convince yourself that the rotation matrices rotate the points in a positive (right-hand) sense around the coordinate axes.

4.5. Relativity and Causality

We have made something of an issue of the invariant interval s^2 in the discussion thus far and it is time now to explain why. One of the bedrock principles upon which modern physics is grounded is **causality**. Causality has its roots in the Newtonian notion that, crudely, things change for a reason; there is cause and effect. That is, a particle's trajectory will only deviate from a straight line if there is some force applied to it. Conversely, observed deviations from straight line motion imply the existence of some applied force.

Suppose that, in your laboratory, some event A occurs at a time t_a: a particle changes its trajectory, for example. Then, then at some subsequent time t_b another event B occurs and subsequently, at a later time t_c yet another event C occurs:

$$A \longrightarrow B \longrightarrow C.$$

In some general sense, we may infer from our observations that A causes B which, in turn, causes C; the order in which the events occur is of critical importance. What we would like to know is this: would other observers all see the events in the same order and thereby infer that A causes B, *et cetera*?

As we can see from figure 4.9, the Lorentz transform preserves the ordering of time-like sequences. If t_a occurs before t_b in your laboratory, then it will occur before t_b in *all* laboratories. All observers will attest to the same sequence order. If your inference is that A causes B, then all observers will make the same inference.[21] Even though, as we shall see, time is not an invariant quantity, the notion of time-ordering of a sequence of events is.

> EXERCISE 4.26. Plot the one-dimensional Lorentz transform of two time-like vectors: $t_a = (1.0, 0.0)$ and $t_b = (1.1, 0.0)$. What happens at large values of the transform parameter ζ? (Do the hyperbolas ever cross?)

FIGURE 4.9. The Lorentz transform of a sequence of time-like vectors preserves the sequence order

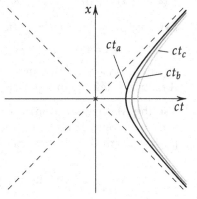

The second bedrock principle of modern physics is **relativity**. As enunciated by Albert Einstein in 1905, the principle of relativity is one of egalitarianism: *any* (astute) observer will derive the same equations of motion to describe the same phenomenon. That is, Einstein suggested that this mathematical enterprise in which we have been engaged: studying the transformation properties of equations, has to be taken seriously. The different transformations are a mathematical representation of different observers. Invariance under the transformations means, quite simply, that different observers will utilize the same form of the equations to represent the same phenomena. The importance of the Lorentz transformations, in particular, is that they are the only linear coordinate transformations that preserve both the wave equation and causality.

The principle of causality is linked to the finite propagation velocity for light. If we consider the spacetime plot shown in figure 4.8, points that are described by space-like vectors cannot be causally related. Suppose that an event P_1 occurs at the point in spacetime $\mathbf{r}_1 = (ct_1, x_1, y_1, z_1)$. We ask when does an observer located at $\mathbf{r}_2 = (ct, x_2, y_2, z_2)$ notice that the event occurred? We note that light will take a time given by the following relation:

---

[21]This is not to say that there will never be scientific debate. Cassini, for example, ultimately did not believe in the finite velocity of light, despite his own and Rømer's data.

$$\Delta t_{21} = [(x_2 - x_1)^2 + (y_2 - y_1)^2 + (z_2 - z_1)^2]^{1/2}/c$$

to travel between the points. So, for times $t - t_1 < \Delta t_{21}$, light has not yet managed to traverse the distance between the points. In this case, the number $c(t - t_1)$ will be less than the spatial distance between the points and, consequently, the quantity $(c(t - t_1))^2 - (x_2 - x_1)^2 - (y_2 - y_1)^2 - (z_2 - z_1)^2$ will be negative. The vector $\mathbf{r}_2 - \mathbf{r}_1$ here will be space-like and thus the spacetime point $\mathbf{r}_2$ cannot be influenced by the event P_1. For times $t > t_1 + \Delta t_{21}$, the converse is true and the vector $\mathbf{r}_2 - \mathbf{r}_1$ will be time-like. For those points, the event P_1 may have, indeed, provoked some other event to occur.

This is, in mathematical terms, precisely the problem that Rømer faced when observing the eclipses of Io. The fact is that Io orbits Jupiter and is subject to the gravitational force that we investigated earlier. Io travels on an elliptical, and thereby periodic, trajectory around Jupiter. Indeed, as Galileo surmised in 1610, Io represents a celestial clock.[22] The earth, however, is located distantly from Io. This means that a significant amount of time Δt_{21} must elapse before events that occurred at Io can be known to observers on earth. Because the earth-Io distance is not a constant, the time Δt_{21} varies. Fortunately, better solutions to the problem of establishing longitude were discovered that eliminated the need for ephemeris tables.

In our modern-day world, we can observe a related phenomenon when airplanes fly overhead. Light takes very little time to propagate from an airplane to the ground and airplanes move very slowly when compared to the velocity of light, so we see the airplane at almost its exact physical position. Sound, which can also be described by a wave equation, travels with a characteristic velocity that is much smaller, $v_c \approx 340 \, \mathrm{m/s}$. As a result, the sound of the airplane appears to lag the apparent position of the airplane.

EXERCISE 4.27. Consider an airplane travelling at 500 km/hr at an altitude of 2 km. If the velocity of sound waves in air is 340 m/s, how long does it take sound to propagate from the airplane to the ground? How long does it take light to propagate from the airplane to the ground?

How far will the airplane travel horizontally in the time that it takes sound to propagate to the ground? Can you explain why the sounds of passing aircraft do not seem to emanate from the airplanes but from somewhere behind the airplanes?

---

[22]There are, of course, small perturbations arising from Io's interactions with the other Jovian moons that affect the period. These are smaller than the precision of Rømer's experiments.

Einstein's physical insight into the importance of what is now called the Special Theory of Relativity has driven much of modern physics.[23] Let us examine what the Lorentz transforms mean physically. If we go back to Equations 4.13 and ignore the constant terms, then dimensionally, we recognize that $x' = a_1 x + a_2 t$ must have the following dimensional equation:

$$(L) = a_1(L) + a_2(T).$$

For this to make sense, a_1 must be dimensionless and a_2 must have dimension of L/T, i.e., a_2 must have the dimension of a velocity. The second equation $t' = a_4 x + a_5 t$ has the following dimensional equation:

$$(T) = a_4(L) + a_5(T),$$

which requires a_4 to have dimension of (T/L) or inverse velocity and a_5 to be dimensionless. Ultimately, we found that the choice of $a_1 = \cosh \zeta$ and $a_2 = c \sinh \zeta$ gave rise to a transformation that preserved the form of the wave equation. If we take the ratio of $a_2/a_1 = c \tanh \zeta$, we find that a_2/a_1 has the dimension of a velocity; let us call it u. The boost transformations are generated if our two observers have a relative velocity u.

Recall now that $\cosh^2 \zeta - \sinh^2 \zeta = 1$ and that $\tanh \zeta = \sinh \zeta / \cosh \zeta = u/c$. Putting these two facts together, we can obtain the following:

$$\cosh^2 \zeta [1 - \tanh^2 \zeta] = \cosh^2 \zeta [1 - (u/c)^2] = 1,$$

whereby we have the two following relations:

$$(4.27) \qquad \cosh \zeta = \left[1 - \left(\frac{u}{c}\right)^2\right]^{-1/2} \quad \text{and} \quad \sinh \zeta = \frac{u}{c}\left[1 - \left(\frac{u}{c}\right)^2\right]^{-1/2}.$$

These quantities u/c and $[1 - (u/c)^2]^{-1/2}$ occur frequently in discussions of relativity and have customary symbols of β and γ, respectively. So, one will frequently see the Lorentz transforms written in terms of $\gamma = \cosh \zeta$ and $\beta\gamma = \sinh \zeta$. The two forms are equivalent but many authors prefer the physical interpretation afforded by the relative velocity u, as opposed to the more geometrical interpretation we have provided with the use of the hyperbolic functions.

The function γ only has real values for the domain $-c \leq u \leq c$. A consequence of the Lorentz transformation is that observers cannot move faster than light. This is, of course, a stunning blow for science fiction writers everywhere. The Milky Way galaxy, depicted in figure 1.4, has a diameter of about 10^5 light-years. Without the capacity for traveling at vastly greater velocities than that of light, it is unlikely that space explorers will ever

---

[23]This theory of relativity is *special* due to the assumption that the transform must be linear. A more general theory of relativity relaxes that restraint.

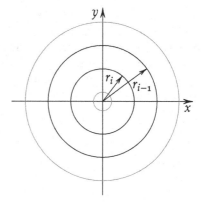

FIGURE 4.10. Waves (in two dimensions) spread in circles, reflecting the rotational symmetry of the wave equation. The distance r_i measures the distance to the wave front from the origin for the ith excitation

visit any sizable portion of the galaxy, much less colonize distant worlds and establish interstellar trade routes.[24]

EXERCISE 4.28. Plot the function γ as a function of $\beta = u/c$ over the domain $-1 \leq \beta \leq 1$. For what values of β does γ differ appreciably from 1? If the magnitude of c is about 3×10^8 m/s, what must u be for γ to be 1.01 or 1.1? Hint: It might be more revealing to plot γ on a log-log plot.

Another major consequence of causality and relativity is that all observers measure the same velocity for light. Einstein, in fact, used the constancy of the velocity of light as a basis for his derivation of the theory of relativity. So, the two propositions can be considered equivalent: causality is preserved is the same as the velocity of light is invariant. That all observers measure the same velocity for light is a consequence of preserving the wave equation. The characteristic velocity v_c (or c in the case of light waves) does not change under Lorentz transformation.

At first this result might seem nonintuitive but let us consider what happens when we launch a wave. If we drop a stone into a pond, disturbances that we call ripples spread across the surface in a series of circles of ever increasing radius. The radius, in fact, is just the characteristic velocity multiplied by the time elapsed since the stone hit the water: $r = v_c t$. If we now consider dropping stones repeatedly, with a constant time interval T between stones, we might see a pattern like that depicted in figure 4.10.

The distance travelled by the wave generated by the ith stone at the time t_1 will be $r_i = v_c t_1$ but this is related to the distance travelled by the previous

---

[24]An obvious way to circumvent this limitation is to alter the human life span. If humans could live to be a thousands of years old, then a two-hundred year journey to a distant star might be an appropriate use of one's time. Such discussions, of course, are more appropriate to classes in molecular biology.

$(i-1)$ stone: $r_{i-1} = v_c(t_1 + T)$. Because more time has elapsed since the previous stone was dropped, r_{i-1} will be larger than r_i. Now, the distance between the two wavefronts is given by $r_{i-1} - r_i = v_c T$. This distance is independent of the time and is given a special name: the **wavelength**. It is usually denoted by the Greek letter λ. For periodic excitations, the wavelength and period are related by the following simple formula:

$$(4.28) \qquad\qquad \lambda = v_c T.$$

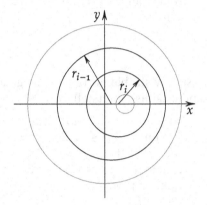

FIGURE 4.11. For a moving source, each wavefront spreads in a circle centered on the point where it was generated. Subsequent wavefronts are shifted along the direction of motion of the source, along the x-axis in this picture

At this point, the astute student might question why we are talking about dropping rocks into a pond when the real subject under discussion is the propagation of light. The answer is that the propagation of both ripples on the surface of a pond and light through the universe are described by the wave equation. The physical quantities, the functions f in Equation 4.25, are different but the equations defining their behavior are the same! Because most students have some familiarity with flinging stones into bodies of water, we can use this previous experience to our advantage.

EXERCISE 4.29. Use the *Mathematica* function `ParametricPlot` to plot a series of circles of radii with a constant difference ($r_{i-1} = r_i + 1$). Now use the `Animate` function to vary the base radius r upon which the other radii were defined. You have now constructed a representation of wave propagation for a periodic source.

Now consider what happens if the source is moving. In figure 4.11, we illustrate the result of a source moving along the x-direction. Each wavefront propagates in a circle as before, but subsequent wavefronts are generated from different points along the x-axis. The result is that the distance between wavefronts along the x-axis is altered from the base wavelength λ. This is a phenomenon known as the *Doppler shift*, after the Austrian physicist Christian Doppler who proposed in 1842 that the motion of distant stars might shift their colors, and is familiar to anyone who has heard the change in pitch of a moving vehicle as it passes by one's (stationary) position.

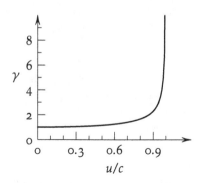

FIGURE 4.12. The function γ has a value of one until the velocity u becomes an appreciable fraction of the velocity of light

To an observer positioned somewhere along the x-axis ahead of the source point, the wavelength appears to be shorter, commensurate with a shorter time interval between excitations. If we define the **frequency** ν to be $\nu = 1/T$, this corresponds to a higher frequency.[25] Conversely, to an observer positioned somewhere along the x-axis behind the source point, the wavelength appears to be longer, commensurate with a longer time interval between excitations or a lower frequency.

Our prediction, therefore, is that light will propagate at its characteristic velocity, independent of the relative motion of any observers, although different observers would characterize the light as having different frequencies. This prediction has been tested numerous times and all experiments to date conclude that light indeed propagates with a constant velocity c. In 1964, for example, experimenters at the Proton Synchrotron accelerator at the *Conseil Européen pour la Recherche Nucléaire* (CERN), slammed protons with an energy of 19.2 GeV into a beryllium target. Amongst the particles created in this process are neutral mesons known as pions (π^0).[26] The π^0 has a mass of 135 MeV/c² and a lifetime of about 8×10^{-17} s. It decays predominantly into two photons, which are known as

---

[25]The inverse period $1/T$ occurs frequently in physics. As a result, it has its own customary name: frequency.

[26]As physicists found more and more excitations of matter, they grouped them by mass. Light particles like electrons were called leptons from the Greek word λεπτός meaning light or thin or delicate. Heavy particles like the proton and neutron were called baryons from the Greek word βαρύς meaning heavy. Mesons were particles of intermediate mass, hence the use of the Greek word μέσον meaning middle.

γ rays for historical reasons. Note that the notation $\pi^0 \to 2\gamma$ has nothing to do with the γ function defined by the Lorentz transform.[27]

The experimenters were able to select π^0s with an energy greater than 6 GeV, which corresponds to a $\gamma > 45$ (See figure 4.12) or a velocity of $u = 0.99975c$ with respect to the laboratory frame of reference. The decay of a π^0 into two γ rays is precisely the sort of moving source that we wish to study. At some brief interval after being created, the pion travelling at a velocity u emits a photon (actually decays into two photons but that doesn't alter the argument) that travels at the velocity c. The experimenters measured the time required for the emitted photons to traverse a distance of 31.45 m, using a methodology not unlike our original flag-waving, timing experiment, albeit with more sophisticated components.

Analysis of the experimental results indicates that the photons produced by relativistic pion decay have a velocity $c_{exp} = (2.9977 \pm 0.0004) \times 10^8$ m/s. The established velocity of light is $c = 2.99792458 \times 10^8$ m/s. The experimental results produce an estimate for the velocity of light that is within one standard deviation of the established value. Considered from a somewhat different perspective, if the velocity of light from the decaying pions were to be of the form $c + ku$, where u is the velocity of the pion and k represents some deviation from the predictions of special relativity, then this experiment sets a limit of $k = (-3 \pm 13) \times 10^{-5}$. Light travels at the velocity of light, even if the source is moving at nearly the velocity of light, as is predicted by the special theory of relativity.

4.6. Eigenzeit

Another major consequence of relativity is that time is no longer an invariant quantity. The Lorentz boosts mix the time and space dimensions in a fashion that Newton did not anticipate. Indeed, as Newton wrote in his *Principia*[28]:

> Absolute, true and mathematical time, of itself, and from its own nature flows equably without regard to anything external, and by another name is called duration: relative, apparent and common time, is some sensible and external (whether accurate or unequable) measure of duration by the means of motion, which is commonly used instead of true time ...

---

[27]The symbol γ is often used to describe the quantum of the electromagnetic field: the photon. Again, one must be cognizant of the meaning of notation and not attempt to simply memorize equations.

[28]This English translation is due to Andrew Motte *circa* 1729.

From this we obtain the notion that time flows like a river. Yet, as we observe in the Lorentz boosts, time and space are interrelated; we must really speak of spacetime (one word) not separately time and space (two words).

For each observer, denoted by the primed variables in the Lorentz transformations, time is not absolute but depends upon the frame of reference (the parameters ζ_i). Einstein used the German word *Eigenzeit* that is formed from the words meaning "own" and "time" to convey this concept that each observer has his own time. The French translation of Eigenzeit would be *temps propre* and in English this concept has come to be known as the *proper time*, which is presumably an unfortunate transliteration of the French. It is probably too late to change what has become common usage but the words "proper time" do not innately convey the concept as well as the German Eigenzeit. Students will have to learn to recognize the meaning of the phrase.

Experimental evidence for this phenomenon can again be found from the realm of particle physics. One member of the lepton family of particles, the muon (μ) has a lifetime of about $\tau = 2.195\,\mu s$.[29] By this, we mean that if one had a bag of N_o muons, the number of muons in the bag would decay according to the rule $N(t) = N_o e^{-t/\tau}$. This lifetime can be determined from cosmic ray experiments but an improved set of measurements were conducted at the CERN Muon Storage Ring in 1977. When muons were accelerated to high velocities ($\gamma = 29.33$), the muon lifetimes were observed to be $\tau = 64.368 \pm 0.029\,\mu s$ for negatively-charged muons (μ^-) and $\tau = 64.419 \pm 0.058\,\mu s$ for positively-charged muons (μ^+). In the primed coordinate system, the Lorentz transformation predicts that $\tau' = \gamma\tau$. The muon lifetime experiments reflect that muons travelling at high velocities do indeed measure time differently than muons (or physicists) that are not travelling at high velocities.

In a remarkable series of experiments conducted in 2010, physicists working with precision atomic clocks have managed to measure the effects of special relativity at velocities as low as 10 m/s.[30] The atomic transitions of a single aluminum atom held in an electromagnetic trap were investigated as a function of the velocity of the atom. By comparing to a second aluminum atom, held in a separate trap, the measurements achieved a precision of approximately one part in 10^{15} in the transition frequencies. This

---

[29]The SI prefix μ and unit s means 10^{-6} seconds and is pronounced microseconds. It is a different entity than the muon, which is also represented by the Greek μ.

[30]The Nobel Prize in Physics in 2012 was awarded to the Moroccan physicist Serge Haroche and the American physicist David J. Wineland "for ground-breaking experimental methods that enable measuring and manipulation of individual quantum systems." Wineland's methods were used by the NIST researchers to study relativistic effects at human scale.

FIGURE 4.13. The relativistic Doppler shift was measured for an aluminum atom confined within an electromagnetic trap. Experimental uncertainties in the relative frequency δf are comparable to the size of the points. Uncertainties in the velocity reflect experimental averaging over a velocity window. The *gray curve* is the relativistic prediction

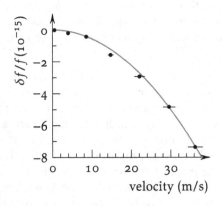

was adequate to resolve the predicted frequency shift $\delta f / f = -v^2 / 2c^2$. As can be seen in figure 4.13, the experimental results are in good agreement with the predicted frequency shift. So, as truly nonintuitive as this concept of Eigenzeit may be, there is substantial experimental evidence that the theory of special relativity provides an accurate description of the nature of spacetime.

EXERCISE 4.30. Instructors are beset with a host of usually lame excuses as to why assignments were submitted after the deadlines. It might seem that this concept of Eigenzeit could prove to be a technically plausible excuse for tardiness. Consider that, in the reference frame of the instructor (considered to be at rest), ten days have elapsed. What would be the requisite γ for a student to have observed an elapsed time of only seven days? What would be the velocity u of the student during that interval?

The NASA New Horizons mission to Pluto left the earth with a velocity of 58,500 km/hr, by some measures the fastest spacecraft ever launched. How does this compare to the necessary velocity u to explain a three-day delay in submitting homework?

Einstein went further than just suggesting that we need to take the Lorentz transformations seriously when discussing the wave equation. Indeed, he proposed that *all* physical theories needed to be formulated in such a way that they do not depend on the observer; mathematically they must be invariant under Lorentz transformation. This was to be a requirement for successful theories and caused physicists to revisit all work done up until that point. This is an important and crucial insight. Einstein recognized that the mathematics has a *physical* interpretation. We are not just conducting messy algebraic manipulations and proving theorems.

The equations we study represent the physical behavior of some system, so we are continually trying to ask the (nontrivial) question: what does it all mean?[31] We must continually rely on experiment to provide guidance in trying to answer that question.

As a consequence of Einstein's interpretation of relativity, one must be quite circumspect in formulating physical theories. In previous chapters, we have examined the kinematic equations that describe the motion of objects. A key element in that formulation was the velocity $\mathbf{v} = d\mathbf{r}/dt$. In a frame of reference that is boosted along the x-direction, another observer would find the following result for the time derivative of the x-component of position:

$$v_x'/c = \frac{dx'}{c\,dt'} = \frac{c\,dt\,\sinh\zeta + dx\,\cosh\zeta}{c\,dt\,\cosh\zeta + dx\,\sinh\zeta}$$

$$(4.29) \qquad = \frac{\tanh\zeta + dx/c\,dt}{1 + dx/c\,dt\,\tanh\zeta} = \frac{\tanh\zeta + v_x/c}{1 + \tanh\zeta\,v_x/c}$$

Note that here we used the Eigenzeit of the observer to compute the derivative. That is the "proper" time for the observer, in that clocks in the observer's frame of reference record t' and not t. The other spatial components transform in similar fashion:

$$(4.30) \qquad v_y'/c = \frac{\mathrm{sech}\,\zeta\,v_y/c}{1 + \tanh\zeta\,v_x/c} \qquad \text{and} \qquad v_z'/c = \frac{\mathrm{sech}\,\zeta\,v_z/c}{1 + \tanh\zeta\,v_x/c}.$$

EXERCISE 4.31. Fill in the missing details in the derivation of Equations 4.30.

EXERCISE 4.32. Plot the functions $\tanh(x)$ and $\mathrm{sech}(x)$ over the domain $-10 \le x \le 10$. What happens to the functions for large values of $|x|$? (What is their asymptotic behavior?)

Plot the function $v_x'(v_x/c, \zeta)$ (in 3D) over the domains $-1 \le v_x/c \le 1$ and $-10 \le \zeta \le 10$. Now plot the (scaled) function $v_y'(v_x/c, \zeta) = \mathrm{sech}\,\zeta/(1 + \tanh\zeta\,v_x/c)$ over the same domains for v_x/c and ζ. Describe the asymptotic behavior of the functions. In particular, in the limit of $\zeta \to \infty$, what is the value of v_y'/c? What does this say about something travelling at the velocity of light?

---

[31]It is to be expected that non-expert students will often ask the (trivial) question: what is the point? Einstein's emphasis on the physical interpretation of our mathematics distinguishes the trivial from the more subtle and insightful.

EXERCISE 4.33.   Consider the limiting case where $\zeta \to \infty$. What is the value of the vector $\mathbf{v}'/c = (v_x'/c, v_y'/c, v_z'/c)$?

Can you use this result to show explicitly that the velocity of light is independent of the frame of reference?

If we recall that the Lorentz factor $\tanh \zeta = u/c$, where u is interpreted as the relative velocity of the observer, then the observer sees a velocity that is not simply the sum of the two velocities. The observed velocity v' is modified by the factor $(1 + uv_x/c^2)^{-1}$.

EXERCISE 4.34.   Rewrite Equations 4.29 and 4.30 in terms of the relativistic parameters β and γ.

The velocity transformation Equations 4.29 and 4.30 are rather messy and, consequently, troubling. The Lorentz transformations that leave the wave equation unchanged do not have the same effect on our kinematic equations. We are led to the unsettling conclusion that the velocity is not a Lorentz invariant quantity. This suggests that the kinematics we have studied in the earlier chapters is not a valid physical theory, given our assertion that relativity must be taken seriously. We are staring at something of an abyss: Newtonian dynamics (in which velocity plays a significant rôle) predicted the behavior of planetary motion and led Rutherford to his nuclear model of the atom. These theories are incompatible with relativity but each separately describes experimental results quite precisely.

Einstein provided an elegant solution to the dilemma. The problem disappears when we consider four-dimensional spacetime and not separately time and three space dimensions. The velocity that we defined in Chapter 1 is a concept valid in three-dimensional space and time. The kinematic theory that we developed in Chapter 2 was successful in describing the time evolution of gravitating masses. What we need to do instead is to think about things four dimensionally. Consider an object at some point in spacetime: $\mathbf{r}_1 = (ct_1, x_1, y_1, z_1)$. What Einstein proposed is that rather than considering the evolution of the object in time, we must consider its evolution in spacetime. We cannot consider quantities like $d\mathbf{r}_1/dt$ because those will not be invariant to Lorentz transformation: each observer has his own proper time. Instead, we need to consider the quantity $d\mathbf{r}_1/ds$, where ds is the Lorentz invariant interval defined by the following relation:

$$
\begin{aligned}
(ds)^2 &= (c\,dt)^2 - (dx)^2 - (dy)^2 - (dz)^2 \\
&= (c\,dt)^2[1 - (dx/c\,dt)^2 - (dy/c\,dt)^2 - (dz/c\,dt)^2] \\
(4.31) \qquad &= (c\,dt)^2[1 - v^2/c^2],
\end{aligned}
$$

where $\mathbf{v} = (dx/dt, dy/dt, dz/dt)$ is the velocity of the object as defined in Chapter 1 and $v^2 = \mathbf{v} \cdot \mathbf{v}$ is the magnitude of the velocity squared. The quantity $d\mathbf{r}_1/ds$, which we can call the four-velocity, is now defined as follows:

$$(4.32) \qquad \frac{d\mathbf{r}_1}{ds} = \frac{1}{c[1 - v^2/c^2]^{1/2}} \frac{d\mathbf{r}_1}{dt} = \frac{\gamma}{c} \frac{d\mathbf{r}_1}{dt},$$

where γ is the relativistic function associated with a boost with velocity v/c.

We know that ds is invariant under Lorentz transformation, so the quantity $d\mathbf{r}_1/ds$ will transform under a boost $\mathbf{B}_i$ like $\mathbf{r}_1$ itself:

$$(4.33) \qquad \frac{d\mathbf{r}_1'}{ds} = \frac{d}{ds} \mathbf{B}_i \mathbf{r}_1 = \mathbf{B}_i \frac{d\mathbf{r}_1}{ds},$$

where $\mathbf{B}_i$ is one of the boosts defined in Equations 4.26. Note here that $\mathbf{r}_1$ is a *four*-dimensional vector in spacetime, where the first component is the time ct_1 and the last three components are the spatial variables.

We can also justify now our choice for the sign of s^2. By choosing the sign as we have, the differential ds defined by Equation 4.31 increases in a positive sense as the time increases in a positive sense. In the limit where $v/c \ll 1$, the factor γ in Equation 4.32 reduces to one and we recover the Newtonian definition of velocity. Additionally, we can note that for $v/c \ll 1$ the invariant interval reduces to the time interval, so that we fully recover Newton's equations of motion when the velocities are small compared to the velocity of light. In this sense, Newtonian mechanics can be thought of as the low-velocity limit of the "correct" relativistic theory.

For students that are concerned we need to revisit the previous chapters to rederive all of the formulas in a relativistic fashion, let us put aside that worry (for now). As we mentioned in the beginning of the text, all physical theories are approximate. Newtonian mechanics provides an adequate description of planetary motion. We now recognize that a relativistic theory will be required to satisfy our principles of causality and relativity but we also recognize that the planetary velocities are small compared to the velocity of light. As a result, we can use Newton's theory of gravitation to study planetary motion and achieve acceptable results. If the velocity of light were much lower, then Newton's approach to computing orbital trajectories would have failed.

EXERCISE 4.35. Use the *Mathematica* function Series to expand the γ function in a power series in $\beta = v/c$. Plot the γ function and the first two terms of the power series. At what value of β does the power series diverge from the actual value by more than 1 %?

EXERCISE 4.36.   Use the data from Table 2.1 and the fact that 1 AU $\approx 1.496 \times 10^8$ km to compute the planetary velocities. The perimeter length L of an ellipse is given by the following:

$$L = 4a\,E(e),$$

where a is the semi-major axis and e is the eccentricity. The function E is the complete elliptic integral of the second kind. When the eccentricity is zero, $E(0) = \pi/2$ and we obtain the usual result for the circumference of a circle. What is the average β for each of the planets? What is the average value of $\gamma - 1$ for each of the planets?

Now Einstein made one further suggestion based on the observation that, in Newtonian mechanics, momentum is a conserved quantity. Rather than considering the time evolution of the velocity, we should consider the spacetime evolution of the momentum. For Newtonian mechanics, one converts between velocity and momentum simply by multiplying velocity by mass; this is, in some sense trivial rescaling of the velocity. In Einstein's relativistic picture, however, we find a somewhat magical outcome.

We first note that Equation 4.32 is dimensionless, because the dimension of the invariant interval ds is (L). To obtain a momentum, which has dimension (ML/T), we need to multiply by mass and something with dimension of a velocity (L/T). The obvious choice for a constant with a dimension of velocity is, of course, the velocity of light c. Multiplying Equation 4.32 by mc, we can define a suitable four-dimensional momentum:

$$(4.34) \qquad \mathbf{p} = mc\,\frac{d\mathbf{r}_1}{ds} = m\gamma\,\frac{d\mathbf{r}_1}{dt} = m\gamma(c, v_x, v_y, v_z).$$

By analogy with the invariant interval s^2, we suggest that the following quantity

$$(4.35) \qquad \gamma^2[(mc)^2 - (mv_x)^2 - (mv_y)^2 - (mv_z)^2]$$

is invariant under Lorentz transformation.

EXERCISE 4.37.   Use the definitions of the boost and rotation matrices in Equations 4.26. Define the following vector

p1 = m{c, vx, vy, vz}/Sqrt[1 − (vx² + vy² + vz²)/c²]

Compute the following transformations and show that the quantity defined in Equation 4.35 does not change. What is the value of the invariant quantity?

(a) R1[a].p1   (b) B1[a].p1   (c) R1[a].B1[b].p1

The relativistic four-momentum vector $\mathbf{p}$ has dimension (ML/T), i.e., it scales like a momentum. Einstein recognized that the first component, the time component, of **p** doesn't contain explicitly the velocity terms like the other components and suggested that we interpret it to be the relativistic energy $\mathcal{E}$ divided by the velocity of light (to make the dimensionality correct). That is, $\mathbf{p} = (\mathcal{E}/c, p_x, p_y, p_z)$ is the relativistic, four-momentum. This has the nice effect of identifying the (scaled) energy as the time component of the vector and harkens back to our earlier assertion that energy conservation is related to time invariance of our equations of motion. We'll revisit this idea in the next section but, for now, let us see the consequences of Einstein's suggestion.

Using the results of the previous exercise, the invariant associated with the four-momentum vector is the following:

$$m^2 c^2 = (\mathcal{E}/c)^2 - p_x^2 - p_y^2 - p_z^2,$$

or, upon solving for the energy,

(4.36) $$\mathcal{E}^2 = p^2 c^2 + m^2 c^4,$$

where we have defined the term $p^2 = p_x^2 + p_y^2 + p_z^2$ to be the squared magnitude of the spatial components of the four-momentum vector **p**. In the case of an object at rest, we obtain the renowned $\mathcal{E} = mc^2$ as the zero-momentum limit of Equation 4.36.

We now have a Lorentz-invariant form of kinematics that makes significantly different predictions about the behavior of systems at high velocities. Recall that we earlier defined the kinetic energy of an isolated mass to be $T = \frac{1}{2}mv^2$. From this we can infer that

$$v_{\text{Newton}} = (T/2m)^{1/2}.$$

From Equation 4.36, the energy is $\mathcal{E} = mc^2$ when the particle is at rest. As a result, we should define the relativistic kinetic energy to be $T = \mathcal{E} - mc^2$. If we use the definition of the momentum provided in Equation 4.35, to determine the velocity, we obtain the following:

(4.37) $$v_{\text{Einstein}} = c \left[1 - \left(\frac{mc^2}{mc^2 + T} \right)^2 \right]^{1/2}.$$

In figure 4.14, we plot the results of an experiment conducted by the American physicist William Bertozzi in 1964 in which electrons were accelerated to different energies T and the time was recorded for the electrons to travel a fixed distance. The experimental velocity was obtained by simply dividing the distance by the elapsed time, which estimates the average velocity over the interval.

From the figure, we see that the prediction of Newtonian mechanics, that the velocity continues to increase as the energy is increased does not fit the data. Instead, the data are described well by Einstein's prediction that the velocity limits to the velocity of light. So, Einstein's theory, which was motivated by the need to develop a kinematic theory that transforms properly under Lorentz transforms, appears to be the correct theory at high energies.

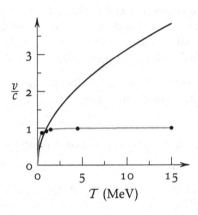

FIGURE 4.14. This experiment measured the time required for electrons to travel a fixed distance. The experimental v/c is obtained from the distance divided by elapsed time. The standard deviation of the measurements is comparable to the size of the points. The prediction of Newtonian mechanics is drawn as the *black line* and the prediction of Einsteinian mechanics is drawn as the *gray line*

EXERCISE 4.38. Use the definition of the relativistic kinetic energy $T = \mathcal{E} - mc^2$ and Equation 4.34 to derive the relativistic velocity v/c in terms of T, as depicted in Equation 4.37.

4.7. Unfinished Business

We have been treading through the shallows of some relatively deep mathematical waters in this chapter. We have encountered partial differential equations, which provides us with the means for discussing functions of more than one variable. The rules for partial differentiation are similar enough to those for ordinary differentiation that our usage should not have proven to be too problematic. We have introduced matrix notation, mostly as a visual aid, but there is a significant body of mathematics underlying linear algebra. Learning more about matrix and tensor algebras will help us to be more proficient in our calculations. We introduced the Lorentz transformations, boosts and rotations, that lead into the mathematical province of group theory. We left unproven our assertion that all Lorentz transformations can be obtained from combinations of the ones we introduced. The proof of this assertion can be established quite readily with some more math under our belts. We began the chapter with a discussion of statistics and probability. In truth, one can usually find several undergraduate (and graduate) courses devoted to each of these subjects and we cannot, of course, delve into much detail here. Suffice to say that there is more math on the agenda for students wishing to pursue physics more earnestly.

After developing the special theory of relativity, Einstein asked, what seemed to him, to be the next obvious question. The Lorentz transforms are linear, by which we mean that second derivatives among the transform variables vanish: $\partial^2 x'/\partial x^2 = 0$, for example. This corresponds to observers in frames of reference that are rotated and boosted, i.e., moving with constant velocities. What would happen if we relax that requirement? What happens to accelerated observers, i.e., would they also write down the same equations of motion?

Einstein consulted with his friend Marcel Grossmann about what mathematics would be required to study such a problem and was told to go study something else, the math was too difficult. Einstein was persistent and, over the course of the next decade and with continual support from Grossmann and others, managed to develop what is now called his General Theory of Relativity. This ultimately proved to be more than just a theory of accelerating observers: the theory Einstein developed provides a natural definition of the gravitational force as an intrinsic property of spacetime. The General Theory of Relativity incorporates Newtonian gravitational theory as the low energy limit of a much more complex description of the universe. It took Einstein the better part of a decade of work to formulate the theory and the differential geometry required to express the theory is usually taught only to very advanced undergraduates or graduate students. We shall not attempt to follow that pathway here.

There are some important lessons here, though, and the first is to remember that all physical theories are *approximate*. Newton's theory of gravitation was enormously successful in explaining the behavior of the solar system. Even with greatly enhanced measurement capabilities over the instruments employed by Galileo, Rømer and Cassini, there are few places in our own solar system where Newton's theory does not agree with experiment. In truth, Einstein had no compelling experimental evidence to launch him on a quest to develop a better theory of gravitation and that was not his original intent. He was just curious about where the thread of accelerating observers might lead and he followed that idea to a new theory. A second lesson is to maintain your scientific curiosity; you cannot predict where it might lead.

Einstein's general theory of relativity extends Newton's gravitational theory into regions where the masses of stars are much larger than that of our own sun and where some observations are in conflict with Newton's theory. For the most part, those deviations are all explained by Einstein's more complex theory, which contains all sorts of concepts like black holes and wormholes in space This does not mean that we should discard Newton's ideas. Instead, we need to recognize the limits of the theory and use it in places where it applicable and not use it in circumstances where it is not going to yield reasonable results.

So, if students are troubled by the fact that somehow all of the work done in the first few chapters is for naught, that it has all been supplanted by a relativistic theory, they need not worry. For most terrestrial applications, relevant velocities are so small compared to the velocity of light that the relativistic corrections can be neglected.

EXERCISE 4.39.   In Einstein's interpretation, the first component of the four-dimensional momentum vector is energy, actually $\mathcal{E}/c = \gamma mc$. For a particle moving by itself, we earlier defined the kinetic energy to be $T = \frac{1}{2}mv^2$, where v is the velocity and m the mass.

Use the *Mathematica* function Series to expand the γ function in a power series in v/c. Show that you recover the Newtonian definition of kinetic energy if you keep only the first term in the power series.

Plot the γ function and the first and second approximations. For what values of v/c is the Newtonian value within 10 % of the Einsteinian value of the energy?

EXERCISE 4.40.   In the previous chapter, we discussed α particle scattering and used the results of those experiments to make a sweeping discovery about the nature of matter. Geiger and Marsden's early experiments used α particles coming from radioactive decay that had kinetic energies on the order of 5 MeV.

If the rest mass of the α particle is 3727 MeV/c², what is γ for 5 MeV α particles? What is the corresponding velocity v/c?

Should we worry about relativistic effects in these experiments?

More on the Nature of Matter

We introduced the wave equation in the previous chapter and focussed much of our discussion on the properties of the equation itself and the consequences of the principles of causality and relativity. Here, we will move on to a discussion of the properties of solutions of the wave equation. As we mentioned previously, waves are a manifestation of energy propagating through a system. By their very nature, waves involve motion, something that is difficult to convey in print.

In figure 5.1, we illustrate the result of rain drops falling on a small body of water. Students can undoubtedly utilize their own experiences and imagination to set the ripples into motion. Over time, the circles expand and eventually fade. As can be seen from the figure, when ripples from two separate drops overlap, a complex pattern arises but each ripple continues to expand independently. This is a situation that is quite different from the one we expect for material objects. If two billiard balls collide, they recoil in directions different from their initial pathways. At least to a first approximation, waves appear to simply pass through one another.

We'll examine the implications of this observation in more detail but for the moment, we shall assert that the ripples satisfy a wave equation in two dimensions like that in Equation 4.17. The ripples are displacements of

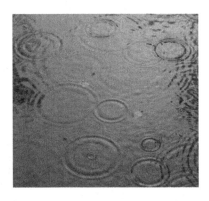

FIGURE 5.1. Rain drops splashing into a pond generate ripples. Part of the kinetic energy of the drops is converted into the waves that propagate across the water surface (Image courtesy of Alan Watson-Featherstone (Executive Director of Trees for Life))

© Mark A. Cunningham 2015
M.A. Cunningham, *Neoclassical Physics*, Undergraduate Lecture Notes in Physics, DOI 10.1007/978-3-319-10647-2_5

the fluid surface above and below the nominal water level in the puddle and so, in this case, the function $f(x, y, t)$ is a representation of the fluid level, or the deviation of the fluid level from its equilibrium position.

The story begins, in large measure, with d'Alembert's publication in 1747 on the vibration of strings.[1] In this work, d'Alembert derives the one-dimensional wave equation and finds solutions in the form of forward- $f(x - t)$ and backward-propagating $f(x + t)$ functions. As a mathematician, d'Alembert was interested in the properties of the potential solutions of the partial differential equation. Subsequently, the problem was also investigated by the Swiss physicist Leonhard Euler who approached the problem in a manner close to the way we would do so today. At the time, the correspondence between mathematics and physics had not yet been established and Euler's physics-driven approach was rebuked by d'Alembert. Today, while we treat the wave equation as an established fact, its origins were quite contentious. This is not an unusual occurrence in the development of scientific theory.

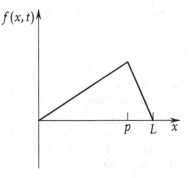

FIGURE 5.2. A plucked guitar string of length L would have an initial shape like that indicated, although the y-axis has been exaggerated

Euler derived the one-dimensional wave equation from physical principles and envisioned that one could find solutions if the initial displacement and velocity of the string were known; in mathematical terms, solutions would be specified by the conditions on the string's position and velocity at some initial time. Euler envisioned that a plucked guitar string might look like the curve depicted in figure 5.2 initially and constructed solutions from that assumption. In a response to Euler's publication, d'Alembert pointed out that, at the point $x = p$, the derivative is discontinuous and, hence, undefined. That is,

$$\left.\frac{\partial f(x, t)}{\partial x}\right|_{x=p^-} \neq \left.\frac{\partial f(x, t)}{\partial x}\right|_{x=p+}.$$

[1] D'Alembert published his "Recherches sur la courbe que forme une corde tendue mise en vibration," in *Histoire de l'académie royale des sciences et belles lettres de Berlin* in 1747, followed by an addendum "Suite des recherches sur la courbe..." in 1750.

From this, d'Alembert concluded that the second derivative must also be undefined at $x = p$ and, therefore, Euler's solutions were invalid.

The debate between d'Alembert and Euler reflects the fundamental differences between mathematicians and physicists. D'Alembert was, perhaps, the preeminent mathematician of his generation and was not particularly interested in the physical applications of the wave equation. Instead, he was interested in the mathematical analysis of such systems. Euler, for his part, was not terribly concerned that his strategy failed only at a single point in the interval $0 \leq x \leq L$. He reasoned that, for infinitesimal displacements, that the discontinuity in the derivative would also be infinitesimal and, thereby, close enough to continuous. Such lack of rigor is known to set mathematicians' teeth on edge.

5.1. Vibrations

Euler's arguments, which sound like many of the plausibility arguments put forward in this text, do not rise to the level of mathematical proof, as d'Alembert was prompt to recognize. In retrospect, Euler was an accomplished mathematician in his own right, so it is somewhat surprising that he did not mount a more successful mathematical defense. We can argue that Euler was more concerned with exposing the workings of vibrating strings than he was in categorizing the solutions of second order partial differential equations. As a result, he was not overly concerned with mathematical rigor. Seen in this light, Euler sits solidly in the physicists' camp in this debate. Euler was convinced that his mathematical treatment of vibrations on a string captured the physics of the system and produced results that agreed with experiment. Euler presumed that any mathematical discrepancies were minor and the theory could be made rigorous if one wanted to expend the effort.

Euler's and d'Alembert's attitudes broadly reflect those of modern-day practicing physicists and mathematicians. Physicists are often accused of being rather glib about mathematical rigor and mathematicians, in turn, are accused of being overly obsessive about rigorous proof. These differences arise, in large measure, because physicists are intent on interpreting the mathematical results in terms of the behavior of some physical system. Knowing beforehand that such an endeavor will only approximate physical reality perhaps contributes to the physicists' glibness. Mathematicians, on the other hand, live in a world of axiom and corollary where, as Gauss remonstrated, two half proofs do not constitute a whole proof. We have attempted to strike a sensible balance in this text without being overly glib: pointing out potential mathematical issues and pitfalls but not dwelling on all of the mathematical nuances.

Mercifully, the mathematics underlying most physics can be made rigorous and physicists have, from time to time, provided impetus for further mathematical insights. The two fields are, generally, mutually beneficial.

The Swiss physicist Daniel Bernoulli also contributed to our understanding of the vibrating string problem, noting in 1755 that the spatial variation in the solution could be provided by the fundamental displacement identified by the English mathematician Brook Taylor: $f(x) = \sin(\pi x/L)$. Time variation could be constructed by multiplying by an appropriate sinusoidal function:

$$f(x,t) = A\sin(\pi x/L)\cos(\pi v_c t/L)$$

and the general solution could be constructed from a series of this fundamental and its harmonics:

$$(5.1)\quad f(x,t) = A_1\sin(\pi x/L)\cos(\pi v_c t/L) + A_2\sin(2\pi x/L)\cos(2\pi v_c t/L) + \cdots.$$

D'Alembert was equally critical of Bernoulli's series solution, expressing concerns that Bernoulli had not demonstrated that such a series would converge.

> EXERCISE 5.1.   Plot the first three spatial functions suggested by Bernoulli: $\sin(n\pi x)$, where n is an integer and we choose $L = 1$. Could these represent the deflections of a guitar string?
>
> Plot the first spatial function multiplied by the cosine function suggested by Bernoulli. Use the `Manipulate` function to investigate the behavior over the domain $0 \le v_c t \le 10$
>
> EXERCISE 5.2.   Use the trigonometric product to sum rule to show that Bernoulli's proposed solution is of the form $f(x+v_c t)+f(x-v_c t)$, i.e., that it is of the form suggested by d'Alembert.

The Italian/French mathematician Joseph-Louis Lagrange published a paper on the nature of sound in 1759 that included a section on the vibrating string.[2] In his approach, Lagrange avoided the wave equation altogether. Instead, Lagrange considered the problem of a series of small masses coupled together by springs. Each mass feels the forces exerted by the springs on either side and the motion of each mass can be computed from the resulting acceleration, using Newton's $F = ma$. Lagrange then took the limit of an infinite number of masses and infinitesimal spring lengths. In this limit, the summations become integrals. Today, we can recognize that Lagrange's approach provides another mathematical

---

[2]Lagrange published his "Reserches sur la nature et la propagation du son" in the *Miscellanea Taurinensia*, the proceedings of the Turin Academy of Sciences that he formed with his students.

representation of the vibrating string. His results are mathematically equivalent to those found through use of the wave equation directly.

The problem of the vibrating was not ultimately resolved until the 1800s when the French mathematician Jean Baptiste Joseph Fourier proved that the series solution proposed by Bernoulli did, indeed, converge and could be shown to provide a unique solution to the wave equation. The integrals that Lagrange derived actually provide the coefficients A_1, A_2, etc. Interestingly, Fourier was not studying the wave equation; he was investigating the diffusion of thermal energy through materials, which was a process also governed by a partial differential equation but in which there was only a first derivative with respect to time. Nevertheless, Fourier's results were applicable to the general problem of partial differential equations and demonstrated that Euler's and Bernoulli's approaches could be made mathematically sound. It is also somewhat remarkable that Lagrange did not discover the Fourier series of his own accord but he was not interested in developing a series solution to the vibrating string problem. Instead, he was seeking an integral solution to a physical problem and was not looking for mathematical theorems.

5.2. Point Sources

Deriving general solutions to the wave equation will require somewhat more mathematical sophistication than we can expect for an introductory course. It happens though, that for some simple cases, solutions can be readily obtained. Consider, for example, the case of a symmetric, point source. Here, we expect the solution to be solely a function of the distance from the source to the point at which the wave is to be measured. In n dimensions, the vector from the source point to the measurement point would have n components: $\mathbf{r} = (x_1, x_2, \ldots, x_n)$. The wave equation in n dimensions can be written as follows:

$$(5.2) \qquad \sum_{i=1}^{n} \frac{\partial^2 f(\mathbf{r}, t)}{\partial x_i^2} - \frac{1}{v^2} \frac{\partial^2 f(\mathbf{r}, t)}{\partial t^2} = 0.$$

We want to obtain solutions for $f = f(r, t)$, where $r = |\mathbf{r}|$. This means that we need to think of f not as a function of the individual components x_i but as a function of $r = (x_1^2 + x_2^2 + \cdots + x_n^2)^{1/2}$. To do this, we will need to change variables by applying the chain rule:

$$\frac{\partial f}{\partial x_i} = \frac{\partial f}{\partial r} \frac{\partial r}{\partial x_i}.$$

The second derivative can be obtained by repeated application of the chain rule:

$$\frac{\partial^2 f}{\partial x_i^2} = \frac{\partial^2 f}{\partial r^2}\left(\frac{\partial r}{\partial x_i}\right)^2 + \frac{\partial f}{\partial x_i}\frac{\partial^2 r}{\partial x_i^2}.$$

EXERCISE 5.3. The partial derivative with respect to the variable x_i (denoted $\partial/\partial x_i$) is computed by holding all variables constant except x_i. Use the definition $r = (x_1^2 + x_2^2 + \cdots + x_n^2)^{1/2}$ to show that the following relations are true:

$$\frac{\partial r}{\partial x_i} = \frac{x_i}{r} \quad \text{and} \quad \frac{\partial^2 r}{\partial x_i^2} = \frac{r^2 - x_i^2}{r^3},$$

The n-dimensional wave equation, in terms of the variable r, can be written as follows:

(5.3)
$$\frac{\partial^2 f}{\partial r^2} + \frac{n-1}{r}\frac{\partial f}{\partial r} - \frac{1}{v^2}\frac{\partial^2 f}{\partial t^2} = 0.$$

Notice that there is now a first-order spatial derivative term. This term will generally complicate the process of obtaining a solution when the dimension n is larger than one. We can utilize a mathematical trick of sorts to simplify the problem.

EXERCISE 5.4. Use the results of the previous exercise and the wave Equation 5.2 to obtain Equation 5.3.

Consider the trial function $g(r,t) = r^p f(r,t)$. We will see what happens when we apply the wave equation to the function g. The partial derivatives of $g(r,t)$ with respect to r can be obtained in a straightforward manner:

$$\frac{\partial g}{\partial r} = r^p \frac{\partial f}{\partial r} + pr^{p-1}f$$

and

$$\frac{\partial^2 g}{\partial r^2} = r^p \frac{\partial^2 f}{\partial r^2} + 2pr^{p-1}\frac{\partial f}{\partial r} + p(p-1)r^{p-2}f.$$

If we substitute these results into the wave Equation 5.3 and collect terms, we obtain the following result:

(5.4) $$r^p\left[\frac{\partial^2 f}{\partial r^2} - \frac{1}{v^2}\frac{\partial^2 f}{\partial t^2}\right] + (2p+n-1)r^{p-1}\frac{\partial f}{\partial r} + p(p+n-2)r^{p-2}f = 0.$$

If we now choose $p = (1-n)/2$, the term multiplying the first order derivative of f will vanish. Using this value of p, the final term in Equation 5.4 becomes

$$p(p+n-2)r^{p-2}f = -\frac{(n-1)(n-3)}{4}r^{p-2}f.$$

Consequently, when $n = 1$ or $n = 3$, this last term vanishes and we are left with just a one-dimensional wave equation for the function f.

As we have seen previously in one dimension, *any* function $f(r - vt)$ or $f(r + vt)$ would be a solution of the wave equation in one dimension. Remarkably, in three space dimensions, any function $f(r - vt)/r$ or $f(r + vt)/r$ will also be a solution of the wave equation.

> EXERCISE 5.5. Fill in the missing algebraic steps in the derivation of Equation 5.4. Show that for $n = 3$ the function $f(r - vt)/r$ will be a solution of the wave equation.

> EXERCISE 5.6. Demonstrate that the function $f(r, t) = \sin(kr - \omega t)/r$ is a solution of the three-dimensional wave equation, provided that $k^2 = \omega^2/v^2$. Here the wave vector k is related to the wavelength λ by $k = 2\pi/\lambda$.

Unfortunately, there are no simple mathematical tricks to provide simple solutions of the wave equation in two dimensions. A simple solution in two dimensions would have been extremely valuable because we can readily visualize ripples on a pond and could then make use of our experiences with ripples to support our understanding of the mathematical representation of those ripples. This is, however, not possible. So, we shall continue our investigations with three-dimensional systems and use the fact that point sources in three dimensions can have relatively simple representations. We will also continue to rely on our experience with two-dimensional ripples to provide us with physical insights but we shall have to add solution of the two-dimensional wave equation to the (growing) list of as-yet unresolved issues.

5.3. Beats

As we have mentioned, one of the characteristics of waves is that they appear not to interact directly; they pass through one another. We certainly observe such behavior in ripples on a water surface, as in figure 5.1. This property has been termed *interference*. Mathematically, we can represent this phenomenon by noting that the result of two waves $f_1(\mathbf{x}, t)$ and $f_2(\mathbf{x}, t)$ propagating through some region of space is characterized by the sum of the two: $f_1(\mathbf{x}, t) + f_2(\mathbf{x}, t)$.

Consider the simple case of a continuous wave of a single frequency ω propagating in one dimension that we can take to be the x-direction. (Alternatively, we could think of a wave propagating through three dimensions but over a short enough distance that we do not need to worry about a factor of $1/r$.) We can represent this wave as a sine function:

$f(x,t) = \sin(kx - \omega t)$, where the wave vector k and the frequency are related by $k = \omega/v_c$. Here v_c is the characteristic wave velocity.

EXERCISE 5.7. Plot the function $\sin(x - 2t)$ for the range $-10 \leq x \leq 10$ and use the Animate function to examine the behavior for the range $0 \leq t \leq 60$. What is the direction of propagation? How does the signal change if you modify the 2 to another value like 2.2 or 2.5? What is the wavelength λ?

Suppose now that there is a second continuous wave propagating in the same region. The sum of the amplitudes of the two waves as a function of time at the position $x = 0$ is illustrated in figure 5.3.[3] If the two waves have different frequencies of oscillation, we observe a curious phenomenon known as *beats*. In the early part of the plot, the two waves are oscillating nearly in phase and so the amplitude of the sum is almost double the amplitude of the individual waves. After a time, the two waves are oscillating out of phase and the amplitudes cancel. Overall, we observe a modulation of the total amplitude that appears at a frequency that is the difference between the frequencies of the two original waves. This phenomenon can be observed readily with two sound sources that oscillate at close frequencies.

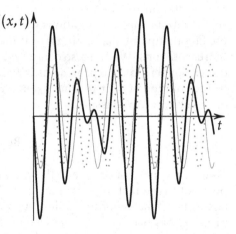

FIGURE 5.3. Two waves of slightly different frequencies are draw in *light gray* and *dotted gray*. The sum of the two waves is drawn in *black*

---

[3]Changing position to some other value of x will shift the trace to the left or right but does not affect the argument. We describe this as a phase shift.

Exercise 5.8. Consider the function

$$f(x,t) = \sin(x - 2t) + a\sin(x - (2 + b)t).$$

For a fixed position in space, say $x = 0$, plot the function for the range $0 \leq t \leq 50$. Use the `Manipulate` function to vary the values of the parameters $0 \leq a \leq 1$ and $0 \leq b \leq 0.5$. (Note: Use the `PlotRange` option of the `Plot` function to prevent the plot from rescaling when you use the sliders to vary the parameters).

What is the maximum value of f when $a = 0$ and $b = 0$? What is the maximum value when $a = 1$ and $b = 0$? What happens when b is non-zero?

Two identical tuning forks generate sound at the same frequency, say 440 Hz.[4] Adding a small mass to one of the arms of one tuning fork will cause it to vibrate at a slightly different frequency. Striking the tuning forks simultaneously will give rise to two waves propagating through space but a distant observer hears essentially one tone that is alternately louder and softer. Hence, detuning the tuning forks gives rise to this amplitude modulation of the wave field that we call beats.

Similarly, tuning one string of one guitar to a slightly different pitch than that of a second guitar (or tuning fork) will also lead to a wavering amplitude if both strings are plucked simultaneously. Readjusting the tuning of the string so that no wavering is heard is an indication that the two strings are vibrating at precisely the same frequency. This modulated amplitude is precisely represented by the behavior illustrated in figure 5.3.

The phenomenon of beats is predicted by the properties of solutions to the wave equation. That we can also confirm that acoustic waves possess this property adds to the evidence that the physical behavior of sound waves is indeed represented by the wave equation. Electromagnetic waves also display the phenomenon of beats; this a general property of any physical process that is described by the wave equation. It is also a phenomenon that can be used for practical purposes. Obviously, tuning a guitar or piano can be expedited by exploiting the phenomenon but one can also construct a precision method for measuring the frequency of electromagnetic waves. One can beat two lasers with similar frequencies against one another to produce a difference signal at a much lower frequency. That difference signal can, in turn, be beaten against a microwave signal to produce a lower frequency difference. This chaining can be conducted down to low enough frequencies that one can simply measure the frequency directly. This bootstrap strategy has enabled the development of extraordinarily precise measurements of time.

---

[4]This is the frequency of concert A.

5.4. Interference

Consider now what would happen if we had two point sources in close proximity. As the wave propagate from the sources, the wave field will be generally quite complex. So, let us look at a simple system in which the waves are represented by the function $f(r,t) = \sin(r - vt)/r$. We put one source at the origin and the second at some distance d along the x-axis. In figure 5.4, we plot the amplitude of the waves in the $z = 0$ plane at the time $t = 0$. This snapshot of the two sources depicts a rather complex pattern. The source locations are clearly indicated by the spikes near the center of the figure. Noticeable at large distances are lines where the amplitudes cancel. These lines of zero amplitude are known as nodal lines.

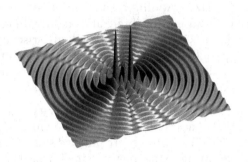

FIGURE 5.4. Two sources are separated by a small distance. The amplitude of the summed waves on the same plane that contains the sources is drawn as the *light gray* surface. At large distances from the sources, there exist nodal lines where the summed amplitude vanishes

Remarkably, the nodal lines are fixed in space. That is, they do not change positions as the waves propagate. What this implies is that for two sources, there will be fixed points in space where the waves cancel exactly. The origin of the nodal lines can be understood from the following analysis. At some distant point $\mathcal{P}$ from the two sources, as depicted in figure 5.5, the waves will have travelled a distance ℓ or $\ell + \delta$. When δ is half of the wavelength λ, the wave amplitudes are opposite in sign and cancel.[5] At very large distances, the interior angles α approach $\pi/2$. In that case, the lower triangle becomes a right triangle and the values of δ and d are related by the trigonometric relation:

$$\frac{\delta}{d} = \cos(\pi/2 - \theta) = \sin\theta.$$

EXERCISE 5.9. Plot the functions $\sin(x)$, $\sin(x+a)$ and $\sin(x)+\sin(x+a)$ for $0 \le x \le 15$. Use the `Manipulate` function to vary the value of the phase over the range $0 \le a \le 10$. For what values of a does the sum vanish?

---

[5]The waves will cancel also when δ is any multiple of a wavelength plus another half wavelength. That is, for $\delta = (2n + 1)\lambda/2$, where n is any integer.

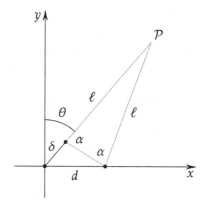

Figure 5.5. Two sources are separated by a small distance d. The path length to some distant point $\mathcal{P}$ from the nearer source is ℓ and from the other source is $\ell + \delta$. At very large distances ℓ, the interior angles α approach $\pi/2$

Exercise 5.10. Let us examine the assertion made in the text that the interior angles α of an isosceles triangle approach $\pi/2$ when the sides ℓ are large. We know that the base of the triangle is less than the distance d, so let us just consider a triangle with base d and sides ℓ. Show that $\tan \alpha = [4\ell^2/d^2 - 1]^{1/2}$, which approaches $\tan \alpha = 2\ell/d$ for $\ell \gg d$.

Plot $\tan^{-1} x$ and the constants $0.99\pi/2$ and $\pi/2$. For what value of $x = 2\ell/d$ is the inverse tangent within one percent of $\pi/2$? Is the assertion a reasonable one?

Exercise 5.11. Let us visualize solutions to the three dimensional wave equation. Define the function $f(x,y,z,t) = \sin(r - 3t)/r$, where $r = (x^2 + y^2 + z^2)^{1/2}$. Use the Plot3D function to plot f over the range $-50 \le x \le 50$ and $-50 \le y \le 50$. It is somewhat challenging to manipulate three dimensional plots in the *Mathematica* program because the plots are computationally intensive. Choose $z = 0$ and $t = 0$ initially. What happens if you now change the time to some other positive value? Now plot the function $f(x,y,z,t) + f(x-8,y,z,t)$. What is the result? How does the plot change if the second source point is moved to $x - 12$?

Exercise 5.12. Repeat the previous exercise using the DensityPlot function. This will produce a different representation of the field amplitudes. At what angle θ should we expect the first nodal line when the second source is at the position $x + 8$? Is this what you observe?

We can now address the question of what would one observe given the wave fields as described above. For an observer standing at some point $\mathcal{P}$ in the far field of the sources, the waves continuously propagate past. Observers do not measure the wave amplitudes directly, though. Without going into any particular detail on the nature of the observation process (photographic film, CCD arrays, microphones, etc.) we can assume that, generally, measurements of wave phenomena are proportional to the wave intensity $I(r)$, where the intensity is defined as the square of the wave amplitude.

To understand why the intensity is important, let us consider the function $f(x,y,z,t) = \sin(r - vt)/r$. If we integrate over one period of oscillation ($T = 2\pi/v$) the integral vanishes. If instead we integrate the square of f, we find the following:

$$(5.5) \qquad I(r) = \int_{0}^{T} dt\, f^2(x,y,z,t) = \frac{\pi}{vr},$$

where the period is defined as $T = 2\pi/v$.

> EXERCISE 5.13. Plot the (scaled) wave intensity from Equation 5.5 in the $z = 0$ plane, i.e., for points $r = (x, 20, 0)$, where $-20 \leq x \leq 20$. How does the intensity change off axis? What happens if you change y to be 40?

What happens now if we have two sources? We measure the *coherent* intensity. That is, we measure the square of the summed amplitudes:

$$I(r) = \int_{0}^{T} dt\, [f_1(x,y,z,t) + f_2(x,y,z,t)]^2.$$

For the case where we have offset one source by a distance d along the x axis (as depicted in figure 5.5) and restrict ourselves to the $z = 0$ plane, we can show that the intensity is given by the following expression:

$$(5.6) \qquad I(x,y) = \frac{\pi[d^2 - 2dx + 2r_1^2 + 2r_1 r_2 \cos(r_1 - r_2)]}{v r_1^2 r_2^2},$$

where $r_1 = (x^2 + y^2)^{1/2}$ and $r_2 = ((x-d)^2 + y^2)^{1/2}$. In figure 5.6, we plot the intensity for two closely spaced sources at a large distance $y = L$.

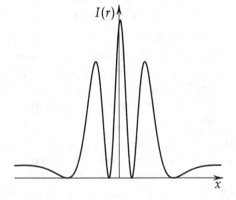

FIGURE 5.6. The intensity $I(r)$ is shown as a function of x for a large value of y. The vertical distance z has been set to zero

It is possible to conduct an interference experiment in the laboratory, where a light source is allowed to strike an opaque barrier that has two thin slits cut into its surface. Measuring the perpendicular distance L from the barrier to a distant screen and the distance Δx between the first two dark spots, it is not difficult to show that $\Delta x/2L = \tan\theta$, where θ is defined as in figure 5.5. Recall that we have also determined that $\lambda/2d = \sin\theta$, where d is the source spacing and λ is the wavelength. Equating the two expressions, we can show that the wavelength of the wave is given by the following relation:

$$(5.7) \qquad \lambda = \frac{d\,\Delta x}{[(\Delta x/2)^2 + L^2]^{1/2}}.$$

Thus, if we know the distance between the two slits d and can measure Δx and L, then we can determine the wavelength of the wave. Alternatively, if we know the wavelength, we could determine the slit spacing.

> EXERCISE 5.14. Define the function $f(x,y,t) = \sin(r - vt)/r$, where $r = (x^2 + y^2)^{1/2}$. Use the Integrate function to demonstrate that the integral of f over one period $T = 2\pi/v$ vanishes. What value do you obtain for the integral of f^2?

> EXERCISE 5.15. Plot the function $I(r)$ defined in Equation 5.6. Use a large value for y (≈ 1000) and a large range for x ($0 \le x \le 5000$). Use the Manipulate function to vary the source distance over the range $0 \le d \le 30$. How does the intensity change as d changes?

It is possible to extend this discussion to the situation in which there are N sources. We can note that for N sources of the form $f(x_i,y,t) = \sin(r_i - vt)/r_i$ that the intensity will have terms of the form $\sin^2(r_i-vt)/r_i^2$ and cross terms of the form $\sin(r_i - vt)\sin(r_j - vt)/r_i r_j$. We find that the following relations hold:

$$\int_0^T dt\,\sin^2(r_i - vt) = \pi \quad \text{and} \quad \int_0^T dt\,\sin(r_i - vt)\sin(r_j - vt) = \pi\cos(r_i - r_j).$$

Using these results, it is possible to show that the N-source intensity can be written as follows:

$$(5.8) \qquad I_N(r) = \sum_{i=1}^{N} \frac{\pi}{r_i^2} + 2\sum_{i=1}^{N} \frac{\pi}{r_i} \sum_{j=i+1}^{N} \frac{\cos(r_i - r_j)}{r_j}.$$

This leads to relatively complicated intensities for small values of N but for large N, there is a remarkable outcome.

If we refer to figure 5.5, we obtain cancellation of the wave amplitudes when δ is half of the wavelength λ. The wave amplitude is a maximum when δ is a full wavelength. If we consider adding more sources, say one at $x = -d$, it is clear that the path length to the point $\mathcal{P}$ will be $\ell + 2\delta$. If $\delta = \lambda$, this source will also add constructively to the amplitude at $\mathcal{P}$. With a large number of sources, we find sharp peaks in the intensity at the angles where $\lambda = d\sin\theta$, as indicated in figure 5.7. A realization of the N source model is known as a **diffraction** grating.[6] The diffraction effect of a large number of slits was originally discovered by the Scottish mathematician James Gregory, who observed in 1673 that a narrow beam of sunlight passing through a bird's feather split into a series of colored ovals.

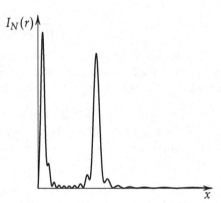

FIGURE 5.7. The N slit intensity for $N = 10$ shows a single sharp peak to the right of the central maximum. The sources were assumed to be separated by the constant value d

EXERCISE 5.16.   Plot the function defined by Equation 5.8 for $y = 1000$ and a source spacing of $d = 12$. Examine the range $0 \le x \le 5000$. Convince yourself that you obtain the two-slit result when $N = 2$ before proceeding. Study the behavior as you set $N = 3, 5$ and 10. (Note: The double sum can be quite numerically intensive. Use caution if you decide to investigate large values of N.)

EXERCISE 5.17.   Redraw figure 5.5 and place a source at $x = -d$. Convince yourself that, if $\alpha = \pi/2$, the path from the source at $x = -d$ to the observation point $\mathcal{P}$ is $\ell + 2\delta$.

5.5. Diffraction

Diffraction refers to the observed phenomenon that waves bend around obstacles (or that a slit in an opaque mask can act like a point source). An example of water waves diffracting through an opening in the breakwater is illustrated in figure 5.8. The wave fronts inside the breakwater are circular and centered on the opening. This behavior is in stark contrast

---

[6]The term diffraction was coined by the Italian Francesco Maria Grimaldi in his 1665 publication *Physico mathesis de lumine, coloribus, et iride, aliisque annexis libri duo.*

to the sharp shadowing that we would expect of material objects. Imagine spray painting a picket fence and holding a plywood sheet behind the fence to catch the overspray. Paint droplets that did not strike the fence would land on the plywood. If we examine the plywood, we would expect a sharp boundary between the painted and unpainted regions reflecting the fact that paint droplets that strike the fence do not reach the plywood. Waves do not behave like paint droplets. As Grimaldi noted in 1665, waves can bend around obstacles.

FIGURE 5.8. Ocean waves incident on the breakwater in the port of Alexandria, Egypt diffract through the opening (Image ©Google Earth 2012)

An empirical explanation for the observed behavior of waves was provided by Christiaan Huygens in 1678.[7] Huygens suggested that one can envision the process of wave propagation by what has come to be known as wavelet construction. If we envision a point source (point A in figure 5.9), the wave fronts will propagate in a spherically-symmetric manner with velocity v_c. After some time t_1, the wave front will have progressed from point A to a spherical shell of radius $r = v_c t_1$ and depicted by the arc HI. If we consider now each point along the arc HI to itself be a point source (four representative points b are identified), then the wave front at time $t = 2t_1$ will have progressed to the radius $r = 2v_c t_1$ depicted by the arc KL. Each point b along HI will have generated a spherically-symmetric wavelet (colored gray) that will have propagated a distance $r = v_c t_1$. The wavelets constructively interfere at the wave front and destructively interfere away from the wavefront. The sum (integral) over all of the points along the arc HI will lead to the new wave front KL.

Similarly, at a subsequent time $t = 3t_1$, the wave front will have progressed to a radius $r = 3v_c t_1$, depicted by the arc DF. Again, the wave front can be constructed from wavelets originating at all of the points d along the arc KL.

Huygens' construction has a geometric simplicity that is quite attractive and provides a geometrical explanation for diffraction. Each point in

---

[7]Huygens' *Traité de la lumière* was published in 1690 and was based on his earlier presentation in 1678 to the French Royal Academy.

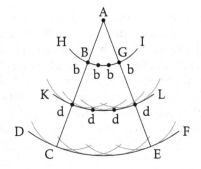

FIGURE 5.9. Huygens wavelet construction of wave propagation. A wave from the source A gives rise to a spherically-symmetric distribution that can be constructed from an infinite series of wavelets

space acts as a point source, generating the next wave front. For waves in free space, the wave propagates along rays (ABC or AGE in figure 5.9), where the rays represent the direction of motion and are locally perpendicular to the wave fronts. This ray-like nature of light is what led Newton to propose his corpuscular theory of light. When a wave encounters a barrier with a hole in it, however, the points in the hole act like individual point sources and the wave diffracts into the space beyond the hole. Hence, a slit in an opaque barrier acts like a point source, like we see in figure 5.8.

EXERCISE 5.18. Let us examine Huygens' assumption that the wavelets will constructively interfere at the wave fronts and destructively interfere elsewhere. Define the function $f(x,y,t) = \sin(r - 3t)$, where $r = (x^2 + y^2)^{1/2}$, that vanishes if $|r - 3t| > \pi$. This represents a single oscillation of a sinusoidal wave. Use the Plot3D function to plot $f(x,y,2\pi)$ over the range $-40 \le x \le 40$ and $-40 \le y \le 40$. What happens when t changes?

Let us consider a line of sources along the x-axis. Use the Sum function to plot the function $f_N(x,y,t)$ where we define

$$f_N(x,y,t) = \frac{1}{2N+1} \sum_{j=-N}^{N} f(x+j,y,t).$$

Plot f_N for $t = 2\pi$ and $N = 15$ for the same x and y range as before. What is the behavior of the summed function? Does the wave front look like (part of) a plane wave? What happens if you set $N = 30$?

EXERCISE 5.19. Assume that you have a geometry where waves propagate through a narrow channel and then enter into a semi-infinite space. Sketch the wave fronts for plane waves in the channel as they emerge into the open space.

Huygens' approach to the problem of wave propagation is, of course, empirical and does not provide quantitative results. Nevertheless, his idea encompasses a number of essential features of wave propagation, to the extent that geometrical constructions with compass and straightedge can represent dynamical phenomena like wave propagation. To obtain quantitative answers, we are faced with the problem of solving a second-order partial differential equation subject to boundary conditions. This is beyond the mathematical skills of most students at this point in their careers, so we shall defer calculation of quantitative solutions to subsequent courses and restrict our discussions to the more empirical approach Huygens suggested.

There are some inconsistencies with Huygens' construction, though, that we should discuss. First, solutions of the wave equation for point sources only take the form of functions $r^{(1-n)/2} f(r \pm vt)$ in one and three dimensions. In two dimensions, the solutions are more complex. This generally isn't a restriction, as we are most often thinking about solutions in three space dimensions (even if forced to plot two-dimensional slices). Second, in figure 5.9, we only plotted the forward propagating section of the wavelets. If we are to really think of the points b and d as point sources, they should radiate in all directions, not just the forward direction. In two dimensions, such backward propagating terms are present but that is not the case in one and three dimensions. So, we can add a prescription for application of Huygens' wavelets that one should only use the forward propagating portion of the wavelet when constructing the wave fronts in free space.

It happens though, that backward propagating wavelets do arise when waves encounter physical obstacles. For wave scattering from small objects, construction of solutions to the wave equation does involve a mathematical process of adding the initial source waves to wave propagating away from the object in all directions and enforcing suitable boundary conditions at the surface of the object. For small (effectively spherical) object, a reasonable approximation is that the total wave field will be given by the sum of the incident wave plus a point source centered on the object.

Remarkably, the German physicist Max von Laue and his experimental collaborators Walter Friedrich and Paul Knipping demonstrated in 1912 that x-rays diffracted from a crystal of copper sulphate. This was remarkable in one sense because x-rays were widely believed to be corpuscular in nature. Diffraction is a property of waves, so the demonstration that x-rays diffract confirms that they are electromagnetic waves, like light and radio waves. The result was also remarkable because it now meant that x-rays could be used to probe atomic structure, at least for crystalline solids. In their apparatus, Friedrich and Knipping sent a narrow beam of x-rays through a crystal and thence onto a photographic plate. They observed a series of bright spots around the central beam spot that they interpreted as due to the diffraction of the x-rays. Laue provided an explanation of why the experiment produced the spots that were observed. His ideas were generalized and extended to more crystal forms by the physicist William Lawrence Bragg. In the summer of 1913, Bragg and his father William Henry Bragg, who had developed an apparatus that detected x-rays by the ionization they produce in a gas cell, measured the diffraction patterns for a number of crystals, including KCl, NaCl and diamond and were able to reconstruct the lattice spacings of each of the crystals. They provided the foundation for the science that we today call crystallography.[8]

To see how Laue and the Braggs resolved the problem of diffraction, consider that atoms in a crystal form some sort of three-dimensional, regular lattice. Then, starting from some atom in the lattice, we can find the nearest neighboring atom some distance a away. Let's define the vector $\mathbf{a}$ by the difference between the position vectors of the two atoms: $\mathbf{a} = \mathbf{r}_{i+1} - \mathbf{r}_i$, where $\mathbf{r}_i$ is the position of some atom in the lattice and $\mathbf{r}_{i+1}$ is the position of the nearest neighbor. For a regular lattice, we would also find atoms at all integer multiples of $\mathbf{a}$ from the initial atom.

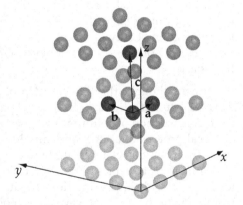

FIGURE 5.10. Atoms occupy positions in a regular lattice defined by the vectors $\mathbf{a}$, $\mathbf{b}$ and $\mathbf{c}$. We can generally align $\mathbf{a}$ with the x-axis of some coordinate system but it is not always the case that the crystal directions $\mathbf{b}$ and $\mathbf{c}$ are perpendicular to $\mathbf{a}$. They will not then align with the y and z directions

Now, let's find the next nearest atom to the initial atom that is not on the line defined by $\mathbf{a}$. We can define this direction to be $\mathbf{b}$, as illustrated in figure 5.10. The vectors $\mathbf{a}$ and $\mathbf{b}$ define a plane in three dimensions. If we now find the nearest neighbor atom that is not on the $\mathbf{a}$-$\mathbf{b}$ plane, we can define the direction $\mathbf{c}$. Together, the vectors $\mathbf{a}$, $\mathbf{b}$ and $\mathbf{c}$ form a basis in three dimensions, although not necessarily an orthogonal basis. The vectors $\mathbf{a}$,

---

[8]Laue won the Nobel Prize in 1914 "for his discovery of the diffraction of x-rays by crystals" and the Braggs shared the Nobel Prize in 1915 "for their services in the analysis of crystal structure by means of x-rays." Laue's work was based on transmitted x-rays and Bragg's apparatus utilized reflected x-rays, causing some to note that the sequential Nobel Prizes were awarded for two sides of the same coin.

b and **c** provide what we might call the natural basis for a crystal, as all atoms in the crystal can be indexed by integer displacements in **a**, **b**, **c** from one another.

Mathematically, by basis, we mean that any vector **r** can be uniquely decomposed into components:

$$(5.9) \qquad \mathbf{r} = r_a\mathbf{a} + r_b\mathbf{b} + r_c\mathbf{c},$$

where the components are real numbers. In our usual Cartesian basis, the unit direction vectors $\hat{\mathbf{x}}$, $\hat{\mathbf{y}}$ and $\hat{\mathbf{z}}$ are orthonormal. That is, they are orthogonal and normalized such that $\hat{\mathbf{x}} \cdot \hat{\mathbf{x}} = 1$, etc. The problem of decomposing a vector into components along some non-orthogonal, non-normalized basis vectors has been solved by the mathematicians by introducing the dual basis.[9]

Consider the vectors $\boldsymbol{\alpha}$, $\boldsymbol{\beta}$ and $\boldsymbol{\gamma}$ defined as follows:

$$(5.10) \qquad \boldsymbol{\alpha} = \frac{\mathbf{b} \times \mathbf{c}}{\mathbf{a} \cdot (\mathbf{b} \times \mathbf{c})}, \qquad \boldsymbol{\beta} = \frac{\mathbf{c} \times \mathbf{a}}{\mathbf{a} \cdot (\mathbf{b} \times \mathbf{c})} \quad \text{and} \quad \boldsymbol{\gamma} = \frac{\mathbf{a} \times \mathbf{b}}{\mathbf{a} \cdot (\mathbf{b} \times \mathbf{c})}.$$

These are the dual vectors to the original vectors **a**, **b** and **c** and they also form a basis in three dimensions.

EXERCISE 5.20.   Define the basis vectors $\mathbf{a} = (a_1, a_2, a_3)$, $\mathbf{b} = (b_1, b_2, b_3)$ and $\mathbf{c} = (c_1, c_2, c_3)$. Show that the following vector identities are true:

$$\mathbf{a} \cdot (\mathbf{b} \times \mathbf{c}) = \mathbf{b} \cdot (\mathbf{c} \times \mathbf{a}) = \mathbf{c} \cdot (\mathbf{a} \times \mathbf{b}).$$

Use this result to prove that the following relations hold:

$$\begin{array}{lll} \mathbf{a} \cdot \boldsymbol{\alpha} = 1 & \mathbf{a} \cdot \boldsymbol{\beta} = 0 & \mathbf{a} \cdot \boldsymbol{\gamma} = 0 \\ \mathbf{b} \cdot \boldsymbol{\alpha} = 0 & \mathbf{b} \cdot \boldsymbol{\beta} = 1 & \mathbf{b} \cdot \boldsymbol{\gamma} = 0 \\ \mathbf{c} \cdot \boldsymbol{\alpha} = 0 & \mathbf{c} \cdot \boldsymbol{\beta} = 0 & \mathbf{c} \cdot \boldsymbol{\gamma} = 1. \end{array}$$

EXERCISE 5.21.   Consider a rectangular solid, with basis vectors $\mathbf{a} = (a, 0, 0)$, $\mathbf{b} = (0, b, 0)$ and $\mathbf{c} = (0, 0, c)$. Construct the dual vectors for this system. Can you explain why the crystallographers would use the terminology reciprocal vectors?

The components of some arbitrary vector **r** can now be decomposed into components along **a**, **b** and **c** as follows:

$$(5.11) \qquad \mathbf{r} = (\mathbf{r} \cdot \boldsymbol{\alpha})\mathbf{a} + (\mathbf{r} \cdot \boldsymbol{\beta})\mathbf{b} + (\mathbf{r} \cdot \boldsymbol{\gamma})\mathbf{c}.$$

Conversely, the vector **r** can be decomposed into components along the dual vectors as follows:

$$(5.12) \qquad \mathbf{r} = (\mathbf{r} \cdot \mathbf{a})\boldsymbol{\alpha} + (\mathbf{r} \cdot \mathbf{b})\boldsymbol{\beta} + (\mathbf{r} \cdot \mathbf{c})\boldsymbol{\gamma}.$$

---

[9]In crystallography, the dual basis is usually referred to as the reciprocal basis.

Exercise 5.22.   Consider the basis vectors $\mathbf{a} = (1, 0, 0)$, $\mathbf{b} = (1, 2, 0)$ and $\mathbf{c} = (0, 2, 3)$. What are the reciprocal vectors?

Consider now the vectors $\mathbf{r}_1 = (5, 9, 12)$ and $\mathbf{r}_2 = (-3, 14, 6)$. What are the components of $\mathbf{r}_1$ and $\mathbf{r}_2$ in the $\mathbf{a}$, $\mathbf{b}$ and $\mathbf{c}$ basis?

The problem of diffraction from a crystal lattice is much like the problem that we solved for N slits. If a crystal is illuminated by electromagnetic waves (x-rays), each atom in the beam will scatter some of the incident energy. Each atom will act like a point source and, in an initial approximation, have the same amplitude. In practice, this assumption of equal amplitudes is not a particularly good choice but it will suffice for now. We will obtain constructive interference only for particular directions where the path lengths from the different sources are multiples of a wavelength λ. The difference here is that the sources are not just equidistant along some x-direction but occupy the lattice positions in three dimensions. If we use the natural coordinates for the lattice sites, then the sum over the source positions as in Equation 5.8, Laue recognized, becomes three sums over the lattice indices.[10]

Consider now an incident wave propagating in the direction $\mathbf{k}_0$. This wave will have a wavelength $\lambda = 2\pi/|\mathbf{k}_0|$. The diffracted wave will have the same wavelength as the incident wave, hence the wave vector $\mathbf{k}$ of the diffracted wave will have the same magnitude as the incident wave. For two atoms at locations $\mathbf{r}_1$ and $\mathbf{r}_2$, constructive interference will be found when $(\mathbf{k} - \mathbf{k}_0) \cdot (\mathbf{r}_2 - \mathbf{r}_1) = 2\pi N$, for N an integer. We know that $\mathbf{r}_2 - \mathbf{r}_1 = h\mathbf{a} + k\mathbf{b} + l\mathbf{c}$, where (h, k, l) are integers, so the condition for constructive interference reduces to the following equation:

$$(5.13) \qquad (\mathbf{k} - \mathbf{k}_0) \cdot (h\mathbf{a} + k\mathbf{b} + l\mathbf{c}) = 2\pi N,$$

for some integer N. The solution is immediate, if we write $\mathbf{k}$ and $\mathbf{k}_0$ in terms of the dual vectors:

$$(\mathbf{k} - \mathbf{k}_0) = k_\alpha \boldsymbol{\alpha} + k_\beta \boldsymbol{\beta} + k_\gamma \boldsymbol{\gamma}.$$

Then, computing the dot product we find:

$$h\, k_\alpha + k\, k_\beta + l\, k_\gamma = 2\pi N.$$

This will be satisfied if each of the components of $\mathbf{k}$ is an integer multiple of 2π. Recall the wave vector is $|\mathbf{k}| = 2\pi/\lambda$, so the result is true if the difference in wave vector is an integral multiple of the dual vectors:

$$(5.14) \qquad \mathbf{k} - \mathbf{k}_0 = h\boldsymbol{\alpha} + k\boldsymbol{\beta} + l\boldsymbol{\gamma}.$$

---

[10]Traditionally, the integers h, k and l are used for the indices in crystallography, in honor of the British mineralogist William Hallowes Miller, whose *A Treatise on Crystallography* was published in 1839. We shall continue the practice but note the potential confusion of the index k with the wave vector $\mathbf{k}$.

The consequence of this result is that information about the lattice positions of atoms in a crystal is directly encoded in the positions of coherent interference. We can use Huygens' principle to discover the structure of materials at an atomic level.

FIGURE 5.11. A beam with wave vector $\mathbf{k}_0$ is directed towards a crystal lattice. Diffracted rays $\mathbf{k}$ from the top layer of atoms (*gray dots*) will coherently interfere at particular angles θ. Diffracted rays from the next layer of atoms will only interfere coherently with those from the top layer when the difference in path lengths 2δ is a multiple of the wavelength λ

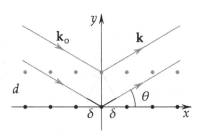

Solving Equation 5.14 will provide us with a direct means of interpreting the diffraction patterns observed by Laue, Friedrich and Knipping. We'll investigate this in more detail presently but a more intuitive approach was pursued by W. L. Bragg, whose idea is indicated in essence by figure 5.11. Bragg's father had devised an instrument that measured x-ray intensity by the ionization current produced in a gas cell, which he considered superior to dealing with photographic plates. He then fashioned a spectrometer, not unlike the apparatus used by Geiger and Marsden to measure α particles (See figure 3.5), where the detector was no longer a microscope but Bragg's gas cell and the α particle source was replaced by an x-ray tube and various crystals replaced the thin metal foils. This apparatus enabled the Braggs, during the summer of 1912, to rapidly measure the currents as a function of scattering angle.

What the younger Bragg recognized, was that if we consider scattering from planes of atoms, there will be special angles for which the diffracted rays originating from the top plane of atoms interfere coherently, as we saw in figure 5.7. If subsequent atomic planes are displaced by a distance d, then the diffracted rays from the lower planes will not generally interfere coherently with those from other planes. The criterion for coherent interference is that the path length 2δ between rays arising from the different planes be a multiple of the wavelength λ. From the geometry of figure 5.11, it is apparent that $\delta = d\sin\theta$, from whence W. L. Bragg deduced that

(5.15) $$n\lambda = 2d\sin\theta$$

would result in coherent interference from the diffracted rays. Bragg's intuitive result paved the way for direct interpretation of the diffraction patterns obtained by Laue and the results he and his father were obtaining with their x-ray diffractometer. There was initial doubt as to the actual wavelengths of the x-rays, so the Braggs reported their initial results in terms of the variable d/λ. We'll return to the issue of determining λ shortly.

Exercise 5.23. W. L. Bragg observed an interference fringe at 10.3° when examining the simple cubic lattice of KCl. What is the ratio d/λ that corresponds to this angle? (For Bragg's geometry, constructive interference occurs when $n\lambda = 2d\sin\theta$.) If $\lambda = 0.11$ nm, what is the lattice spacing in KCl? What do we mean by the size of a potassium atom?

It is a bit tricky to see how to obtain solutions to Equation 5.14. We'll utilize a graphical illustration of the problem devised by the German physicist Paul Peter Ewald. Suppose, for example, that we have managed somehow to align the beam with the **a** direction. Where will we observe constructive interference? If we plot the $\mathbf{k_o}$ vector and a sphere of the same radius, coherent diffraction will occur when the point $\mathbf{k} = \mathbf{k_o} + h\boldsymbol{\alpha} + k\boldsymbol{\beta} + l\boldsymbol{\gamma}$ lies on the surface of the sphere. This is illustrated in figure 5.12, where the reciprocal lattice is displayed as a series of gray dots. Lattice points that are on (or close to) the spherical surface are colored black. This pattern will extend off to infinity from the origin and, hence, a photographic plate would record spots that correspond to the spots on the surface of the Ewald sphere. Calculating the angles between the spots permits us to use the previous result that $n\lambda = d\sin\theta$, where here d can be mapped to one of the crystal dimensions a, b or c.

If the beam is rotated around the crystal, or the crystal within the beam, the pattern of dots changes. In figure 5.13, the beam is directed diagonally across the unit cell of a cubic lattice. Notice the three-fold symmetry in the diffraction pattern. This mirrors the three-fold symmetry that you would observe if you look diagonally across a cube. The Braggs exploited the natural crystal symmetries in their experiments, directing the x-ray beams normal to the crystal planes, where alternately, h, k and l would be constant. Determining structures of more complex crystals was aided by W. L. Bragg's observation that atoms need not be ordered in a simple cubical lattice but could also be arranged where adjacent planes were shifted laterally in face-centered cubic or body-centered cubic forms. Potassium and chlorine are close in size—giving rise to KCl crystals that are cubic—and a quick resolution of the structure of KCl. Sodium, on the other hand, is significantly smaller than chlorine. As a result, the NaCl crystals are ordered in a face-centered cubic form that modifies the diffraction pattern.

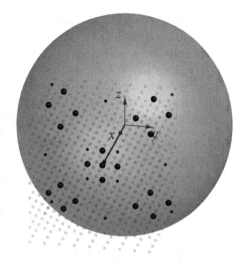

FIGURE 5.12. A beam with wave vector **k** is directed along the positive *x* axis, as indicated by the *arrow*. Diffracted rays from the cubic crystal could lie anywhere on the sphere of radius |**k**|. Coherent diffraction only occurs when the reciprocal lattice points (*gray dots*) lie on the spherical surface (*black dots*)

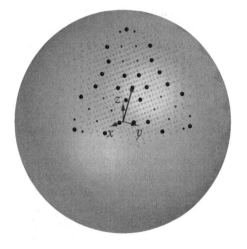

FIGURE 5.13. A beam with wave vector **k** is directed as indicated by the *arrow*. Diffracted rays from the cubic crystal could lie anywhere on the sphere of radius |**k**|. Coherent diffraction only occurs when the reciprocal lattice points (*gray dots*) lie on the spherical surface (*black dots*)

One of the first observations that we can make from figures 5.12 and 5.13 is that the wave vector |**k**| needs to be reasonably large compared to the size of the dual lattice. Otherwise, there will generally be no points that satisfy the conditions of Equation 5.13. As a practical matter, it means that the wavelength λ should be comparable in size or smaller than the lattice spacing. This is why we see no diffraction patterns from crystals using visible light. The lattice spacings for most simple compounds is in the range 0.1–0.4 nm, where the visible spectrum of light spans the range from 400 to 700 nm. It was not until the advent of x-ray tubes that diffraction experiments could yield information about the nature of matter.

EXERCISE 5.24.   Use the Table function (along with Partition and Flatten) to construct a rectangular lattice with $a = 1$, $b = 1$ and $c = 1$ and with 5 points along each direction. Plot the lattice with the ListPointPlot3D function. What symmetry of the lattice do you observe when looking along the **a** axis? What symmetry do you observe when viewing across the diagonal of the lattice? What happens if you change b and/or c to values larger than 1?

EXERCISE 5.25.   It is possible to visualize the Ewald sphere using the Graphics3D function.

$$\text{Graphics3D}[\{\text{Sphere}[\#,0.2]\&/@\text{Tuples}[\text{Range}[-2,2],3],$$
$$\text{Sphere}[\{-6,0,0\},6]\}]$$

will generate a lattice of small spheres (representing the reciprocal lattice) and a larger sphere (representing $|\mathbf{k}|$). Diffraction spots will occur when the Ewald sphere intersects the lattice points. You can focus on the reciprocal lattice portion of the plot by adding the PlotRange->{{-3,3},{-3,3},{-3,3}} directive to the Graphics3D function. What happens if you change the Ewald sphere radius to 2 or 4? What happens if you move the center of an Ewald sphere of radius R to the point $(-R/\sqrt{3}, -R/\sqrt{3}, -R/\sqrt{3})$?

5.6. Spectra

As we mentioned previously, diffraction of light by a collection of thin slits was originally discovered by the Scottish mathematician James Gregory, who observed that a narrow beam of sunlight passing through a bird's feather split into a series of colored ovals. This observation coincided with Isaac Newton's 1666 observations of white light passing through prisms. Newton concluded that white light was composed of a number of different components that we would today describe as having different wavelengths. The amount of light at each wavelength is known as the **spectrum**, a terminology coined by Newton. We can assign a function $I(\lambda)$ that characterizes the intensity as a function of wavelength.

The diffraction effect observed by Gregory provides a means for determining the spectral intensity in a quantitative manner. Newton's prismatic method depended upon another property of waves known as refraction. Waves passing from one medium to the next bend at the interface according to the rule $n_1 \sin\theta_1 = n_2 \sin\theta_2$, where n_1 and n_2 are known as the refractive indices of the materials and the angles θ_1 and θ_2 are measured from the normal to the interface. The index of refraction n depends upon the type of glass used to form the prism, so it was quite difficult to compare the results of different researchers.

A major advance in the new field of spectroscopy was provided by the English scientist Henry Grayson, who developed a series of devices that could inscribe lines in glass microscope slides. Grayson's original interest was in providing a means of measuring the sizes of biological objects viewed through a microscope but he quickly realized that his *ruling engine* could also prepare high quality diffraction gratings. By 1911, Grayson had succeeded in constructing gratings with 4700 lines in each millimeter, or a line spacing of $d = 2.13 \times 10^{-7}$ m. Such a level of precision provided tremendous opportunities to quantify our knowledge of light. It was now possible to measure the wavelengths of visible light to great precision.

EXERCISE 5.26. Suppose that we utilize a diffraction grating with 500 lines/mm. At what angles would we observe diffracted light of wavelength $\lambda = 600$ nm? We have $m\lambda = d \sin\theta$, where m is an integer. What orders m are visible?

If we assume that we have an angular resolution of 1°, what would be our resolution in wavelength in the vicinity of $\lambda = 600$ nm for the first order ($m = 1$)? (What wavelength corresponds to a 1° change in θ?) What would be the wavelength resolution for an angular resolution of one arc-minute?

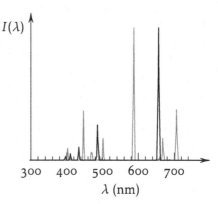

FIGURE 5.14. The spectra of hydrogen (*black*) and helium (*gray*) are plotted as a function of wavelength. The spectral intensity is arbitrarily normalized

The German physicist Gustav Kirchhoff established in 1859 that the alkali and earth-alkali metals sodium, potassium, strontium, lithium, calcium and barium each has a characteristic spectrum that is different from that of all the others. This was the first evidence that the spectrum provides a unique atomic signature. Kirchhoff's measurements depended upon his ability to produce exceptionally pure samples of the materials. His experiments used a flame to generate the spectral lines from the samples and he used a prism to determine the spectra. Use of diffraction gratings subsequently enabled the quantitative measure of the spectral intensity $I(\lambda)$. The visible spectra of hydrogen and helium are illustrated in figure 5.14. There are four prominent lines for hydrogen in the (visible) wavelength range from 400 to 700 nm and more lines for helium in the same range. With the advent of sensors that were sensitive to wavelengths beyond the visible range, into the shorter (ultraviolet) and longer (infrared) wavelengths, it was found that there are many additional lines for each element.

We note that the hydrogen and helium spectra illustrated in figure 5.14 are composed of a series of discrete lines, with different relative intensities for the different lines. Explaining the elemental spectra became the signature problem for physicists that was only resolved with the development of quantum theory. This subject lies beyond our current purview but we can state that the lines composing the spectra are now understood to be generated by transitions between two atomic energy states. The energy associated with the line at a wavelength λ is given by $\mathcal{E} = hc/\lambda$, where h is a constant first introduced by the German physicist Max Planck. A spectral line therefore represents the energy difference between two states and its intensity is a measure of the probability that such a transition will take place.

> EXERCISE 5.27. What is the energy associated with a transition where the wavelength is 397 nm? Use energy units of eV. What is the energy associated with a transition where the wavelength is 656 nm? How do these energies compare to the several MeV energies that Rutherford observed for α particles?

This brings us to the resolution of one problem of x-ray diffraction that we avoided previously. As the Braggs discovered, diffraction occurs when the condition $n\lambda = 2d \sin\theta$ is satisfied. So, the question arose as to why their diffraction experiments (and Laue's photographic plates) produced narrow dots. If the x-ray tubes were producing a wide spectrum of x-rays, then the x-ray interference patterns should be smeared by the existence of more than one wavelength in the x-ray spectrum. Yet, distinct diffraction spots were observed.

At the time, the spectral output of the x-ray tubes had not been determined. X-rays were generated in the tubes when an energetic beam of electrons struck a metal target.[11] Today, we understand that the spectral content of the x-ray tubes is dominated by individual transitions like those depicted in figure 5.14 but at much shorter wavelengths. The transitions that give rise to x-rays involve electrons occupying the lowest energy levels in the metal targets (known for historical reasons as K-shell electrons).

---

[11]The German physicist Wilhelm Conrad Röntgen discovered penetrating radiation that emanated from high voltage cathode ray tubes in 1895. He was awarded the first Nobel Prize in physics in 1901 for his discovery of what he termed x-rays.

As a result, the x-ray spectrum produced by the x-ray tubes is dominated by a single (or a few) individual wavelengths. Hence, the early experimental x-ray studies produced well-defined diffraction patterns. Modern x-ray diffractometers employ specialized optics (monochromators) to ensure that only a single wavelength from the source strikes the target.

5.7. Unfinished Business

Our studies thus far have provided significant insights into the nature of matter and the structure of spacetime. We now know that atoms have a nuclear center that comprises the vast majority of the mass of the atom and that atoms, at least in crystals, occupy regular lattice sites. The nuclear dimension is of the order of a few femtometers (10^{-15} m) and the interatomic spacing in a crystal lattice is of the order of a few tenths of a nanometer (10^{-10} m).[12] As a result, there is a factor of 10^5 between the nominal size of the nucleus and what we might call the size of the atom.

This large difference makes generating any sort of reasonable representations of atoms impossible. Consider representing the nucleus by a circle of radius 1 cm, about the size of a visible dot on the blackboard at the front of the classroom. If scaled correspondingly, neighboring atomic lattice sites would be about 1 km away. No sensible instructor is going to run a kilometer across the campus to mark the neighboring spot in the atomic lattice. Reasonable instructors will perhaps mark the lattice dimensions as a few centimeters or perhaps a meter in size but we know that this scaling is wholly incorrect. Everyone has seen the representation of atoms as a nucleus (depicted by a small cluster of grapes) surrounded by orbiting electrons. We shall not reproduce such a representation here because it is completely misleading. Not only is the scale of such representations incorrect but we now know that it makes no sense whatsoever to think of the atomic constituents as particles in the classical sense.

In 1927, Clinton Davisson conducted a series of experiments at Bell Laboratories in the United States (with Lester Germer) that examined the results of scattering electrons from nickel crystals. In Scotland, meanwhile, George Paget Thomson (son of J. J. Thomson) examined the results of sending beams of electrons through thin metal foils. The results of these experiments turned out to be identical to the results of the Braggs and Laue in scattering x-rays! The electrons diffracted from the atomic

---

[12]To avoid decimals, atomic dimensions are often provided in units of angstroms (Å), named after the Swedish physicist Anders Jonas Ångström, whose 1868 chart of the solar electromagnetic spectrum was graded in units of 10^{-7} mm. This unit became known as the Å in his honor.

lattice in precisely the same way that x-rays did.[13] The interpretation of this result is inexorably that electrons are waves.

The Davisson/Thomson result strikes home the fact that the microscopic world is wholly different than the macroscopic world that we observe with our eyes. We have used the word *particle* to mean a small bit of matter and have undoubtedly visualized particles in our mind's eye as small spheres of some sort. Yet, the experimental evidence dictates that electrons are waves. Fundamentally, the picture of electrons circling about a small cluster of grapes cannot really represent an atom, despite the widespread popular usage of such images.

Envision that somehow we could build a super microscope that would enable us to resolve the world at ever smaller scales: from grains of sand to microorganisms like bacteria. If we were to increase the resolution of our microscope even further, to the level of molecules and atoms, the images would never come into focus. Atoms are not spheres of a particular size and atomic nuclei are not clusters of proton and neutron grapes. The electron clouds that surround the nuclei do not have sharp edges. In some sense, electron cloud is probably a good choice of nomenclature. Clouds in the sky have discernible shapes from afar but, up close, clouds do not have definable edges.

At yet smaller scales, the components of the nuclei are also waves: neutrons also diffract from crystals, as was demonstrated by Ernest Wolland in 1945.[14] So, the nucleus inside of the atom also does not possess a sharp edge. The ideas that the word "particle" may conjure in one's mind are not relevant in discussing the microscopic structure of the universe. Nothing in our perceptible, macroscopic universe, it seems, is truly applicable at the microscopic scale.

> EXERCISE 5.28. The French physicist Louis Victor Pierre Raymond duc de Broglie postulated in 1924 that the electron wavelength λ was related to its relativistic momentum p by the formula $\lambda = h/p$, where h is Planck's constant.[15] Given that atomic crystals have a lattice spacing in the range of 0.3–0.4 nm, what would be the necessary momenta for electrons to display diffraction effects?

---

[13]Davisson and Thomson were awarded the Nobel Prize in 1937 "for their experimental discovery of the diffraction of electrons by crystals."

[14]Wolland was joined in his experiments by Clifford Shull in 1946. Shull received the Nobel Prize in 1994 "for the development of the neutron diffraction technique" (with Bertram N. Brockhouse "for the development of neutron spectroscopy") but Wolland had died in 1984 and was therefore ineligible.

[15]De Broglie was awarded the 1929 Nobel Prize in Physics "for his discovery of the wave nature of electrons."

VI

Terrestrial Mechanics

Newtonian mechanics epitomizes the triumph of the process of physics. One can, with two guiding principles and a handful of mathematical equations, describe the behavior of planets orbiting the sun and α particles scattering from nuclei. As a result of these successes, it is only natural to ask can we not use the same ideas to describe phenomena on our own planet? As the astute student can probably recognize, given that the text does not end at this point, the answer is a resounding yes. Unfortunately, motion in our terrestrial environment is significantly more complicated than the cases that we have studied to this point.

As it happens, essentially the *only* force acting between the earth and the sun is the gravitational force. The sun radiates a constant stream of high energy particles, mostly protons and electrons, into space along with the electromagnetic waves that we call sunlight but the force exerted on the earth by this radiation is miniscule compared to the gravitational force. The sun also has a large magnetic field but it does not extend to any significant degree to the earth's orbital distance. So, to a very good approximation, gravity is the only force acting on the earth due to the sun.

This is not the case for motion of physical objects in our personal experience. Consider, for example, the flight of a golf ball. In order for the ball to move forward, it must displace the air in front of it. This resistance to forward motion can be expressed as a force acting on the ball but this resistive force is not a fundamental force like gravity.[1] Instead, we shall have to empirically determine the nature of the force.

In addition, it is possible to impart spin to a golf ball. Professional golfers do so in a strategic fashion to "shape" their shots to avoid obstacles or tailor their approaches to the greens. Novice golfers often impart spin inadvertently, resulting in hooks or slices where the trajectory of the ball

---

[1]Physicists have thus far identified four fundamental forces: gravity, the electromagnetic force, the strong nuclear force and the weak nuclear force. Other forces with which the student might be familiar, such as capillary forces or frictional forces, are macroscopic manifestations of the electromagnetic force.

© Mark A. Cunningham 2015
M.A. Cunningham, *Neoclassical Physics*, Undergraduate Lecture
Notes in Physics, DOI 10.1007/978-3-319-10647-2_6

diverges dramatically from the intended direction. Apparently, the fact that the golf ball is spinning about its center of mass also gives rise to a force that we shall have to empirically determine.

So, to discuss motion of objects in terrestrial environments, we shall have to modify Newton's Equation 1.7 slightly to account for the fact that objects may be subject to multiple forces. We can write the following succinct relation:

$$(6.1) \qquad\qquad \sum_i \mathbf{F}_i = m\mathbf{a},$$

where the large Greek sigma indicates that one must sum over all of the forces $\mathbf{F}_i$ acting on the object.

Equation 6.1 provides us with a means for quantifying the behavior of terrestrial objects. Consider, for example, a golf ball that is sitting on the ground. The ball is, of course, subject to the gravitational force that is directed toward the center of the earth. Yet, as it is not moving, it is clearly not accelerating. From Equation 6.1, we can infer that the (vector) sum of the forces acting on the ball must be zero, as $\mathbf{a} = 0$. As a result, there must be some *other* force acting on the ball, in opposition to the gravitational force. This force is provided by the matter that makes up the earth itself, which resists any subsequent motion of the golf ball towards the earth's center.

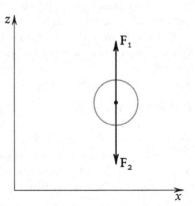

FIGURE 6.1. A free-body diagram represents the forces (*arrows*) acting on an object (*gray circle*). A coordinate system is necessary, as the forces are vectors

Physicists have developed an abstract, pictorial means of representing Equation 6.1 known as a **free-body diagram**. In a free-body diagram, an object is represented by a circle or rectangle. The object may be an airplane or a baseball or a battleship but it is not important to draw a faithful representation of the object itself. Forces acting on the (center of mass of the) object are depicted as arrows, as indicated in figure 6.1. While we have been discussing a golf ball sitting on the ground, the free body diagram illustrated in figure 6.1 would also represent situations in which a helicopter was hovering motionless above the ground and a tug-boat was floating in a harbor. In both those situations, the gravitational attraction of the earth provides a downward force on the objects and, in each case, there is some other force opposing the gravitational force. For the helicopter, we call that force "lift" and, for the tugboat, we call it the "buoyant force."

For the case of the golf ball sitting on the ground, the resulting equation for the acceleration is $F_1 + F_2 = 0$, where if we assume from figure 6.1 that z is directed upward from the earth's surface, the gravitational force acting on the golf ball would be F_2. The force opposing the gravitational force must be equal and opposite, in order for the acceleration to vanish: $F_1 = -F_2$. We shall make further use of the free-body diagram as we continue to examine the motion of terrestrial objects.

6.1. Motion Near the Earth's Surface

In the vicinity of the earth's surface, gravitational acceleration of objects can be considered to be constant, at least to a first approximation. For golf balls, baseballs or other objects that do not travel far in the vertical direction compared to the earth's radius of 6400 km, we can construct a local Cartesian coordinate system in which the gravitational force acts in the vertical (z) direction, as depicted in figure 6.2. In this regard, motion in the vertical direction will always be subject to a constant acceleration, whereas motion in the horizontal directions may or may not be accelerated.

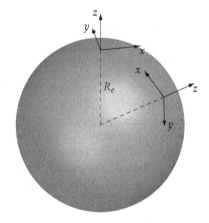

FIGURE 6.2. At the surface of the earth, one can define local coordinate systems. The gravitational force on objects near the earth's surface acts along the vertical ($-z$) direction

Let us begin by considering relatively slow-moving objects, where we can neglect resistive forces that arise due to motion through the atmosphere.

The gravitational field **G**, as we defined in Equation 2.40, depends on the distance of an object from the center of the earth. For motion in the vicinity of the earth's surface, a difference of even a kilometer in the vertical direction is a small fraction of the earth's radius of 6400 km. As a result, the acceleration due to gravity can be considered constant and vertically directed. This is, of course, only approximately true and the approximation fails if we consider rockets or other projectiles that traverse a significant distance across the earth's surface, but is a generally good approximation for golf balls and the like.

So, neglecting air resistance, motion in the vicinity of the earth's surface is subject to an acceleration **a** $= (0,0,-g) = -g\hat{z}$, where we assume the z axis to be upwardly directed. The acceleration is the time derivative of velocity, so because **a** is a constant, this can be integrated immediately:

$$\int_{t_1}^{t_2} dt\, \frac{d\mathbf{v}}{dt} = \mathbf{a} \int_{t_1}^{t_2} dt.$$

Thus, we find the following relation[2]:

$$\mathbf{v}(t_2) - \mathbf{v}(t_1) = \mathbf{a}(t_2 - t_1).$$

Recall that velocity is the time derivative of position. If we choose t_1 to be some fixed, initial time, then after rearranging and integrating the equation again, we obtain the following:

$$\int_{t_1}^{t_2} dt\, \frac{d\mathbf{r}}{dt} = \int_{t_1}^{t_2} dt\, [\mathbf{v}(t_1) + \mathbf{a}(t - t_1)].$$

Finally, we have

(6.2) $$\qquad \mathbf{r}(t_2) - \mathbf{r}(t_1) = \mathbf{v}(t_1)(t_2 - t_1) + \frac{1}{2}\mathbf{a}\,(t_2 - t_1)^2.$$

This is the same kinematic equation that we obtained originally (Equation 1.6) as an approximation to motion. In the case of constant acceleration, the kinematic equation is exact.

> EXERCISE 6.1. Let define the initial time to be $t_1 = 0$ and set the coordinate origin to be $\mathbf{x}(t_1) = (0,0,0)$. Suppose that an object has an initial velocity $\mathbf{v}(t_1) = (v_1 \cos\theta, 0, v_1 \sin\theta)$ and is subject to the gravitational acceleration $\mathbf{a} = (0,0,-g)$. What is the free-body diagram for the object? What is the position of the object at a time t_2? How long does it take the object to return to the height $z = 0$? How far has the object travelled horizontally in that time?

---

[2] Again, there is a potential notational ambiguity here. On the left hand side of the equation, the velocity **v** is a function of time and is measured at two specific times. On the right hand side, the time difference multiplies the constant acceleration.

EXERCISE 6.2. In the *Mathematica* program, define the functions $x(t, x_1, v_1) = x_1 + v_1 t$ and $z(t, z_1, v_2) = z_1 + v_2 t - 10 t^2$. These functions represent the motion of an object in the x-z plane as a function of time, initial position and initial velocity. Use the `ParametricPlot` function to plot the position of the object as a function of time over the range $0 \le t \le 2$. Choose $x_1 = z_1 = 0$ and $v_1 = v_2 = 5$.

Use the `Plot` function to plot the functions x and z as a function of time. Calculate and plot the derivatives of x and z. Use the `Manipulate` function to vary the initial velocities over the range $0 \le v_1 \le 5$ and $0 \le v_2 \le 5$. How do the plots change as a function of initial velocity?

Constant acceleration in the vertical direction gives rise to parabolic motion; the position depends quadratically on the time and thus a plot of the vertical component of position versus time sweeps out a parabola. Moreover, without acceleration in the horizontal direction, the trajectory of an object sweeps out a parabola in space also. (The horizontal position changes at a constant rate, proportional to the velocity in the horizontal direction. A plot of vertical position versus horizontal position instead of time results in a simple rescaling of the horizontal axis.)

For many physical objects, neglecting the resistive force that arises from displacing air (or fluid) in the path of travel is a poor approximation. Watching the trajectories of baseballs or golf balls, it is evident that the path is not parabolic. The trajectory is not symmetric; the final portion of the trajectory is much steeper than the initial portion.

Unfortunately, there is no universal law for the resistive force that affects flying objects. Indeed, the motion of objects through fluids or fluids around objects is generally quite complex and depends strongly on the properties of the fluid like density, viscosity and compressibility and the relative velocity of the object. For example, a chunk of rock entering the earth's atmosphere at a relative velocity of 17 km/s leaves a fiery trail across the sky: we call such objects meteors or, more romantically, shooting stars.[3] The passage of the rock through the atmosphere transfers so much energy to the air as it travels that the air glows and the rock partially melts, leaving a bright trail across the sky. We observe no such dramatic behavior of rocks hurled into a pond. They generate splashes of water upon striking the pond surface (and ripples) but leave no bright trails through the atmosphere.

---

[3]In 2013, a 17–20 m diameter meteor plunged through the sky over Chelyabinsk, Russia. Its trajectory was recorded by now-ubiquitous cameras and, for a time, was brighter than the sun. The shock wave generated by the meteor's passage through the atmosphere damaged over 7200 buildings and injured thousands, mostly due to flying glass from broken windows.

So, in truth, the study of the flight of golf balls involves rather complex fluid mechanics that we shall not pursue here. Rather, we shall investigate simpler models that can capture some of the behavior of more complex phenomena. As we have stated previously, this is a common tactic in physics: study a problem for which you can generate analytic solutions before venturing into deeper mathematical waters. Such a strategy allows us to study (simplified) systems and gain some understanding of how the various parameters affect the behavior of the system.

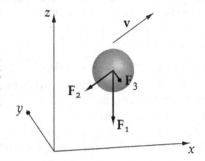

FIGURE 6.3. The free-body diagram for an object (*gray sphere*) moving with a velocity **v** includes the gravitational force $\mathbf{F}_1$ and other, aerodynamic forces $\mathbf{F}_2$ and $\mathbf{F}_3$

If we consider the flight of a physical golf ball that is initially travelling in the x-z plane, the ball may ultimately hook or slice into the transverse y direction. This implies that there must be some force acting on the ball in that direction: without a component of acceleration in the y-direction, the y-component of velocity will remain zero and, hence, the y-component of position will not change. In figure 6.3, we have illustrated the free-body diagram for a golf ball. Using our conventional choice for coordinate systems, there is a constant gravitational force $\mathbf{F}_1$ acting on the ball in the vertical z direction. We anticipate that there will be a force $\mathbf{F}_2$ oppositely directed to the velocity **v** that opposes the forward motion of the ball and there may also be some component of force perpendicular to the direction of motion $\mathbf{F}_3$ that accounts for hooks and/or slices.

Let us ignore, for the time being, other forces and focus on the force that opposes the motion. We propose to investigate a resistive force $\mathbf{F}_2$ that is proportional to the velocity: $\mathbf{F}_2 = -\alpha \mathbf{v}$, where α is a constant of proportionality and the sign is chosen to produce a force in opposition to the direction of motion. We note that α must have dimension. The dimensional equation is $(M \cdot L/T^2) = \alpha(L/T)$, from which we can infer that α has units of (M/T). One question that arises is can we infer any physical properties of the parameter α? The physical characteristics of the system include the size of the object (L) or its cross-sectional area (L^2), the density of the air (M/L^3) and the viscosity of the air (M/LT). In 1851, the British physicist George Stokes analyzed the problem of a small sphere drifting through a fluid and proposed that the constant α can be interpreted on dimensional

grounds as the product of the size of the sphere and the fluid viscosity.[4]
The result Stokes obtained turns out to have limited applicability but we
shall see where the idea leads.

The kinematic equation describing this simplified system can be obtained
from Equation 6.1 (dividing by the mass of the object m) and can be written as follows:

(6.3) $$\mathbf{g} - \frac{\alpha}{m}\mathbf{v} = \mathbf{a} = \frac{d\mathbf{v}}{dt},$$

where m is the mass of the object and $\mathbf{g} = -g\hat{\mathbf{z}}$ is the gravitational acceleration. If we assume that motion takes place in the x-z plane, Equation 6.3
separates into two component equations:

$$-\frac{\alpha}{m}v_x = \frac{dv_x}{dt} \quad \text{and} \quad -g - \frac{\alpha}{m}v_z = \frac{dv_z}{dt}.$$

These can be solved quite readily with a change of variables. Let $u = -g-(\alpha/m)v_z$, then $du = -(\alpha/m)dv_z$ and the equation for v_z can be rewritten
as follows:

$$\int \frac{du}{u} = -\frac{\alpha}{m} \int dt.$$

Integrating both sides, we find $\ln u = -(\alpha/m)t + c$, for some constant of
integration c. Exponentiating both sides of this result yields the following
result:

$$u = -g - (\alpha/m)v_z = c'\,e^{-(\alpha/m)t},$$

where the original constant of integration c has become a multiplicative
factor c'. If we define an initial condition that $v_z(t = 0) = v_2$, then this will
fix the multiplicative factor. We find the following result for the velocity
in the z-direction:

(6.4) $$v_z = v_2\,e^{-(\alpha/m)t} - \frac{mg}{\alpha}\left[1 - e^{-(\alpha/m)t}\right].$$

If we assume that v_x has some initial value v_1, then we can also show that
the following relation holds for v_x:

(6.5) $$v_x = v_1\,e^{-(\alpha/m)t}.$$

We have plotted the velocities for different values of α in figure 6.4. With
no resistive force ($\alpha = 0$), we can see from figure 6.4 that the velocity in the
x-direction is unchanged over time and that the velocity in the z-direction
changes linearly under the influence of the constant gravitational acceleration. When a resistive force is present (gray curves in figure 6.4) the
velocity in the x-direction decreases. Initially, the rate of change of velocity in the z-direction is larger than that of the gravitational acceleration.
At later times, the velocity in the z-direction changes less.

---

[4]Stokes derived the result $\alpha = 6\pi\eta R$, where η is the viscosity and R the sphere radius.

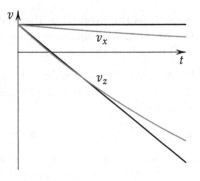

FIGURE 6.4. The time evolution of the velocity depends on the value α/m. The *black curves* represent the case $\alpha/m = 0$ and the *gray curves* represent a case where $\alpha/m \neq 0$

EXERCISE 6.3. Complete the missing steps in the derivation of Equation 6.4. Use these results to derive Equation 6.5.

EXERCISE 6.4. Use the Plot and Manipulate functions to plot the velocity of an object as a function of time, using Equations 6.4 and 6.5. Assume that $v_1 = v_2 = 5$ m/s, $g = 10$ m/s², and consider the interval $0 \leq t \leq 2$. What happens as α/m changes from 0.1 to 0.5? What happens at very long times ($t \to \infty$)?

One problem you may find in plotting the solutions to v_z is the fact that the representation in Equation 6.4 depends inversely on α. Actually, the vertical velocity has a finite limit when α vanishes. Note that the exponential function has a series expansion:

$$e^{-x} = 1 - x + \frac{x^2}{2!} - \frac{x^3}{3!} + \cdots$$

whereby

$$1 - e^{-x} = x - \frac{x^2}{2!} + \frac{x^3}{3!} - \cdots$$

Hence, as α limits toward zero, the vertical velocity becomes $v_z = v_2 - gt$, which is the result we obtained previously for constant acceleration.

For very long times, the exponentials in Equations 6.4 and 6.5 tend to zero. Consequently, at long times, there will be no horizontal velocity. The vertical velocity tends to a constant value $v_z = -mg/\alpha$, that is often called the *terminal velocity*. This occurs when the velocity-dependent resistive force is equal to the gravitational force. At that point, there is no further acceleration of the object.

Now to examine the trajectories of objects subject to the resistive force, we need to integrate Equations 6.4 and 6.5 to obtain the positions. In the horizontal direction, we have

$$\int dt \frac{dx}{dt} = v_1 \int dt\, e^{-(\alpha/m)t}.$$

If the time integrals extend from $t = 0$, then we find the following:

(6.6)
$$x - x_0 = \frac{mv_1}{\alpha}\left[1 - e^{-(\alpha/m)t}\right],$$

where x_0 is a constant of integration that defines the initial position in the x-direction. The vertical direction is a bit more tedious, as there are more terms, so we shall state the result:

(6.7)
$$z - z_0 = \frac{m}{\alpha}\left\{-gt + \left(v_2 + \frac{mg}{\alpha}\right)\left[1 - e^{-(\alpha/m)t}\right]\right\},$$

where we have again assumed an initial time of $t = 0$ and an initial vertical position z_0.

> EXERCISE 6.5. Use the function D to demonstrate that Equations 6.6 and 6.7 yield the velocities found in Equations 6.5 and 6.4.

> EXERCISE 6.6. Integrate Equation 6.4 to obtain Equation 6.7.

> EXERCISE 6.7. Take the limit of Equations 6.6 and 6.7 as α tends to zero. Are your results reasonable?

If we now plot the results of Equations 6.6 and 6.7, we can observe directly the influence of the resistive force. When the resistive force vanishes, as we can see in figure 6.5, we obtain a parabolic trajectory. For increasing values of α, the object travels shorter and shorter horizontal distances. Additionally, the maximum vertical distance obtained is also reduced as α increases.

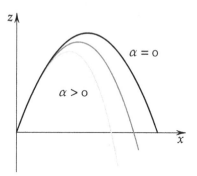

FIGURE 6.5. The parameter α defines the resistive force. For $\alpha = 0$ (*black curve*), one obtains a parabolic trajectory. For increasing values of α (*gray* and *light-gray curves*) the trajectory deviates from parabolic. The curves were generated for the same total time interval

> EXERCISE 6.8. Use the ParametricPlot function to plot the position as a function of time, as defined by Equations 6.6 and 6.7. Use $g = 10\,\text{m/s}^2$ and initial velocity $\mathbf{v} = (5., 0., 5.)\,\text{m/s}$. What happens as α/m varies from 0.1 to 0.7? At what times does the object return to the $z = 0$ plane? What happens at very long times?

6.2. Energy Conservation

We can define mechanical **work** W as follows:

$$(6.8) \qquad W = \int d\mathbf{s} \cdot \mathbf{F},$$

where the integral extends over some (three-dimensional) path. For gravitating systems, we found that the force was directed along the vector
$\mathbf{r}_2 - \mathbf{r}_1$. From Equation 2.6, if we consider the work done on an object by
moving it in a gravitational field, only the component of the path along
the direction defined by $\mathbf{r}_2 - \mathbf{r}_1$ will contribute to the integral. In a spherical coordinate system centered at $\mathbf{r}_1$, this direction corresponds to the
radial direction.

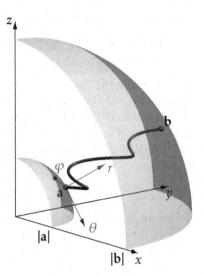

FIGURE 6.6. If we set the coordinate
origin at $\mathbf{r}_1$, the motion of an object
($\mathbf{r}_2$) from **a** to **b** can follow a tortuous path. The potential energy is
constant on surfaces of constant radius and thus depends solely on the
magnitudes $|\mathbf{a}|$ and $|\mathbf{b}|$

We have depicted a particular path in figure 6.6, where the vector $\mathbf{r}_2 - \mathbf{r}_1$
evolves from point **a** to **b** over some time. Recalling the original definition of the gravitational force from Equation 2.4, we can state that the
work involved in moving a mass M_2 from **a** to **b** is given by the following
expression:

$$W = -GM_1M_2 \int_a^b d\mathbf{s} \cdot \frac{\mathbf{r}_2 - \mathbf{r}_1}{|\mathbf{r}_2 - \mathbf{r}_1|^3}.$$

The dot product selects only the component of $d\mathbf{s}$ that is parallel to $\mathbf{r}_2 - \mathbf{r}_1$,
i.e., the radial direction. So, if the point **b** is located somewhere on the
sphere of radius $|\mathbf{a}|$, the integral will vanish due to the fact that the dot
product vanishes. If **b** is not on the spherical surface, the integral can

be solved most readily if we utilize spherical coordinates. An arbitrary infinitesimal step in spherical coordinates can be written as follows:

$$d\mathbf{s} = dr\,\hat{\mathbf{r}} + r\,d\theta\,\hat{\boldsymbol{\theta}} + r\sin\theta\,d\varphi\,\hat{\boldsymbol{\varphi}},$$

where the polar angle θ is measured from the z-axis and the azimuthal angle φ is measured from the x-axis. The unit vectors $\hat{\mathbf{r}}$, $\hat{\boldsymbol{\theta}}$ and $\hat{\boldsymbol{\varphi}}$ are orthogonal.

EXERCISE 6.9.   The unit vectors in spherical coordinates can be resolved in Cartesian coordinates as follows:

$$\hat{\mathbf{r}} = (\sin\theta\cos\varphi, \sin\theta\sin\varphi, \cos\theta)$$

$$\hat{\boldsymbol{\theta}} = (-\cos\theta\cos\varphi, -\cos\theta\sin\varphi, \sin\theta)$$

$$\hat{\boldsymbol{\varphi}} = (\sin\varphi, -\cos\varphi, 0).$$

Show that the following results are true:

$$\begin{array}{lll} \hat{\mathbf{r}}\cdot\hat{\mathbf{r}} = 1 & \hat{\boldsymbol{\theta}}\cdot\hat{\mathbf{r}} = 0 & \hat{\boldsymbol{\varphi}}\cdot\hat{\mathbf{r}} = 0 \\ \hat{\mathbf{r}}\times\hat{\mathbf{r}} = 0 & \hat{\boldsymbol{\theta}}\cdot\hat{\boldsymbol{\theta}} = 1 & \hat{\boldsymbol{\varphi}}\cdot\hat{\boldsymbol{\theta}} = 0 \\ \hat{\mathbf{r}}\times\hat{\boldsymbol{\theta}} = \hat{\boldsymbol{\varphi}} & \hat{\boldsymbol{\theta}}\times\hat{\boldsymbol{\varphi}} = \hat{\mathbf{r}} & \hat{\boldsymbol{\varphi}}\times\hat{\mathbf{r}} = \hat{\boldsymbol{\theta}} \end{array}$$

Using our compact notation, where $\mathbf{r}_2 - \mathbf{r}_1 = r_{12}\,\hat{\mathbf{r}}$, the only non-vanishing component of the integral over the path $d\mathbf{s}$ is the radial component over the scalar dr_{12}:

$$W = -GM_1M_2 \int_{|\mathbf{a}|}^{|\mathbf{b}|} \frac{dr_{12}}{r_{12}^2}$$

$$= GM_1M_2\left[\frac{1}{|\mathbf{b}|} - \frac{1}{|\mathbf{a}|}\right].$$

If we now refer to the definition of the potential energy, given in Equation 2.16, we find that the difference in potential energies between positions **a** and **b** is given by the following:

$$\mathcal{U}_b - \mathcal{U}_a = -GM_1M_2\left[\frac{1}{|\mathbf{b}|} - \frac{1}{|\mathbf{a}|}\right],$$

which is identical to the work done to move from **a** to **b** (up to a sign). Notice that the work performed to move the object from **a** to **b** does not depend upon the path, only on the endpoints. We describe such a force as a *conservative* force.

We interpret this result to mean that if we perform work on an object to change its position in a gravitational field, then this work is manifested by a change in the potential energy of the system. Work is, consequently, a form of energy.

In Chapter 2, we defined (in Equation 2.16) the mechanical energy $\mathcal{E}$ of the gravitational system to be the sum of kinetic T and potential $\mathcal{U}$ terms. In our present discussions about motion near the surface of the earth, we can adopt a simplified form of the equation. First, if we take M_1 to be the mass of the earth and M_2 to be the mass of our object, then the sum $M_1 + M_2$ is effectively the same as the mass of the earth itself. Hence, we can approximate the kinetic energy of our object as follows:

(6.9) $$T = \frac{M_1 M_2}{2(M_1 + M_2)} v^2 \approx \frac{1}{2} M_2 v^2.$$

where v is the object's velocity. Similarly, near the earth's surface, the potential energy can be simplified. Using our local Cartesian coordinate system, the vector $\mathbf{r}_2 - \mathbf{r}_1$ is given by $\mathbf{r}_2 - \mathbf{r}_1 = (R_{earth} + z)\hat{\mathbf{z}}$, where here z measures the distance from the earth's surface. In this case, the potential energy can be written as follows:

(6.10) $$\mathcal{U} = -GM_1 M_2 \frac{1}{R_{earth} + z} \approx -\frac{GM_1 M_2}{R_{earth}}(1 - z/R_{earth}),$$

where in the last step, we have made use of the approximation $(1 + x)^{-1} \approx 1 - x$ when x is a small number. If we consider changes in the potential energy at two different heights z_1 and z_2, we can write the following result:

(6.11) $$\Delta \mathcal{U} \equiv \mathcal{U}(z_2) - \mathcal{U}(z_1) = M_2 g(z_2 - z_1),$$

where we have defined the gravitational acceleration $g = GM_{earth}/R_{earth}^2$. Equations 6.9 and 6.11 define the energy state for objects near the earth's surface. The kinetic energy is quadratically dependent upon the velocity and the potential energy is linearly dependent upon the height above the earth's surface.

> EXERCISE 6.10. Plot the functions $f(x) = 1/(1 + x)$ and $f_1(x) = 1 - x$. For what values of x do the two differ by more than 1 %?. If the earth's radius is 6400 km and the highest point in a golf ball trajectory is approximately 100 m, is the approximation we make with f_1 a reasonable one?

> EXERCISE 6.11. If an object is held at rest at an altitude h above the earth's surface and then released, what would be the velocity just before impact? Hint: use the fact that energy is conserved.

If we now want to focus on the motion of objects in the vicinity of the earth's surface, we have a relatively simple prescription for defining the potential energy: we can equate it to the negative of the work required to move the object from **a** to **b**. The gravitational force, in our usual coordinate system, has the form $\mathbf{F}_1 = -mg\,\hat{\mathbf{z}}$. As a result, the work integral projects out only the z-components of any trajectory. We have then

$$\mathcal{U}_b - \mathcal{U}_a = -W = mg \int_a^b dz = mg(b_z - a_z),$$

where a_z and b_z are the z-components of $\mathbf{a}$ and $\mathbf{b}$, respectively.

Consider now what happens when we include the resistive force. We have plotted the mechanical energies for the trajectories for different values of the parameter α (shown in figure 6.5) in figure 6.7. For the case where $\alpha = 0$, we observe that the total mechanical energy $\mathcal{E}$ is constant. The kinetic T and potential $\mathcal{U}$ energies vary in such a way that the sum is preserved, as we have come to expect.

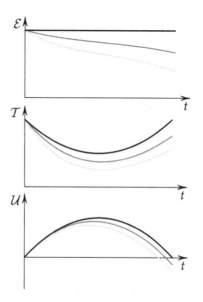

FIGURE 6.7. A resistive force causes the loss of energy from the moving object. When $\alpha = 0$ (*black curve*), the total mechanical energy $\mathcal{E}$ remains constant, where the kinetic T and potential energies $\mathcal{U}$ do not. For increasing values of α (*gray* and *light-gray curves*) the total energy decreases. All curves were generated for the same total time interval

Such is not the case for values of α greater than zero. The total energy decreases over time. In some sense, this loss of energy would be a crisis: we have previously linked energy conservation to time invariance through Noether's theorem. Energy conservation is a basic principle of physics and a violation of this basic principle would be devastating.[5]

From a physical perspective, however, there is no real mystery. Some of the kinetic energy of the object is simply transferred to molecules in the air, albeit in a fashion that makes quantitative analysis difficult. One can

---

[5]Experimental verification of a violation of energy conservation would, most likely, lead to award of a Nobel Prize in physics. So, it would not be completely devastating to everyone. Experimentalists are always quietly looking for such results but, thus far, no exceptions have been found.

imagine that, as the object plows through the atmosphere, a wake of disturbed air is generated, much like the wake trails that boats generate as they traverse the surface of the water. Unfortunately, the air is not visible to our eyes, so we cannot directly visualize this behavior.

If we consider now the work performed by the resistive force $\mathbf{F}_2$, we find the following:

$$W = -\alpha \int_a^b d\mathbf{s} \cdot \mathbf{v} = -\alpha \int_a^b [dx\, v_x + dz\, v_z].$$

We know x and z as functions of the time t and, so, we can rewrite the integral in terms of an integral over time:

$$W = -\alpha \int_0^{t_2} dt\, [v_x^2 + v_z^2],$$

where we have utilized the fundamental theorem of calculus and the fact that $dx/dt = v_x$. and $dz/dt = v_z$. As before, we have also set the initial time to be $t_1 = 0$.

Substituting in our results for Equations 6.4 and 6.5, we can obtain the following result for the work performed:

(6.12)
$$\begin{aligned}
W = &-\frac{m}{2}\left[v_1^2 + \left(v_2 + \frac{mg}{\alpha}\right)^2\right]\left[1 - e^{-2(\alpha/m)t}\right] \\
&+ 2\frac{m^2 g}{\alpha}\left(v_2 + \frac{mg}{\alpha}\right)\left[1 - e^{-(\alpha/m)t}\right] - \frac{m^2 g^2}{\alpha} t
\end{aligned}$$

If we compute the sum $T + U - W$, we find that it is a constant, $m(v_1^2 + v_2^2)/2$, as illustrated in figure 6.8.

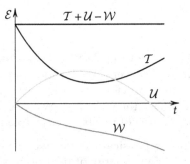

FIGURE 6.8. The resistive force causes the loss of energy from a moving object. The kinetic T (*dark gray curve*) and potential energies U (*light gray curve*) do not sum to a constant. If we include the work W (*gray curve*) performed on the external world, we can account for all of the energy

We can interpret this result as the total energy of the object plus air system is conserved. The total energy includes the mechanical energy of the object, as we defined previously, plus the work done by the object on the external world. The resistive force is an example of a *non-conservative* force; the work depends upon the path, not just the end points of the path. Such a force is also described frequently as a *dissipative* force, as energy is dissipated from the mass M_2 as time passes.

EXERCISE 6.12. Show that integrating Equations 6.4 and 6.5 results in Equation 6.12.

EXERCISE 6.13. Define functions for T, U and W in the *Mathematica* program. Show that $T + U - V = m(v_1^2 + v_2^2)/2$. Use the Plot function to demonstrate this graphically for the case where $a/m = 0.4$, $g = 10 \, \text{m/s}^2$ and $\mathbf{v} = (5, 0, 5) \, \text{m/s}$. Examine the range $0 \le t \le 1$.

EXERCISE 6.14. Clouds, we know, are made of small droplets of water. Drops of water spraying from a lawn sprinkler fall to the ground in less than a second. Rain drops fall from the sky, so why don't clouds? Use Stokes's formula for $a = 6\pi\eta R$ and compute the terminal velocity of a water droplet with $R = 10 \, \mu\text{m}$. Assume that $\eta = 1.75 \times 10^{-5} \, \text{kg/m·s}$.

Now at this point, the alert student may ask what would happen if we introduce the aerodynamic force $\mathbf{F}_3$ that accounts for deflection of the object from the plane of motion. We have associated this force with spinning objects. As it happens, though, a force that acts perpendicular to the direction of motion does no work; the dot product between the force and the path vanishes. Consequently, incorporating such a force does not affect our discussions of energy conservation in our model system.

6.3. Physics of Baseballs

The simple velocity-dependent force that we have investigated thus far possesses some of the properties that we need to describe the motion of objects like golf balls and baseballs. Unfortunately, actual trajectories of these objects are not well described by a linear dependence on velocity but appear, instead, to depend upon the square of the velocity. In his 1851 paper, Stokes introduced a dimensionless ratio that we now call the **Reynolds number**, that measures the relative magnitude of the inertial forces, that we have discussed previously, and the viscous forces at work in the fluid. For motion of an object through a fluid, the Reynolds number can be written as follows:

$$(6.13) \qquad\qquad R_e = \frac{\rho v L}{\eta},$$

where ρ is the fluid density, η is the fluid viscosity, v is the relative velocity of the object in the fluid and L is a characteristic size of the object. Technically, the motion of objects the size of baseballs through air is associated with a high Reynolds number and Stokes's model that depends linearly on velocity is not valid. The difference in trajectories is illustrated in figure 6.9.

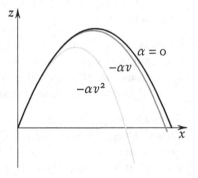

FIGURE 6.9. The resistive force valid for the flight of baseballs scales like the square of velocity. The *black curve* is the trajectory obtained with no resistive force and the *gray curve* is that obtained with a force that scales linearly with velocity. The *light gray* curve is the more realistic result obtained when the resistive force scales like v^2

EXERCISE 6.15. What is the Reynolds number for a $10\,\mu m$ cloud droplet? What is the Reynolds number for a $7.5\,cm$-diameter baseball? Assume $\eta = 1.75 \times 10^{-5}\,kg/m{\cdot}s$ and $\rho = 1.3\,kg/m^3$.

In 1916, Lord Rayleigh developed an equation for the motion of a sphere through compressible fluids, subject to a resistive force that scaled like the square of the velocity.[6] If we again assume that motion is restricted to the x-z plane, then the equations of motion for the sphere can be written as follows:

$$m\frac{d^2x}{dt^2} = -\alpha\frac{dx}{dt}\left[\left(\frac{dx}{dt}\right)^2 + \left(\frac{dz}{dz}\right)^2\right]^{1/2}$$

(6.14)
$$m\frac{d^2z}{dt^2} = -mg - \alpha\frac{dz}{dt}\left[\left(\frac{dx}{dt}\right)^2 + \left(\frac{dz}{dz}\right)^2\right]^{1/2}.$$

The coefficient α in this case also has dimension. The dimensional equation associated with Equations 6.14 is $ML/T^2 = \alpha L^2/T^2$. This implies that α has dimension of M/L. If we again consider the physical characteristics of the fluid: density (M/L^3) and viscosity (M/LT) and the size of the object (L), then the proportionality factor α cannot depend upon the viscosity as there is no time dependence in α.

Rayleigh suggested that the appropriate form for α is $\alpha = \frac{1}{2}C_D\rho L^2$, where C_D is some dimensionless number, ρ is the fluid density and L^2 can be interpreted as the cross-sectional area of the object along the direction

---

[6]John William Strutt became the third Baron of Rayleigh of Terling Place, Witham upon the death of his father in 1873. Rayleigh's interest in science was an unusual preoccupation for the landed gentry but his scientific career was quite remarkable. Rayleigh was awarded the Nobel Prize in physics in 1904 for his discovery of the element argon.

of travel. For a sphere, this is just πR^2, independent of the direction of travel. The drag coefficient C_D will depend on other physical characteristics of the object itself. We can envision that a smooth metal ball will slip through the air more readily than a rough, nearly spherical, rock.

Unfortunately, while we have the equations of motion, it is not possible to solve the equations analytically. The dependence on the magnitude of the velocity $|v|$ couples the x- and z-components of the equations in a non-trivial manner. If we exclude, for the moment, motion in the horizontal direction, then the one-dimensional problem of a sphere falling vertically can be solved. In this case, the velocity in the z-direction satisfies the following equation:

$$\frac{dv_z}{dt} = -g + \frac{\alpha}{m}v_z^2.$$

If we rearrange terms and integrate, we obtain the following result:

$$\int \frac{dv_z}{g - \alpha v_z^2/m} = -\int dt$$

$$\left(\frac{m}{\alpha g}\right)^{1/2} \tanh^{-1}\left[\left(\frac{\alpha}{mg}\right)^{1/2} v_z\right] = -t + c$$

where c is a constant of integration. Solving now for v_z, we find that, using the fact that $\tan(-x) = -\tan(x)$:

(6.15) $$v_z = -\left(\frac{mg}{\alpha}\right)^{1/2} \tanh\left[\left(\frac{\alpha g}{m}\right)^{1/2} t\right] + c',$$

where now c' is some constant that will be determined by the initial conditions. Comparing this result with our previous result (Equation 6.4) for a linear resistive force, we see that the terminal velocity scales like the square root of mg/α instead of linearly.

> EXERCISE 6.16.   Consider the case of an object dropped from rest ($v_z = 0$), subject to either linear or quadratic resistive forces. For the same value of $\alpha g/m$, how does the time evolution of the velocity differ? Plot v_z for each of the two models using the Plot function. Assume $g = 10 \text{ m/s}^2$ and $\alpha/m = 0.4$.

If we really want to pursue the study of physical objects, we are now faced with something of a dilemma. We possess the equations of motion but cannot obtain an analytic solution for the two-dimensional problem. To proceed, we can simplify the system further or just solve the problem numerically. Physicists often follow either or both pathways. In the limit of very shallow trajectories, or nearly horizontal motion, an approximation can be developed that leads to an analytic expression for the trajectory. If the problem that you want to study involves nearly horizontal motion, then that approach would be satisfactory. If instead, you want to consider the problem more generally, then a numerical approach will provide the answers necessary but with the loss of an explicit, functional dependence on the parameters. One has to run a number of numerical cases to extract information on how each of the parameters affects the result.

At this point, we shall make use of the new computational tools at our disposal to study the problem numerically. This will free us from any restrictions on the applicability of our results. We shall utilize the *Mathematica* function NDSolve to construct solutions to the system of equations with a quadratic resistive force. The syntax associated with using this function is not intuitive, so we shall provide an example in the following Exercise.

EXERCISE 6.17. In the *Mathematica* program, type the following:

```
Manipulate[ Module[
    {soln=NDSolve[{x'[t]==vx[t],z'[t]==vz[t],
    vx'[t]==-a*vx[t]Sqrt[vx[t]^2+vz[t]^2],
    vz'[t]==-10-a*vz[t]Sqrt[vx[t]^2+vz[t]^2],
    x[0]==0,z[0]==0,vx[0]==10,vz[0]==10},
    {x,z,vx,vz},{t,0,2}]};
    ParametricPlot[Evaluate[{x[t],z[t]}/.soln],{t,0,2},
    PlotRange->{{0,20},{-4,6}}]],{a,0,0.5}]
```

This will result in a plot of the trajectory of an object as a function of the manipulatable parameter $a = \alpha/m$.

Technically, the NDSolve function returns a structure. To visualize the results of a computation, we must utilize the Evaluate function. To be able to use the Manipulate function on the plotted values, we have to recompute the solutions to the differential equations that also depend upon the parameter a. This is achieved by enclosing the calls to NDSolve and ParametricPlot inside of the Module directive. The ' notation is a shorthand for the D function.

First, how does the trajectory change as a increases? Now, plot the velocities as a function of time. How do the velocities change as a increases?

Now compute the energies $T/m = (v_x^2 + v_z^2)/2$ and $U/m = gz$. Plot T/m, U/m and their sum. Is this constant if a is greater than zero? Where did the energy go?

We can now utilize the ability to solve the equations numerically to study the behavior of some systems in detail. The approach is general but we shall focus on the flight of baseballs as an example. Actually, baseballs are very complex objects, so we shall again recognize that our results are approximate. Equations 6.14 and Rayleigh's suggestion for the form of α provide us with a framework for studying the motion of baseballs but measurements on real baseballs indicate that the drag coefficient C_D is a function of velocity and can depend upon the orientation of the ball. So, the true description of baseballs in flight is more complex that we shall admit here.

Consider then, the flight of a batted ball subject to the quadratic resistive force proposed by Rayleigh. A well-struck ball has an initial magnitude of velocity of approximately $v_o = 45\,\text{m/s}$ (100 mph).[7] Let us first address the question of what is the maximum horizontal distance the ball can travel? If we define the angle θ to be the angle the initial velocity vector makes with the horizontal direction, then $v_1 = v_o\cos\theta$ and $v_2 = v_o\sin\theta$, where we again use $v_x(t = 0) = v_1$ and $v_z(t = 0) = v_2$. From Rayleigh's formula, the parameter α is defined by the cross-sectional area of a baseball ($L^2 = 0.00441\,\text{m}^2$), the density of air ($\rho = 1.22\,\text{kg/m}^3$) and the drag coefficient C_D. At a velocity of $45\,\text{m/s}$, the drag coefficient has a value of about $C_D = 0.3$. As we are now utilizing somewhat more precise values of constants, we should use a better value for the gravitational acceleration $g = 9.8\,\text{m/s}^2$. The mass of baseball can be assumed to be 0.145 kg.

> EXERCISE 6.18.   Baseballs in a vertical wind tunnel will hold their position when the air velocity is $42.5\,\text{m/s}$. At this velocity, the gravitational and aerodynamic forces are equal. This implies that the terminal velocity of a baseball is $42.5\,\text{m/s}$. What value of C_D does this represent?

> EXERCISE 6.19.   Using the information above, compute the trajectory of the baseballs as a function of initial angle θ and find the angle for which the maximum distance is greatest. The distance will be maximum when $z = 0$. If you have a numerical solution to NDSolve saved as the variable soln, then the time at which z vanishes can be found with FindRoot[Evaluate[z[t]/.soln],{t,1}], where the final 1 is a guess as to the position of the root. (If you choose to guess 0, then the FindRoot function will return the first root it finds. As the vertical position is zero initially, you need to guess a later time.) The horizontal distance is simply the value of x at that time. It is well known that, with no resistive force, the maximum distance is obtained when $\theta = \pi/4$ (45°). What happens to your result if C_D is

---

[7]Traditionally, baseball utilizes British units, with velocities measured in miles/hour and distances measured in feet. One can certainly compute solutions in those units but we shall utilize SI units for consistency.

changed from $C_D = 0.3$ to $C_D = 0.25$? What happens to your result if the initial velocity is 49 m/s?

EXERCISE 6.20.   The other parameter affecting the trajectory is the air density ρ, which depends upon a number of factors, including the temperature, humidity and altitude of the stadium. Air density decreases with increasing temperature and also decreases with increasing humidity. (The molecular weight of water is 18 Da, where the molecular weight of air, which is composed of diatomic nitrogen (28 Da) and diatomic oxygen (32 Da), is greater.) On a hot, humid afternoon the air density may be as low as 1.145 kg/m³, whereas on a cold evening in October, the air density may be as high as 1.295 kg/m³. Suppose that a ball is batted with an initial angle of 35° with the horizontal and an initial velocity of 45 m/s. How does the air density affect the distance travelled?

EXERCISE 6.21.   The resistive force depends upon the parameter α/m. What is the range of α/m values that we obtain for baseball trajectories, if the baseball diameter can range from 73 to 76 mm, the air density can range from 1.145 to 1.295 kg/m³, C_D can take on values from 0.2 to 0.3 and the baseball mass can range from 0.142 to 0.149 kg?

At this point, we should make an attempt to deal with the fact that baseballs not only translate through space but also generally rotate about their center of mass. A complete treatment of the phenomena associated with baseball trajectories still eludes scientists but a reasonable approximation to their behavior was originally postulated by Newton in 1762, who was considering the flight of tennis balls.[8] Today, the approximation is usually associated with the German physicist Heinrich Magnus and is known as the Magnus force.

If the baseball is rotating about some axis, it possesses an angular momentum **L** (that will be conserved) and an angular velocity ω, as depicted in figure 6.10. As the center of mass of the ball translates with a velocity **v**, portions of the ball below the rotational axis have a relative velocity with respect to the fluid that is increased by as much as ωR, where R is the radius of the ball. On the opposite side of the ball, the relative velocity is decreased by the same amount. As the force on the ball depends on the relative velocity through the fluid, the rotation gives rise to a differential force perpendicular to the direction of motion. As illustrated in the figure,

---

[8]Mythically, baseball was not invented until 1839 by Abner Doubleday. Nevertheless, the similar English game of rounders was certainly being played in the 1700s. From an abstract perspective, the equations are just as applicable to tennis or any other sport involving spherical balls.

the force on the bottom portion of the ball will be greater than the force on the top of the ball, giving rise to a force that points generally upward.

In studying the nature of this differential force, the velocity dependence has been experimentally determined to be proportional to the velocity squared. As a result, the Magnus force is generally written in the following form:

$$(6.16) \qquad \mathbf{F}_3 = \frac{1}{2}C_L\rho L^2 v(\hat{\boldsymbol{\omega}} \times \mathbf{v}),$$

where C_L is a dimensionless coefficient, ρ is the fluid density, L^2 is interpreted as the cross-sectional area of the object and $\hat{\boldsymbol{\omega}}$ is a unit vector that points in the direction of the angular velocity. Note that this representation of the force is essentially identical to the one proposed by Lord Rayleigh for resistive motion. This is not entirely coincidental. Note also that we have not provided any sort of derivation for Equation 6.16; none exists. It is, instead, a relatively simple, phenomenological representation of the motion of objects derived from experimental data. A detailed understanding of the complex fluid dynamics problem underlying the flight of baseballs has not yet been developed but, mercifully, Equation 6.16 appears to capture the lion's share of the physics of spinning balls.

Equation 6.16 bears some scrutiny. We note first that the force does not seem to depend upon the magnitude of the angular velocity ω. It seems intuitively obvious that, if the ball is spinning more rapidly, the aerodynamic force would be greater. This is, indeed, the case. Experimental results in wind tunnels and with pitching machines that can independently control the spin on the ball and the linear velocity indicate that the angular velocity does affect the force; however, the angular velocity turns out not to be independent of the linear velocity v. For thrown or batted baseballs, the mechanical aspects of propelling the ball couple the angular and linear momenta. Aerodynamics studies of thrown and batted balls utilize what is termed the spin ratio $S = \omega R/v$, which the ratio of the tangential velocity ωR of the surface of the ball to its linear center-of-mass velocity v. For a ball that is rolling, without slipping, along a horizontal

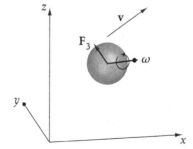

FIGURE 6.10. The angular velocity $\boldsymbol{\omega}$ of the ball gives rise to a resistive force $\mathbf{F}_3$ that is proportional to the cross product $\boldsymbol{\omega} \times \mathbf{v}$

surface we find $S = 1$. In principle, the rotational motion of the baseball flying through the air is not strictly coupled to the linear motion, so S could have any value and need not be restricted to the range $0 \leq S \leq 1$. Yet, thrown baseballs invariably have a spin ratio in the range $S = 0.1–0.3$, suggesting that the spin and velocity are coupled. As it happens, over this range of S, the drag coefficient C_L is nearly constant. So, even though we have not explicitly included the spin rate into the equation for the Magnus force, spin is implicitly included through its correlation with velocity.

Second, Equation 6.16 assumes that the angular velocity of the ball is a constant. If you drop a spinning object into a tub of water, the rotational motion rapidly dissipates. The viscosity of air is significantly less than that of water but we should anticipate that there will be dissipative torques acting on the ball even in air. The experimental evidence at this point is not definitive but suggests that the relaxation time for such torques is of the order of twenty to twenty-five seconds. That is, if we consider the time dependence of the angular velocity to be approximated by the relation $\omega = \omega_0 e^{-t/\tau}$, the relaxation time τ is quite long compared to the times that balls are actually in flight. Pitched balls reach the catcher (or the bat) within half of a second of their release. Fly balls are in the air for just a few seconds, certainly not twenty seconds. As a result, the fact that the representation of the Magnus force in Equation 6.16 ignores dissipative torques likely has little effect on its predictive value. This is not likely to be true for other sorts of projectiles but it seems to be a reasonable approximation for baseballs.

We shall proceed to make use of Equation 6.16 to study the aerodynamic effects of the Magnus force. We note that the Magnus force does no work on the ball: $\mathbf{F}_3 \cdot d\mathbf{s} = 0$. What this means is that the magnitude of the velocity of the ball is unaffected by the Magnus force. The direction of the velocity will change but not the magnitude.

6.4. Spin on Baseballs

Addition of another force on the baseball of course complicates matters. We can no longer restrict ourselves to motion in a plane. Even if the motion starts in one plane, the Magnus force will generally deflect the motion into the third dimension, as is illustrated in figure 6.11. As a result, we shall have to consider the problem as an intrinsically three dimensional one.

We can begin by defining the equations of motion of a ball with spin. Utilizing the coordinate system that we have employed thus far, we can write the following relations:

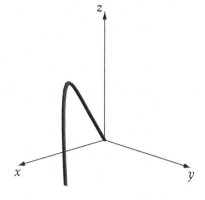

FIGURE 6.11. The Magnus force in addition to the resistive force of Rayleigh gives rise to three dimensional trajectories. Motion initially started in the x-z plane but bends significantly into the y-direction

$$m\frac{d^2x}{dt^2} = \left[-\alpha\frac{dx}{dt} + \beta\left(\omega_y\frac{dz}{dt} - \omega_z\frac{dy}{dt}\right)\right]v(t)$$

$$m\frac{d^2y}{dt^2} = \left[-\alpha\frac{dy}{dt} + \beta\left(\omega_z\frac{dx}{dt} - \omega_x\frac{dz}{dt}\right)\right]v(t)$$

(6.17) $$m\frac{d^2z}{dt^2} = -mg + \left[-\alpha\frac{dz}{dt} + \beta\left(\omega_x\frac{dy}{dt} - \omega_y\frac{dz}{dt}\right)\right]v(t)$$

where

(6.18) $$v(t) = \left[\left(\frac{dx}{dt}\right)^2 + \left(\frac{dy}{dt}\right)^2 + \left(\frac{dz}{dz}\right)^2\right]^{1/2}$$

and we have used the notation $\alpha = C_D\rho L^2/2$ and $\beta = C_L\rho L^2/2$. The spin direction is represented as $\hat{\omega} = (\omega_x, \omega_y, \omega_z)$. The vector $\hat{\omega}$ has unit magnitude, so its components can also be presented in terms of the polar angle θ and azimuthal angle φ that we have used previously in the spherical coordinate system. In this case we would find $\omega_x = \cos\phi\sin\theta$, $\omega_y = \sin\phi\sin\theta$ and $\omega_z = \cos\theta$. One sometimes finds Equations 6.17 presented using the angular form.

The obvious result of incorporating the Magnus force into our discussion is that each of the components of velocity is coupled into all three of the equations. Where we might have been disappointed that there were no analytic solutions of Equations 6.14, a brief examination of Equations 6.17 and 6.18 will lead us to conclude that a numerical approach is going to be necessary. Even with the rather drastic simplifications in the physics associated with the Magnus force, we have not generated a system of equations that is separable, certainly not for arbitrary values of the parameters α and β. Let's begin with the following exercise.

Exercise 6.22.   First, let us define the general system of equations in the *Mathematica* program. Type the following:

```
v[t]=Sqrt[vx[t]^2+vy[t]^2+vz[t]^2]
eqns={x'[t]==vx[t],y'[t]=vy[t],z'[t]==vz[t],
    vx'[t]==(-a vx[t] + b(wy vz[t]-wz vy[t]))v[t],
    vy'[t]==(-a vy[t] + b(wz vx[t]-wx vz[t]))v[t],
    vz'[t]==-g+(-a vz[t] + b(wx vy[t]-wy vx[t]))v[t],
    x[0]==0,y[0]==0,z[0]==0,
    vx[0]==v1,vy[0]==v2,vz[0]==v3}
soln1=NDSolve[eqns/.{a->0.4,b->0.2,
    v1->10,v2->0,v3->10,wx->0,wy->0,wz->1,g->10},
    {x,y,z,vx,vy,vz},{t,0,2}]
ParametricPlot3D[{x[t],y[t],z[t]}/.soln1,{t,0,2}]
```

This will generate a parametric plot of the trajectory for a specific set of values of the parameters. Note here we have used the parameters $a = \alpha/m$ and $b = \beta/m$ but this is a choice not a necessity. We have made use of the function ReplaceAll (via its shortcut /.) to provide values of the parameters before passing the system of equations to NDSolve. In this example, the resulting trajectory is plotted in three dimensions. The initial velocity was chosen to be in the *x-z* plane. Does the trajectory remain in that plane?

Let us now consider the flight of a batted ball with an initial velocity of 45 m/s. Suppose that the ball initially departs at an angle of 35° with respect to the horizontal. Depending upon how the bat and ball collide, the ball may initially have topspin or backspin. That is, if we choose the usual coordinates, the angular velocity vector will either point along the $+\hat{y}$ or $-\hat{y}$ directions.[9] The force arising from topspin or backspin is directed in the *x-z* plane, meaning that motion that originally started in the plane will remain in the plane.

Exercise 6.23.   Assume that $C_L = 0.23$. Using the parameters for baseballs from the previous exercises, how does the trajectory change when $\omega_y = -1, 0, 1$ and $\omega_x = \omega_z = 0$? In particular, what is the change in the maximum horizontal distance (measured when $z = 0$)?

Exercise 6.24.   Batters that are late on a pitch impart significant side spin on the ball. For right-handed batters, this implies $\omega_z = -1$ and

---

[9]We utilize the right-hand convention. If you curl the fingers of your right hand along the direction of rotation of the ball, the thumb points in the direction of the angular velocity.

the x-direction lies along the first base line. For left-handed batters, $\omega_z = 1$ and the x-direction would lie along the third base line. Use an initial velocity of 45 m/s and an initial angle of $\theta = 35°$. What happens to the balls that initially start out along the base lines for right- and left-handed batters?

Pitchers utilize spin to affect the trajectory of thrown balls. Here, we consider that the pitch is thrown with an initial velocity that is predominantly x-directed. A normal, fastball delivery imparts backspin to the ball but very few pitchers release the pitch such that the angular velocity is purely in the $\pm y$-direction. Generally, the angular velocity vector is in the y-z plane. The $\boldsymbol{\omega} \times \mathbf{v}$ spin-dependence on the force indicates that when the angular velocity is parallel to the linear velocity, there is no Magnus force. There is, therefore, no competitive advantage to produce a pitch with a spin component in the x-direction. In fact, spin along the direction of motion adds gyroscopic stabilization to the ball. Pitchers call such pitches "mistakes" and batters call them home runs.[10]

Major league pitchers typically throw fastballs with an initial velocity in the range of 40–45 m/s. Such pitches typically have the angular velocity vector predominantly in the y-direction but "cutting" fastballs have the angular velocity vector tilted more into the z-direction. Breaking pitches are released to provide topspin but the mechanics required to do so reduces the initial velocity to the range of 30–37 m/s. Changeups and split-fingered fastballs are released with velocities in a similarly low range of velocities but have little or no spin, as do knuckeballs.

The nominal distance from the pitching rubber to home plate is 18.44 m but major league pitchers release the ball from a point that is nearly two meters in front of the rubber. Depending on individual pitchers, the ball is released from a height of approximately 1.5 m above the ground towards a target varies in height but can be considered to be, on average, about one meter above the ground.

EXERCISE 6.25. At what angle φ from the horizontal must a pitcher release a 42 m/s fastball with backspin to hit the target described above if the travel distance is 16.5 m? Consider that the angular velocity makes an angle $\theta = 65°$ with respect to the z-axis (nearly horizontal). What is the result of varying the angle φ by $\pm 1°$? How does the velocity of the ball change over the course of the trajectory?

EXERCISE 6.26. At what angle φ from the horizontal must a pitcher release a 35 m/s curve ball with topspin to hit the target described

---

[10]Such mistakes are readily recognized by batters. The spinning laces produce a noticeable circle on the leading face of the ball.

above if the travel distance is 16.5 m? Consider that the angular velocity makes an angle $\theta = 35°$ with respect to the z-axis (nearly vertical). What is the result of varying the angle φ by $\pm 1°$? How does the velocity of the ball change over the course of the trajectory?

EXERCISE 6.27. Compare the trajectories of the fastball and curveball from the two previous exercises. How do the trajectories differ? What happens if the curve ball is delivered along the same initial direction as was required for the fastball to hit the target? What can you say about the comment that pitchers have "late movement" on their pitches?

EXERCISE 6.28. Denver is a notoriously bad location for pitchers. The lower air density, as we have seen, adds significantly to the travel distance for batted balls. How does the change in ρ affect curve balls? Consider the change in trajectory that occurs when a curve ball is thrown with $\rho = 1.225 \, \text{kg/m}^3$ (sea level) and $\rho = 1.145 \, \text{kg/m}^3$. Can you explain why pitchers have less success in Denver?

The models that we have constructed enable us to investigate a number of aspects of the flight of baseballs They are based on reasonable physical insights but significant approximations from the much more complex fluid dynamics problems that underlie the models. An alternative approach used by professional baseball utilizes a nine-parameter model of the trajectory. The parameters are the initial position $\mathbf{x}_0$ and velocity $\mathbf{v}_0$ and three components of a constant acceleration vector $\mathbf{a} = (a_x, a_y, -g + a_z)$.

EXERCISE 6.29. Construct a model in which the accelerations are constant in each direction. Compare the trajectory obtained from this model to the trajectory obtained from numerical models. In particular, examine the results from the fastball and curveball exercises above. Can you obtain a trajectory for the nine-parameter model that is close to that obtained from the more sophisticated model? (To quantify your results, consider the difference between the two models over the course of the trajectory. How much does the maximum difference deviate from zero?)

6.5. Wind

The final aspect of baseball physics that we shall discuss here is the effect of wind on the baseball trajectory. This has a pronounced effect on batted balls and a lesser effect on pitched balls. (Pitched balls are in flight for less time.) Incorporating the wind velocity into the equations of motion for the baseball is actually a relatively simple accomplishment if the wind can be considered to have a constant velocity $\mathbf{v}_w$. For real baseball stadiums, this may be a rather dubious approximation. There are numerous videos of hapless, albeit professional, athletes trying to field balls caught in the swirling winds in the old Candlestick park that was the home field for many years of the San Francisco Giants. Nevertheless, if the wind velocity is constant, it simply adds to the velocity $\mathbf{v}$ of the baseball.

Essentially, the aerodynamic force is proportional to the relative velocity of ball through the air: $\mathbf{v}_{rel} = \mathbf{v} - \mathbf{v}_w$. Once the ball is in flight, it is subject to aerodynamic forces that depend on $\mathbf{v}_{rel}$, so incorporating the effects of wind involves replacing $\mathbf{v}$ with $\mathbf{v}_{rel}$ in our previous equations of motion.

EXERCISE 6.30. Write the equations of motion for a spinning ball in the presence of wind.

EXERCISE 6.31. Consider the trajectory of a 45 m/s batted ball in the presence of a crosswind with velocity $\mathbf{v}_w = \hat{\mathbf{y}}7$ m/s. How does the crosswind affect the trajectory?

6.6. Friction

One difficulty students find with the typical statement of Newton's laws of motion are that they are simply contrary to human experience. For example, the statement that objects in motion tend to stay in motion is wholly opposite to the observation that, if you have the misfortune to run out of gas in your automobile, you will coast to a stop an uncomfortably long distance away from any gas stations. Similarly, a book pushed across a tabletop slides to a stop in a short distance. Neither object remains in motion. As a result, it would seem that the best statement of terrestrial motion is that objects in motion tend to stop and then stay stopped.

We now recognize, of course, that the complete statement of Newton's first law is that objects in motion tend to stay in motion unless acted upon by an external force. It is the last phrase that is often overlooked but explains the observation that terrestrial objects tend to be stationary. We ascribe the observed behavior to be due to a dissipative force called **friction**, that acts like the resistive forces we have been discussing that oppose the motion of objects in air. For objects resting or moving along a surface, like the book on the tabletop, the frictional force is an empirical construct that seems to explain the gross behavior of terrestrial objects.

At a microscopic level, neither the table surface nor the cover of the book are smooth. Indeed, at an atomic scale, the surfaces in contact resemble mountain ranges more that they do flat planes, as sketched in figure 6.12. As a result, when the surfaces are moving, the mountain ranges scrape

FIGURE 6.12. The surfaces of the the table and book can be considered to be smooth at a macroscopic level. At a microscopic (atomic) scale, the surfaces are not smooth

past one another, with some material being transferred across the interface in a complex process. Even for objects that are rolling, there are attractive interactions where the mountain ranges contact one another that must be disrupted for the object to move. Thus, we observe dissipative, frictional forces for rolling objects as well. At a macroscopic level, it often suffices to use a very simple model for friction of objects resting on a horizontal surface: that the frictional force $\mathbf{F}_f$ is a constant force proportional to the gravitational force acting on the object, i.e., the weight of the object:

$$(6.19) \qquad\qquad |\mathbf{F}_f| = \mu m g,$$

where the proportionality factor μ is called the coefficient of friction. The direction of the force is opposite to the direction of motion.

Assuming that the frictional force is constant is obviously a tremendous simplification of the complex interactions that are involved in the physical processes at work when two objects slide past one another. Nevertheless, for modest-sized objects that are not moving too rapidly initially, this approximation appears to describe the behavior of sliding objects reasonably well.

EXERCISE 6.32.   Consider a rectangular block of mass m_1 sliding along a horizontal surface, where the coefficient of friction is μ_1. Draw the free-body diagram for this system.

If the block initially has a velocity $\mathbf{v} = v_1 \hat{\mathbf{x}}$, how far will the block travel before stopping? How much time will it take the block to stop?

What was the initial mechanical energy of the system? How much work was done by the frictional force?

The question now arises as to what happens when the surface is no longer horizontal? Consider the situation illustrated in figure 6.13. If the block initially has a velocity v_1 and then comes to rest after travelling a distance

FIGURE 6.13. A block slides a distance d down a surface that is inclined an angle θ from the horizontal before stopping

d, what can we conclude about the frictional force? Well, we can state that the block initially possessed mechanical energy $\mathcal{E}_1 = \frac{1}{2}mv_1^2 + mgz_1$, where z_1 is the block's initial vertical position (center of mass). At the final position, the mechanical energy is just $\mathcal{E}_2 = mgz_2$, where z_2 is the final vertical position of the block. We recognize that $z_1 - z_2 = d\sin\theta$, so the difference in mechanical energies must be due to the work done by friction:

$$(6.20) \qquad \mathcal{E}_1 - \mathcal{E}_2 = \frac{1}{2}mv_1^2 + mgd\sin\theta = |\mathbf{F}_f|d.$$

Here, we have assumed that the frictional force is a constant, so the work done is simply proportional to the path length d.

Determining the nature of the frictional force in Equation 6.20 is an experimental task. We can envision sliding blocks down a plane, measuring their initial velocities and stopping distances. Our independent variables are the angle θ and initial velocity v_1 and, of course, the materials from which the blocks and planes are constructed. We have suggested that the frictional force does not depend upon velocity, at least for objects like blocks sliding on planes. Furthermore, we have stated that there is a coefficient of friction that depends upon the surfaces in contact. We would not expect that coefficient to depend on the angle. Given our previous statement of the frictional force for horizontal surfaces (Equation 6.19), we should expect the frictional force obtained through our series of experiments to also depend upon the object mass and gravitational acceleration and some form of trigonometrical dependence on the angle θ.

EXERCISE 6.33. Draw the free-body diagrams for the block sliding down the inclined plane:

(a) When the block has a velocity v_1.
(b) When the block is at rest.

We can make a hand-waving argument about the nature of the frictional force as follows. When the surface on which the block rests is inclined from the horizontal, the gravitational force that is vertically directed can be resolved into two components: one component that is directed perpendicular to the surface and a second that is directed tangential to the surface: $\mathbf{F} = -mg\hat{z} = \mathbf{F}_{\text{tang}} + \mathbf{F}_{\text{perp}}$. It is the tangential component of the gravitational force that causes a ball to roll down an inclined plane; the perpendicular component is directed into the surface of the plane. The tangential component is proportional to the gravitational acceleration and the sine of the angle θ: $|\mathbf{F}_{\text{tang}}| = mg\sin\theta$. The perpendicular component is proportional to the cosine: $|\mathbf{F}_{\text{perp}}| = mg\cos\theta$.

So, we can predict that the frictional force on an inclined surface to be proportional to the component of the gravitational force that is perpendicular to the surface of the plane[11]:

$$(6.21) \qquad\qquad |F_f| = \mu mg\cos\theta.$$

There are thus two factors entering into dissipative motion on an inclined surface. First, changing the elevation changes the gravitational potential energy, resulting in a lengthening of the distance d an object would slide. Second, the frictional force is also reduced, further lengthening the distance d an object would slide. A staple of undergraduate physics laboratories is the conduct of just such a series of experiments as we have described to elucidate the nature of the frictional force.

> EXERCISE 6.34.   Given our prediction for the form of the frictional force (Equation 6.21), use Equation 6.20 to predict the outcomes of experiments. Suppose the mass is $m = 0.2$ kg and the initial velocity is in the range $v_1 = 0.2$–0.5 m/s. Consider a coefficient of friction of $\mu = 0.3$ and that the angle θ is in the range $0 \le \theta \le 25°$. How does the stopping distance vary? What precision would you need to determine whether the cosinusoidal dependence on θ is a reasonable representation of the frictional force?

> EXERCISE 6.35.   Suppose that we conduct experiments where the block initially is sliding *up* the plane before coming to rest. How does the analysis change? Using the parameter values from the previous exercise, how does the stopping distance vary?

---

[11]The perpendicular vector to a surface is often called the normal vector. As a result, the force opposing the perpendicular component of the gravitational force is often referred to as the "normal" force. There are no "abnormal" or "paranormal" forces.

VII

Celestial Mechanics

Isaac Newton's notable success in providing a theoretical explanation for the motion of planets around the sun was followed quickly by his realization that the gravitational problem involving three bodies was immensely more difficult than the two-body problem. Where the two-body problem, as we have seen in Chapter 2, can be solved exactly, Newton's attempts to provide a concise mathematical description of the earth-sun-moon system were not successful.[1] This, of course, is not due to Newton's lack of mathematical skills. No one has ever found a general solution to the three-body problem. Indeed, a series of notable mathematicians all applied their skills to the problem but without success, although it depends somewhat on how one defines success. It is true that no general solutions of the three-body problem have been constructed but the assault on the problem led to powerful new mathematical methods for understanding dynamical systems.[2]

We can begin to see the difficulties that vexed Newton and others when examining the three-body problem. In principle, if we have N gravitating bodies, there are $6N$ functions of position and velocity that are required to define the system. We found, though, that the principle of conservation of momentum led to the observation that the center of mass motion is constant. That is, the position of the center of mass at a time t can be written as $\mathbf{r}_{cm}(t) = \mathbf{x}_{cm} + \mathbf{v}_{cm}(t - t_1)$, where $\mathbf{x}_{cm}$ and $\mathbf{v}_{cm}$ are constants. Hence, the equations of motion have only $6N - 6$ independent variables. In addition, the conservation of angular momentum means that $\mathbf{L}$ is another constant of the motion and conservation of mechanical energy $\mathcal{E}$ implies that there are, in all, only $6N - 10$ independent variables. In studying the two-body problem, we were faced ultimately with finding a solution in two variables. That is, the initial twelve degrees of freedom could be reduced to

---

[1] Newton's colleague Edmond Halley remarked that Newton claimed the problem "made his head ache, and kept him awake so often, that he would think of it no more."

[2] The methods due to the Italian/French mathematician Joseph-Louis Lagrange (originally Giuseppe Luigi Lagrangia) and the Irish physicist William Rowan Hamilton form the basis of the next class in mechanics.

© Mark A. Cunningham 2015

M.A. Cunningham, *Neoclassical Physics*, Undergraduate Lecture Notes in Physics, DOI 10.1007/978-3-319-10647-2_7

studying motion in a plane: if we place M_1 at the origin, then we need only find the coordinates of M_2 as a function of time and we explicitly constructed such solutions in polar coordinates (r_{12}, ψ).

For the three-body problem, we have eight independent degrees of freedom. Where the two-body problem was restricted to a plane, it is obvious that the three-body problem is not. A number of approximate theories were developed based on simplifications of the general problem. We shall investigate one such problem here in which the mass of the third object is vastly smaller than that of the other two. This is known as the *restricted* three-body problem. It is not of great utility for most astronomical problems. For example, the moon's mass is not negligible compared to the mass of the earth and sun. Yet, it is a perfectly reasonable approximation if we talk about artificial satellites.

7.1. Restricted Three-Body Problem

In the case where the third mass M_3 is much smaller than the other two masses, we can approximate the behavior of the system by assuming that the two larger bodies occupy the orbits that we discussed in Chapter 2. The most interesting of these are, of course, the elliptical or circular orbits. Indeed, the restricted three-body problem presumes that the two heavy masses occupy circular orbits. For the two-body problem that we investigated previously, consider the potential energy surface defined by Equation 2.27. We have depicted the potential energy surface in figure 7.1.

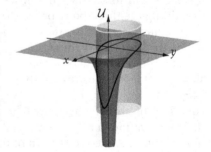

FIGURE 7.1. A body moving on an elliptical trajectory (*light gray ellipse*) sweeps out a path (*dark line*) on the potential energy surface. The potential energy surface is symmetric; circular orbits would have constant potential energy

One is tempted to think about trajectories in terms of the mass M_2 moving along the potential energy surface. For the two-body problem, the potential energy surface is symmetric. So, contours of constant potential energy are circles, corresponding to the circular orbits that we discussed previously. Elliptical orbits trace out complex curves on the potential energy surface due to the fact that it is the total mechanical energy $\mathcal{E}$ that is conserved, not just the potential energy $\mathcal{U}$. As the mass M_2 moves along the ellipse, the decrease in potential energy near the distance of closest approach is compensated by an increase in kinetic energy: a greater velocity.

With two centers, the potential energy surface is no longer symmetric, as is depicted in figure 7.2. The dark lines on the surface represent contours of constant potential energy and one might be tempted to think that these contours represent potential trajectories. The problem with that assumption is that mass M_2 is moving, so the potential energy surface is not static. Consequently, mass M_3 sees a complex, time-varying potential energy surface and any hopes that we might try to interpret the contours in the figure as trajectories are naïve.

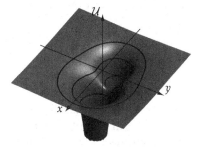

FIGURE 7.2. The potential energy surface is more complex with two centers. Lines of constant potential energy are no longer described by *circles*

EXERCISE 7.1. In the *Mathematica* program, define the function

$$\mathcal{U}(x,y,M) = -M/(x^2 + y^2).$$

Use the Plot3D function to plot $\mathcal{U}(x-0.3, y, 1) + \mathcal{U}(x+0.5, y, M_2)$ over the range $-2 \le x \le 2$ and $-2 \le y \le 2$. How does the surface change as M_2 changes over the range $0 \le M_2 \le 1$?

So, in trying to gain some understanding of the three-body problem, we shall make the same assumption that the initial investigators made: the mass M_3 is sufficiently small that is does not perturb the trajectories of the other two masses. In this case, the problem separates into two parts: the motion of the masses M_1 and M_2 and the motion of the mass M_3 in the presence of those two. The equations of motion for the third mass can be written as follows:

$$\frac{d\mathbf{r}_3(t)}{dt} = \mathbf{v}_3(t)$$

(7.1)
$$\frac{d\mathbf{v}_3(t)}{dt} = -GM_1 \frac{\mathbf{r}_3(t) - \mathbf{r}_1(t)}{|\mathbf{r}_3(t) - \mathbf{r}_1(t)|^3} - GM_2 \frac{\mathbf{r}_3(t) - \mathbf{r}_2(t)}{|\mathbf{r}_3(t) - \mathbf{r}_2(t)|^3}$$

Here, the positions $\mathbf{r}_1(t)$ and $\mathbf{r}_2(t)$ are those that we have previously determined.

EXERCISE 7.2.   Write down the complete set of equations for three masses, using the non-compact notation, as in Equation 7.1. Rewrite them using the concise notation.

In examining Equation 7.1, we shall diverge from the path taken by the early researchers and rely on the technology at our disposal. Let us begin by first making some further simplifying assumptions beyond that M_3 is greatly smaller than either M_1 or M_2. Let us also assume, for now, that the mass M_3 has initial position and velocity in the same plane as M_1 and M_2. If we think about the solar system, the planets occupy orbits that are nearly coplanar and the eccentricities are small, so our assumptions will limit the space of solutions but should permit us to make some sensible inferences. (In addition, plotting the results will be simpler.)

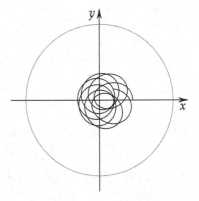

FIGURE 7.3. One mass is located at the origin of the plot and the second occupies a circular orbit (*gray line*). The third mass occupies a crudely elliptical orbit that precesses around the first mass

For example, in figure 7.3, we illustrate a bounded trajectory of M_3. The heaviest mass M_1 is located at the origin of the plot and the second mass ($M_2 = 0.4M_1$) has a circular orbit. We observe that the trajectory of M_3 is nearly ellipsoidal but does not actually repeat. Instead, the ellipse precesses around the mass M_1. The mass M_3 initially begins with an orbital radius of $0.45\ r_{21}$ and, at the initial time, was collinear with masses M_1 and M_2 (all along the positive x-axis). We have plotted only one orbit of M_2 in figure 7.3, so one could well question if the trajectory of M_3 will remain stable for very long times.

The trajectory of M_3 in this example reflects what we could term a *bounded* trajectory. The orbit of M_3 around M_1 is not periodic but it does occupy a finite space, at least for the elapsed time illustrated. Such a bounded solution is actually not common. Far more prevalent are unbounded solutions.

EXERCISE 7.3. In the *Mathematica* program, type the following:

```
x2[t_]:=Cos[t]
y2[t_]:=Sin[t]
r31=Sqrt[x3[t]^2+y3[t]^2]
r32=Sqrt[(x3[t]-x2[t])^2+(y3[t]-y2[t])^2]
eqs={x3'[t]==vx3[t],y3'[t]==vy3[t],
  vx3'[t]==-GM1 x3[t]/r31^3-GM2(x3[t]-x2[t])/r32^3,
  vy3'[t]==-GM1 y3[t]/r31^3-GM2(y3[t]-y2[t])/r32^3}
ics={x3[0]==xo,y3[0]==yo,vx3[0]==vxo,vy3[0]==vyo}
soln=NDSolve[Join[eqs,ics]/.{GM1->1,GM2->0.4,xo->0.45,
  yo->0.0,vxo->0.0,vyo->0.9},
  {x3,y3,vx3,vy3},{t,0,2Pi}]
ParametricPlot[{{x2[t],y2[t]},
  Evaluate[{x3[t],y3[t]}/.soln]},{t,0,2Pi},
  PlotRange->{{-3,3},{-3,3}}]
```

This will create a plot of a circular orbit of M_2 around M_1, scaled such that the radius is one and the orbital period of M_2 is 2π. The third mass is placed initially on the x-axis, with a velocity in the positive y-direction. How do the solutions change as M_2 changes from 0.4 to 0.1? How do the solutions change as the initial velocity of M_3 changes?

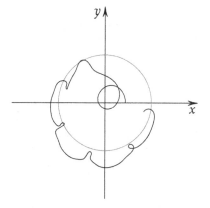

FIGURE 7.4. One mass is located at the origin of the plot and the second occupies a circular orbit (*gray line*). The third mass is eventually kicked out of the system but orbits both M_1 and M_2 at different times

Indeed, compare the trajectory observed in figure 7.4 with that in figure 7.3. In this case, the mass M_3 started at the same initial position but with a different velocity. The initial portion of the trajectory is crudely elliptical but quite rapidly becomes distorted. In this system, M_3 is initially orbiting M_1 but moves for a time to a crudely elliptical trajectory around M_2 and then back into an orbit around M_1. Eventually, the mass scatters from M_2 and is ejected from the system. In these results, for which $M_2 = 0.4 M_1$, we observe that finding stable orbits for the third mass can be challenging. In the solar system, we observe that there are relatively few planets orbiting the sun and they are widely separated. In the subsequent exercises, we'll investigate the inference that gravitational interactions will tend to deplete the interior regions of the solar system of lighter constituents.

> EXERCISE 7.4. Using the parameters from the previous exercise, set the initial velocity of M_3 to be 1.27. Add a plot of the vector $\mathbf{r}_3 - \mathbf{r}_2$. This effectively sets the coordinate origin to the position of M_2. What does the trajectory look like in this coordinate system?

> EXERCISE 7.5. Consider the same equations as defined in the previous exercises but set the initial point for M_3 to be at $x_0 = 5$. Can you find an initial velocity that leads to a circular orbit? If so, what is the orbital period? (Note: you will have to extend the calculations to times longer than that specified in the previous exercise.)

As noted earlier, analysis of the sun-earth-moon system drove Newton to distraction; it is actually quite complex to analyze and does not fall within the assumptions that we have made in defining the restricted three-body problem. We can illustrate one aspect of the complexity by simply calculating the gravitational force on the moon due to the earth and the sun. If we utilize the data in the table below, we can compute the relative forces acting on the moon due to the earth and the sun, respectively. Somewhat surprisingly, the gravitational force of the sun on the moon is significantly larger than the gravitational force of the earth on the moon. We would deem this result surprising because, at least naïvely, one might expect that, if the force on the moon due to the sun is larger than that due to the earth, the moon would move in the direction of the sun. As we have discovered previously, however, the gravitational force is a centripetal force that conserves angular momentum. As a result, the moon does not fly off to orbit the sun independently, although as we have seen from the previous exercises, complex orbits are possible.

> EXERCISE 7.6. Using the values in Table 7.1, what is the ratio of the gravitational forces acting between the sun-moon and the earth-moon systems? Compare that to the ratio of forces for the International Space Station (ISS), which is in a low-earth orbit. Note that,

TABLE 7.1. Orbital parameters

	a (AU)	T (yr)	M (M_{earth})
Sun	—	—	333060
Earth	1	1	1
Moon	0.00257	0.0748	0.0123
ISS	4.5×10^{-5}	1.75×10^{-4}	7.5×10^{-20}

while the ISS is a sizable structure, its mass is quite small in comparison to that of the earth.

Students are encouraged to investigate the variety of different solutions of the restricted three-body problem that can be produced with the equations of Exercise 7.3. What you will find is that very few initial conditions lead to stable orbits, where by stable we mean that the third mass does not fly off to infinity. Most often, the trajectories can hardly be termed orbits, at least in the sense of resembling ellipses.

EXERCISE 7.7.   Study the case of an object orbiting a small mass, which is in turn orbiting a larger mass. Use the example from Exercise 7.3 as a starting point. For numerical reasons, it will prove best to make M_1 the small mass, so let $M_1 = 0.001$. Initially, set $M_2 = 0$ and choose the initial x position to be $x_0 = 0.0025$. This is close to the earth-moon distance (in AU) but the ratio of masses for the earth and sun is 3×10^{-6}, not 10^{-3}, so the model is not a particularly good representation of the sun-earth-moon system.

What initial velocity is required to achieve a circular orbit around M_1? Note that you will need to use the PlotRange option of the ParametricPlot function in order to see this small orbit.

Using the initial velocity that generates a circular orbit with no other mass present, what happens now if M_2 is set to be 0.1 or 1.0?

7.2. Lagrange Points

One of the first problems tackled by early researchers was to identify orbits in which the third mass rotated synchronously with the other two. This was termed stationary orbits, in the sense that, in a coordinate system centered on M_1 that rotated with M_2, the mass M_3 would occupy a single point. The Swiss physicist Leonhard Euler first identified three points that were collinear with the two large masses and, subsequently, Lagrange identified two additional points that were equidistant from the two large masses. The five points in all are usually termed the Lagrange points of the orbit.[3] A depiction of their relative positions is illustrated in figure 7.5.

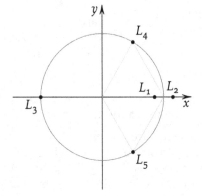

FIGURE 7.5. The mass M_1 is at the origin and the mass M_2 is at the point $(1,0)$. The five Lagrangian points for the case $M_2 = 0.01M_1$ are plotted as *dots*

EXERCISE 7.8. As we saw previously, the rotation matrix defined as follows:

$$\begin{bmatrix} x' \\ y' \end{bmatrix} = \begin{bmatrix} \cos\theta & \sin\theta \\ -\sin\theta & \cos\theta \end{bmatrix}\begin{bmatrix} x \\ y \end{bmatrix},$$

will rotate the coordinate axes by an angle θ. (See Equation 4.24.) Suppose that we now choose θ to be a linear function of time $\theta = \omega t$. Show that the point $\mathbf{r}_2 = (a\cos(\omega t), a\sin(\omega t))$ is mapped to the point $\mathbf{r}_2' = (a,0)$.

EXERCISE 7.9. Derive the equations of motion for the third mass in the rotating coordinate system. Start with Equations 7.1 and use the rotation matrix defined in the previous exercise to define $\mathbf{r}_3'$ in terms of $\mathbf{r}_3$.

The points L_1 and L_2 are approximately equidistant from M_2, at a distance that is given by $d = (M_2/3M_1)^{1/3}$. The point L_3 is located on the opposite side of mass M_1, at approximately the same radius as M_2. The points L_4 and L_5 have the same orbital radius as M_2 but are at an angle of $\pi/3$ ahead and behind M_2, respectively. Hence, the L_4 and L_5 points form equilateral triangles with the masses M_1 and M_2.

Dynamically, it turns out that the collinear points L_1, L_2 and L_3 are not stable but the L_4 and L_5 points can be if M_2 is significantly smaller than M_1. This is certainly the case in the solar system, if we assume M_1 to be the

---

[3]As we have seen previously, the historical record cannot be deduced from the current attribution. While it is important to understand the historical circumstances of discoveries, science is never conducted in isolation. There are generally several individuals who deserve some credit for each discovery.

sun. Indeed, there is a cluster of asteroids in the orbit of Jupiter, known as Trojan asteroids, that are found close to both the L_4 and L_5 points. Recent observations have also found a small asteroid (2010 TK_7) in the earth's orbit, close to the L_4 point. The L_4 and L_5 points have, of course, also been the source of significant interest in the science fiction literature, with numerous suggestions of constructing various structures and observatories at these points.

EXERCISE 7.10. For $M_2 = 0.01 M_1$, compute the orbits of objects starting at the L_1, L_2 and L_3 points. Note that, using the units of Exercise 7.3, that the initial velocity will have the same magnitude as the initial position. (The orbital period for all will be $T = 2\pi$.) Do you find stable orbits? What happens if $M_2 = 0.001 M_1$?

EXERCISE 7.11. For $M_2 = 0.01 M_1$, compute the orbits of objects starting at the L_4 and L_5 points. What must be the initial velocities? (The magnitude of the velocity is one.) What happens if $M_2 = 0.001 M_1$?

7.3. Rocket Equation

Up until this point, we have generally assumed that the masses involved in our equations were invariants. That is, M is not a function of time. For the problems that we have studied until now, that was a good assumption but we would now like to discuss the possibility of traveling between two different gravitating bodies and this will inevitably involve rocket propulsion. In the earth's atmosphere, airplanes can generate lift through aerodynamic forces. In addition, the oxygen required to burn the fuel can be obtained from the atmosphere itself. As a result, air transport vehicles can be designed in a wholly different fashion than space vehicles. In the vacuum of space, there will be no aerodynamic lift and no oxygen to burn fuel: spacecraft fall under different design rules.

To understand how rockets function, consider that you are suspended in space far from anything else. Momentum conservation requires that whatever momentum you possess will be constant over time. So, if you are at rest, you will remain at rest forever. Consider the consequence of throwing some object in your possession: a shoe, for example. The equation that describes this situation is defined as follows:

$$(M_{you} + M_{shoe})\mathbf{v}_0 = M_{you}\mathbf{v}_{you} + M_{shoe}\mathbf{v}_{shoe}.$$

If we assume, for the moment, that $\mathbf{v}_0 = 0$, then we obtain the following result:

$$M_{you}\mathbf{v}_{you} = -M_{shoe}\mathbf{v}_{shoe}.$$

What has happened here is that you have obtained a nonzero velocity such that your momentum is opposite to that of your shoe. Rocket propulsion utilizes this basic strategy to propel objects through space.

What complicates matters in rockets is that the total mass decreases as exhaust gases are expelled. If we think of firing out some infinitesimal mass dM, then we can write a differential equation governing the motion of rockets as follows. In the center of mass frame, the rocket initially has no momentum. If an infinitesimal mass dM is expelled with an exhaust velocity $\mathbf{v}_e$, then we must have

$$(M - dM)d\mathbf{v} = -\mathbf{v}_e dM,$$

where $d\mathbf{v}$ is the infinitesimal change in velocity of the remaining mass $(M - dM)$. If we divide both sides of the equation by M and make the approximation that $(M - dM)/M = 1$, we can integrate both sides to find the following:

(7.2) $$\Delta\mathbf{v} = -\mathbf{v}_e \ln(M_f/M_o),$$

where $\Delta\mathbf{v}$ is the (macroscopic) change in velocity obtained when the mass changes from M_o to M_f; that is, when an amount of mass $M_f - M_o$ has been exhausted. Equation 7.2 is known as the rocket equation.

> EXERCISE 7.12.   Fill in the details of the derivation of Equation 7.2. How does the equation differ if we do not set $\mathbf{v}_o = 0$?

One of the first questions that arises when dealing with rockets is what is the largest $\Delta\mathbf{v}$ that is achievable for a given fuel load. While science fiction spacecraft fly all across the universe with generally modest-sized vehicles, it cannot have escaped the attention of diligent students that NASA rockets are enormous vehicles that return only very small capsules. This is, of course, not due to oversight on the part of NASA engineers but is a direct consequence of dealing with reality.

First, rocket exhaust velocities are rather modest. Liquid hydrogen/liquid oxygen propulsion systems generate exhaust velocities in the neighborhood of 4.5 km/s, where solid rockets have exhaust velocities around 3 km/s.[4] Second, the rocket initially can be thought of as a payload (M_{PL}), fuel (M_f) and structural elements such as tanks, pumps and motors, etc. The mass of the structural elements crudely scales with the mass of fuel, so we can approximate the total mass of the rocket as $M = M_{PL} + (1+\alpha)M_f$, where α is the scaling factor that accounts for the structural elements. (In

---

[4]It is common practice to define the specific impulse $I_{sp} = v_e/g$, where g is the gravitational acceleration at the earth surface. Then the rocket equation is written in terms of I_{sp} in place of v_e. This practice led to the loss of the Mars Climate Orbiter in 1999 due to a units conversion issue: NASA presumed I_{sp} to be defined in SI units but the vendor utilized British units.

NASA parlance, this is the tankage factor.) In this case, the maximum velocity change will be $\Delta \mathbf{v}_{max} = -\mathbf{v}_e \ln((M - M_f)/M)$. If we exponentiate both sides, we find the following:

$$\frac{M - M_f}{M} = e^{-\Delta \mathbf{v}_{max}/\mathbf{v}_e}.$$

If we use the scaling factor α to define the mass of the structural elements, then we can show the following:

(7.3)
$$\frac{M_{PL}}{M} = (1 + \alpha)e^{-\Delta \mathbf{v}_{max}/\mathbf{v}_e} - \alpha.$$

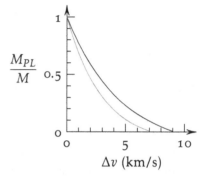

FIGURE 7.6. The ratio of payload mass over total mass at launch is an exponential function of Δv. A liquid fueled rocket (*black curve*) is characterized by a larger $\mathbf{v}_e$ but a scaling factor of $\alpha = 0.15$. A solid fueled rocket (*gray curve*) has a smaller $\mathbf{v}_e$ but a scaling factor of $\alpha = 0.1$

The scaling factor for liquid-fueled rockets is approximately $\alpha = 0.15$, for typical NASA rockets. For solid-fueled rockets, which don't require pumps for fuel transfer, the scaling factor is typically $\alpha = 0.1$. Notice that the curves in figure 7.6 both fall below zero at some value of $\Delta \mathbf{v}$, reflecting the finite values of α. What you can immediately notice from the figure is that the relative fraction of payload is a steeply decreasing function of maximum $\Delta \mathbf{v}$. For example, the Command Module of the Apollo missions to the moon had a mass of about 6000 kg (the Service Module had a mass of 25000 kg) while the Saturn V rocket had a mass of 2.8×10^6 kg, for a ratio of $M_{PL}/M = 0.01$ if you include the Service Module as part of the payload or $M_{PL}/M = 0.002$ if you don't. To escape the earth's gravitational pull, the Apollo vehicles had to obtain a velocity of over 11 km/s. The small mass ratios for the Apollo missions were exacerbated by the need to land on the moon and then return but the exponential behavior of the rocket equation means that the old science fiction vision of rockets leaving the earth and returning intact is simply not achievable with chemical rockets. Not even liquid-fueled rockets with their larger exhaust velocities are capable of single-stage to orbit, much less interplanetary travel.

EXERCISE 7.13. Show that you can obtain Equation 7.3 and then plot the values of M_{PL}/M. Use values for $\mathbf{v}_e$ typical for solid- and liquid-fueled rockets.

EXERCISE 7.14.   One often sees multistage rockets employed. Consider that the payload of the first rocket is another rocket with the final payload. The total $\Delta\mathbf{v}$ will be the sum of the two stages. What is the ratio M_{PL}/M? Plot your results and compare to figure 7.6. For a desired $\Delta\mathbf{v}$, what is the difference in M_{PL}/M for one- and two-stage rockets? Assume that both stages are characterized by the same $\mathbf{v}_e$ and α and that each stage provides half the total $\Delta\mathbf{v}$.

To see the constraints on rocket travel more explicitly, let us include the gravitational field of the earth in our calculations.[5] Let us take the center of the earth to be the point $\mathbf{r}_1$, and the rocket will be at some position $\mathbf{r}_3$. The equations governing the motion of the rocket are given by the following:

$$\frac{d\mathbf{r}_3(t)}{dt} = \mathbf{v}_3(t)$$

$$(7.4) \qquad (M_3 - dM_3)\frac{d\mathbf{v}_3(t)}{dt} = -\mathbf{v}_e\frac{dM_3}{dt} - G\frac{M_1 M_3}{|\mathbf{r}_3 - \mathbf{r}_1|^3}(\mathbf{r}_3 - \mathbf{r}_1),$$

where M_1 is the mass of the earth and M_3 is the mass of the rocket. This looks just like the gravitational force law that we've investigated previously but with the addition of a force term due to the rocket thrust.

The direction of the rocket thrust is defined by the vector $\mathbf{v}_e$. This direction is not necessarily constant in time. The magnitude of the thrust is the product of the exhaust velocity and the mass flow rate. In general, this is a complex problem, so let us make some simplifying assumptions. First, let us assume that the rocket is initially in a circular orbit, which we can take to have a unit radius and a period of $T = 2\pi$. Next, let us assume that the mass flow rate is a constant and that the rocket is fired only for a relatively short period of time (short compared to T). We can assume then that the function $\mathbf{v}_e(t)$ is some constant value (magnitude and direction) during the interval $0 \leq t \leq t_{\text{burn}}$.

EXERCISE 7.15.   In the *Mathematica* program, define the x- and y-components of the thrust vector to be $v_e \cos\theta$ and $v_e \sin\theta$, respectively. Use the If function to set the exhaust velocity to be zero outside the burn interval of $0 \leq t_{\text{burn}} \leq 0.25$. If you choose $GM_1 = 1$, $r_3 = 1$ and $v_3 = 1$, as in the previous exercises, numerically integrate the equations of motion with $v_e = 0$ and show that you obtain a circular orbit. Now set $v_e = 1$. What happens to the orbit when θ takes on values 0, $\pi/2$, $-\pi/2$ and π?

---

[5]Actually, we will utilize a model system in which we take $GM_1 = 1$ and do not deal with the earth's mass explicitly. This will provide us with a test system that embodies many of the issues we wish to study but without the complexity associated with more realistic calculations.

EXERCISE 7.16.   Consider now the case of elliptical orbits.   What orbit results if $v_3 = 0.8$ initially (and $v_e = 0$)?   Now set $v_e = 1$ and $\theta = -\pi/2$.   What happens to the orbit for $t_{\text{burn}} = 0.1$ and $t_{\text{burn}} = 0.2$?

EXERCISE 7.17.   What orbit results if $v_3 = 1.2$ initially (with $v_e = 0$)? Consider now firing the rocket ($v_e = 1$) at the largest orbital radius. (This is the point $\mathbf{r}_\pi$ from figure 2.7.)   This will require that you change the initial burn time to π instead of 0.   What value of t_{burn} is required to circularize the orbit?

The previous exercises demonstrate that when the rocket is fired for a relatively short period of time, an originally circular orbit becomes elliptical (or potentially hyperbolic).   The change in velocity alters the orbit but does not do so in a simple, linear fashion owing to the dominant gravitational force exerted by the earth.   The exercises also provide a clue as to how one might go about visiting other planets.   The idea is due originally to the German physicist Walter Hohmann.[6]   If we begin with a circular orbit, changing the velocity generates an elliptical orbit, as illustrated in figure 7.7.   A second velocity change at the furthermost distance of the elliptical orbit will generate a circular orbit at the larger radius.

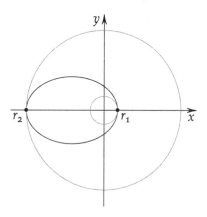

FIGURE 7.7. Changing the orbital radius from r_1 to r_2 can be accomplished via a Hohmann transfer orbit (*black*).   An increase in velocity at the point $\mathbf{r}_1$ will modify the circular orbit into the ellipse.   A second change in velocity (decrease) at the point $\mathbf{r}_2$ will convert the elliptical orbit into a circular orbit

Modifying an orbit can be accomplished with a series of velocity changes. We can envision the alteration of a low earth orbit to a higher, potentially geosynchronous, orbit using this strategy.   It also provides a strategy for visiting other planets, as Hohmann originally proposed.   If we think of the inner circle as the earth's orbit around the sun and the outer circle as the orbit of Mars around the sun, then a journey to Mars would require that the spacecraft be launched from earth at such a time that the journey from

---

[6]Hohmann's *Die Erreichbarkeit der Himmelskörper* was published in 1925.

point $\mathbf{r}_1$ to $\mathbf{r}_2$ along the elliptical trajectory is the same time that it takes Mars to travel along its orbit and arrive at $\mathbf{r}_2$ when the spacecraft arrives.

As a practical matter, the problem is a bit more complicated than just described. The presence of Mars, or some other gravitating body, alters the trajectory from that illustrated in figure 7.7. Nevertheless, Hohmann's strategy forms the basis of the interplanetary travel conducted thus far. It provides an exceptionally fuel-efficient strategy for travel.

> EXERCISE 7.18. Consider the case of an hyperbolic trajectory, which we can arrange using the previous examples if the initial spacecraft velocity is $\mathbf{v}_3 = 2$. If you again use $\mathbf{v}_e = 1$, what length of engine firing (t_{burn}) is required to capture the spacecraft into an elliptical orbit? What must be the direction of $\mathbf{v}_e$?

As an example of a trajectory-changing operation, consider that there is a gravitating body M_2 in orbit around M_1. Let M_2 have an orbital radius of $r_2 = 10$. If we consider M_3 to be initially in an orbit of radius $r_3 = 1$ with an orbital period of $T_3 = 2\pi$, then M_2 must have an orbital period of $T_2 = 2\pi\sqrt{1000}$. A sketch of one potential pathway is illustrated in figure 7.8, in which a second rocket engine firing was performed as the mass M_3 crossed the orbital path of M_2. The following exercise will reconstruct that trajectory.

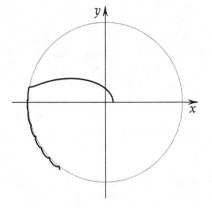

FIGURE 7.8. Two rocket firings, timed appropriately, can move the mass M_3 from an orbit around M_1 (coordinate origin) to an orbit around M_2 (*gray circle*). The *black line* is the trajectory of M_3

> EXERCISE 7.19. Again assume that the mass M_3 is in orbit with an initial position $\mathbf{r}_3 = (1,0)$ and velocity $\mathbf{v}_3 = (0,1)$. Values of $M_1 = 1$ and $M_2 = 0.01$ correspond approximately to the earth/moon mass ratio. Use the Piecewise function to define two engine firings. That is $\mathbf{v}_e$ is nonzero in the intervals $0 \leq t \leq t_{b1}$ and $t_{\text{toi}} \leq t \leq t_{\text{toi}} + t_{b2}$, where t_{toi} is the start time of the second (trans-orbit insertion) rocket firing. For simplicity, the direction of $\mathbf{v}_e$ can be taken to be the same

for both firings. Use $v_e = 1$ and an initial $t_{b1} = 0.36$. The second mass M_2 has a relative phase shift on its orbit of $\phi = 0.67\pi$.

At what time does the mass M_3 cross the orbit of M_2? Use that time as an initial guess for t_{toi}. What values of t_{b2} lead to capture of the mass M_3 into orbit around M_2?

Plot $\mathbf{r}_3 - \mathbf{r}_2$ and convince yourself that the curious path observed in figure 7.8 actually corresponds to an elliptical orbit around M_2.

7.4. Launching from the Earth's Surface

We can now expand our investigations to include the problems that arise from launching a rocket from a planet surface. As we have mentioned, there is one fundamental problem associated with the exponential relationship between the mass ratio and velocity change embodied in the rocket Equation 7.2. In addition to the need to achieve orbital velocity, launching from the surface of the planet requires pushing through an atmosphere that will create drag. The aerodynamic forces affecting a rocket during atmospheric flight are quite considerable. One need only view some of the early films of attempts to launch rockets to appreciate just how difficult the process actually is.

We have certainly paid short shrift to the design and reliability of the pumps that power the rocket engines. We have not addressed the issue of the transition to supersonic flight or even aerodynamic forces at all. Coping with all of these problems will be challenging, so let us make some simplifying assumptions that will provide us with at least some realistic features of rocket flights. In the mathematician's sense, let us reduce the problem to a previously solved result. Suppose that we investigate the problem of achieving a circular orbit of radius $r_3 = 1$ and orbital velocity $v_3 = 1$. This represents the orbit that we have utilized in the previous section, so that we can construct a mission to Mars, for example, by first launching into a low earth orbit and then initiating the Hohmann transfer. For the moment, we shall also ignore atmospheric effects. This is a tremendous simplification, of course, but we shall continue our strategy of attempting simpler problems and analyzing the results before tackling the most realistic problems. In this particular case, this strategy will also enable us to illustrate some other issues associated with space travel.

We shall also not explicitly cope with rocket stages. This verges on making the problem completely unrealistic but, in true hand-waving fashion, let us argue that we can probably fix up those details subsequently. In determining the initial conditions for our simulation, let us first consider that low earth orbits are typically a few hundred kilometers above the earth's surface and the earth's radius is about 6400 km. So, the ratio of orbital radius to earth's radius is about 1.05. Such orbits have an orbital period of about 90 minutes, so a day lasts about 16 orbital periods. We also need to define the initial rocket velocity, which is not zero due to the fact that the earth is rotating. So, we shall investigate launching from a rotating surface where the initial radius is 0.95 and the velocity is $v_0 = 1/16$ and we intend to achieve our canonical orbit, where $r_3 = 1$ and $v_3 = 1$.

If we consider the flight of NASA rockets, we notice that the rockets initially fly upward but then tilt over in the direction of the eventual orbital trajectory. This is, of course, due to the requirement that the velocity $\mathbf{v}_3$ be perpendicular to the position vector $\mathbf{r}_3$ in orbit. We first note that the vertical velocity achieved by the rocket is completely useless: this is actually a radial component of the velocity that is not part of our canonical trajectory. In practice, the vertical segment of the rocket flight primarily serves to reach altitudes where the atmospheric density and, hence, the aerodynamic drag is significantly smaller. Without an atmosphere in our model or worrying about nearby trees and buildings, we could hypothetically just launch the rocket tangential to the earth's surface but, in the interests of retaining a semblance of reality, let us retain the vertical segment. So, to make a modestly realistic approximation of the flight path of a real rocket, let us assume that the rocket exhaust $\mathbf{v}_e$ is directed vertically for a time t_1 and transitions to horizontal during the time interval $t_1 \leq t \leq t_2$ and then continues to burn for a time t_3.

EXERCISE 7.20. Use the `Piecewise` function to define the x- and y-components of $\mathbf{v}_e$ such that

$$\mathbf{v}_e \cdot \hat{\mathbf{x}} = \begin{cases} -v_e & t \leq t_1 \\ -v_e \cos(\pi(t-t_1)/2(t_2-t_1)) & t_1 < t \leq t_2 \\ 0 & t > t_2 \end{cases}$$

and

$$\mathbf{v}_e \cdot \hat{\mathbf{y}} = \begin{cases} 0 & t \leq t_1 \\ -v_e \sin(\pi(t-t_1)/2(t_2-t_1)) & t_1 < t \leq t_2 \\ -v_e & t_2 < t \leq t_3 \\ 0 & t > t_3 \end{cases}$$

Plot the velocity profiles that you define to ensure that you have implemented the `Piecewise` function properly. Define the initial position to be $\mathbf{r}_3 = (0.95, 0)$ and the initial velocity to be $\mathbf{v}_3 = (0, 1/16)$ and choose $v_e = 1.2$. Try initial values of $t_1 = 0.25$, $t_2 = 1.25$ and $t_3 = 1.5$. Plot the trajectory, the orbital distance and the magnitude of the velocity. Adjust the parameters to obtain an orbital height of $r_3 = 1$ and velocity $v_3 = 1$. How does the trajectory change if the engine firing times are changed?

At what time does the trajectory intersect the planet surface (crash)? How far has the planet rotated during that interval?

Not surprisingly, the orbit produced by the example is an ellipse. Unfortunately, the trajectory results in a collision with the planet surface, as depicted in figure 7.9, at approximately the opposite side of the planet from the launch site. Fortunately, we know how to circularize an elliptical orbit. An additional engine firing is required: in the *opposite* direction of travel.

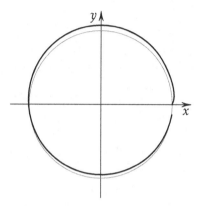

FIGURE 7.9. Launching from the surface of a planet (*gray circle*) with $R_1 = 0.95$ and $GM_1 = 1$. An initial thrust direction in the x-direction lifts the rocket from the planet surface. Subsequently the thrust direction is changed smoothly to the y-direction. The result is an elliptical orbit but, unfortunately, one that intersects the planet surface

EXERCISE 7.21. Add an additional engine firing to the x-component of $\mathbf{v}_e$ from the previous example. Define $\mathbf{v}_e = v_e/2\,\hat{\mathbf{x}}$ for the interval $t_4 \leq t \leq t_5$. At what time does the trajectory from the previous example reach the point $(0, 1)$? Define that time to be t_4. What value of t_5 results in a (nearly) circular orbit?

Plot the trajectory, the orbital radius and velocity. Do you find values close to the desired orbital values?

In retrospect, we should have expected this sort of behavior. When launching a craft from the planet surface, the trajectory is still dominated by the gravitational force exerted by the planet on the spacecraft. As we have seen previously, trajectories are ellipses with different eccentricities and changing the velocity alters the eccentricity. Changing the vehicle position from one radius to another requires that the velocity will contain a radial component. As a result, the orbit will necessarily be elliptical (or potentially hyperbolic), as we have seen in the studies of Hohmann transfer orbits. To achieve a circular orbit, one must apply a second engine firing opposite to the direction of travel. Then, one can achieve the (nearly) circular orbit illustrated in figure 7.10. As a result, we can begin

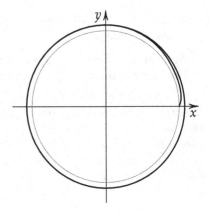

FIGURE 7.10. A second engine firing at the appropriate time will circularize the orbit. The direction of the engine firing must be in the direction of travel, slowing the rocket

to understand why the NASA launch vehicles are so large. The rocket equation greatly restricts the amount of payload that one can place into an orbital trajectory and, moreover, launching from the surface of the planet results in an elliptical orbit. In order to modify that initial orbit into one that does not intersect the planetary surface, requires additional expenditures of fuel to actually slow the spacecraft into a (more) circular orbit. Achieving a stable orbit requires a good deal more sophistication than one might first suspect.

Our analysis has, thus far, ignored the problems of staging and atmospheric drag, so our model represents a very simplistic version of spacecraft launch. We could construct more realistic models by incorporating each of these effects sequentially. At present, our model of atmospheric drag could be represented by a velocity-dependent force, as we have used previously. As we can recall, the drag coefficient depended upon the air density, which we took to be a constant for the problem of baseball flight. For rocket launches, the air density depends exponentially upon the altitude: $\rho = \rho_0 \exp(-\alpha r)$, where ρ_0 is the density at sea level and α is a proportionality factor that describes the rate at which the density decreases as a function of height r. For transonic and supersonic flight, our simple model does not really describe the atmospheric drag particularly well. So, while we can incorporate additional corrections to our simple model, our improvements would not be adequate to address realistic launch issues. We shall consider, instead, another aspect of travel around the solar system.

EXERCISE 7.22. Design a model of engine firings that can achieve a geosynchronous orbit for the model system studied in the previous exercises. A geosynchronous orbit is one that has the same orbital period as the rotational period of the planet. Consider also the requirement that the final orbit be aligned radially with the initial launch point.

EXERCISE 7.23.   Consider that your spacecraft is in the canonical orbit with $r_3 = 1$ and $v_3 = 1$ and that you want to reach a space station that is in a circular orbit with $r_4 = 1.1$. Suppose that the space station has an initial phase on its orbit of 0, i.e., that it is initially collinear. What happens if you try to fly straight toward the station? Would be the direction of the vector v_e? Can you identify a possible trajectory to rendezvous with the space station?

7.5. Gravity Assists

The realities of motion under the influence of the gravitational force bear little resemblance to the portrayal of interplanetary travel in film and print. Chemical rockets simply cannot provide sufficiently large momentum changes to enable a rocket ship to leave the earth's surface, travel to distant planets and return intact. Indeed, as we have seen, less than ten percent of the total mass on the launch pad can be delivered to low earth orbit, even with staged rockets. As a result, exploration of the solar system has proceeded at a pace that lags far behind the imagination of science fiction writers who have already established colonies on distant worlds. In fact, the pace of solar system exploration has been significantly hastened by the advent of digital computers and the insights of Michael Minovitch.[7]

One problem that we encountered in our study of designing a transfer orbit from earth to the moon (Exercise 7.19) was that the elliptical orbit we generated by the rocket firing was significantly altered by the presence of the target planet. In fact, early interplanetary missions incorporated engine firings to compensate for the deviation from the ideal Hohmann transfer orbit, in order to keep the spacecraft on the unperturbed path. While such trajectories were acceptable for lunar missions and voyages to Venus and Mars, the limitations of rocket technology left exploration of the outer solar system out of reach, requiring voyages of unreasonable duration. A Hohmann orbit to Jupiter requires almost three years and a similar transfer to Saturn requires six years. Similar transfer orbits to Uranus and Neptune would require sixteen and thirty years, respectively.

EXERCISE 7.24.   Revisit Exercise 7.19 and examine the behavior of the trajectory when $GM_2 = 0$ and 0.01.

---

[7]Minovitch was a graduate student at the University of California Los Angeles who spent a summer internship working at the Jet Propulsion Laboratory. His work on exploiting gravitational interactions to change spacecraft velocities were detailed in a series of JPL technical reports in 1961.

While we have access to sophisticated computational resources, early mission planners did not. The approach taken at the beginning of the space era was to construct approximations to a proposed trajectory by piecing together conic sections. As we have seen in our previous investigations, a hyperbolic trajectory is approximately two straight paths connected by a short section in which the trajectory bends around the scattering center. We can think of the influence of the gravitating body as having a rather short duration (or short length) despite the fact that the influence extends to infinity. An encounter with a gravitating body can even be crudely approximated by an instantaneous change in direction by some angle θ.

One could then try to plan a mission to the moon by considering the effect of the moon's gravitational force separately from that of the earth. The spacecraft would follow an elliptical trajectory around the earth, like those that we have discussed previously, until it came into close proximity to the moon. For a brief interval, the spacecraft's trajectory would be subject to the moon's gravitational influence, resulting in an hyperbolic trajectory. The modified trajectory would then revert to an elliptical trajectory, modified from the initial ellipse, around the earth. Clearly, gravity does not turn on and off like we are suggesting but this does provide a means for constructing an approximation to the actual path that a spacecraft might take. The approximation is relatively easy to calculate and, presumably, the actual trajectory would be close to that of the estimated path.

Minovitch was tasked with an investigation into improving some of the initial estimates of the trajectory but found a significant insight when he investigated the problem in vector form.[8] The geometry of an hyperbolic trajectory is depicted in figure 3.2. What we found in Chapter 3, where we discussed the two-body problem with masses M_1 and M_2, was that the asymptotic velocity $|v_2 - v_1|$ was a constant and the encounter served to change the direction of the velocity.

Consider now what happens when we investigate the encounter of a small mass M_3 with the target mass M_2 in a three-body problem. From this perspective, the center of coordinates illustrated in figure 3.2 is now placed on the target mass M_2 and the branch of the hyperbola that passes through the point r_0 represents the coordinates $r_3 - r_2$. The velocity of M_3 with respect to M_2 is $v_3 - v_2$. Here, we should expect that the asymptotic values of $|v_3 - v_2|$ will be constant, with only the direction of the velocity changing. This is, as we have demonstrated previously, the result of energy conservation.

---

[8]We have consistently used this representation throughout the text. Astronomers historically utilize a different set of coordinates for their measurements that we shall not describe here.

As we have defined earlier, the kinetic energy of the spacecraft is proportional to the square of velocity: $T \propto \mathbf{v} \cdot \mathbf{v}$. Relative to M_2, we find that

$$(7.5) \qquad (\mathbf{v}_3 - \mathbf{v}_2) \cdot (\mathbf{v}_3 - \mathbf{v}_2) = |\mathbf{v}_3|^2 - 2\mathbf{v}_3 \cdot \mathbf{v}_2 + |\mathbf{v}_2|^2$$

We should note, however, that this reflects the velocity with respect to the target mass M_2, not the velocity with respect to the other mass M_1. If we consider the velocity of M_3 with respect to M_1, something unexpected occurs. Let us calculate the change in kinetic energy of the spacecraft before and after the encounter with M_2, we find

$$
\begin{aligned}
\Delta T_3 &\propto |\mathbf{v}_3(t_0 + \Delta t)|^2 - |\mathbf{v}_3(t_0 - \Delta t)|^2 \\
&= 2[\mathbf{v}_3(t_0 + \Delta t) - \mathbf{v}_3(t_0 - \Delta t)] \cdot \mathbf{v}_2,
\end{aligned}
$$

(7.6)

where we have assumed that $|\mathbf{v}_3|$ is large enough that $\mathbf{v}_2$ is constant over the interval $t_0 - \Delta t \le t \le t_0 + \Delta t$.

> EXERCISE 7.25.   Fill in the missing steps of the derivation of Equation 7.6.

Because the velocity vector $\mathbf{v}_3$ changes direction before and after the encounter, the apparent change in kinetic energy ΔT_3, from Equation 7.6, will not be zero. What Minovitch discovered is that the kinetic energy of the spacecraft relative to M_1 can be increased (or decreased) by its encounter with the target M_2! This struck most researchers of the day as an apparent violation of energy conservation, so Minovitch's suggestion was generally dismissed. Nevertheless, there is no contradiction here. Energy is conserved but there is a distinct velocity change in M_3 that seemed unphysical to early investigators. Our derivation of Equation 7.6 relies, in fact, solely on the assumption that the spacecraft is travelling rapidly on its trajectory compared to the motion of the target planet M_2. We can remove that approximation by, once again, considering the problem numerically.

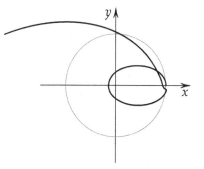

FIGURE 7.11. Two masses M_2 and M_3 orbit the central mass M_1. M_3 is initially on an elliptical trajectory before encountering M_2

To illustrate the effect, let us consider an elliptical orbit that crosses the x axis at the radius $x_3 = 1.01$, just outside our canonical orbit of $r_2 = 1$.

Here, as usual, the mass M_1 is located at the origin. Such an orbit is depicted in figure 7.11, where we have adjusted the phase of M_2 so that the encounter of M_2 and M_3 occurs after M_3 has completed one orbit. After the encounter, M_3 is diverted to a much larger elliptical orbit. If we now plot the velocity (relative to M_1) of M_3 as a function of time on orbit, we observe that prior to the planetary encounter, the velocity peaks near the point of closest approach to M_1 and then would fall back to the value it had initially. At the time t_0, the spacecraft encounters M_2 on its orbit and the velocity of M_3 rises sharply, as depicted in figure 7.12.

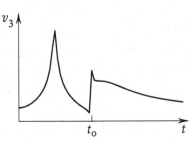

FIGURE 7.12. The spacecraft is initially on an elliptical trajectory ($t < t_0$). After the encounter with the planet, the velocity of the spacecraft is greatly increased

The nagging thought that somehow energy is not being conserved in this process kept Minovitch's insights from being immediately accepted. Yet, we can presume that momentum is conserved in what is an elastic collision between the two bodies M_2 and M_3. This would require that $M_2\Delta\mathbf{v}_2 + M_3\Delta\mathbf{v}_3 = 0$, where $\Delta\mathbf{v}_2$ and $\Delta\mathbf{v}_3$ are the changes in velocities of the two masses before and after the encounter. Hence, $\Delta\mathbf{v}_2 = -(M_3/M_2)\Delta\mathbf{v}_3$. As we saw in Exercise 7.6, the ratio of the mass of the International Space Station to that of the earth is 7.5×10^{-20}. As a result, the velocity change of a planet due to the encounter with even a large spacecraft is going to be inconsequential.

EXERCISE 7.26. Modify the equations from Exercise 7.3 to include a relative phase for M_2: $x_2(t) = \cos(t+\phi)$ and $y_2(t) = \sin(t+\phi)$. Start M_3 at the point $\mathbf{r}_3(t = 0) = (1.01, 0)$ with velocity $\mathbf{v}_3(t = 0) = (0, 0.5)$. For $GM_1 = 1$ and $GM_2 = 0$, plot the trajectory of M_3. Plot the magnitude of velocity $|\mathbf{v}_3|$.

Set the phase angle ϕ so that M_2 and M_3 approach the point $\mathbf{r}_2 = (1, 0)$ simultaneously. (This will be at the end of the first orbit of M_3.) With $GM_2 = 0.01$, plot the new trajectory of M_3 and the magnitude of velocity $|\mathbf{v}_3|$.

How is the trajectory of M_3 altered as you change the phase angle ϕ? That is, if the mass M_3 reaches the point $(1, 0)$ before or after M_2?

Despite the early skepticism, the strategy of utilizing gravity assists to improve the flight times to distant planets was rapidly adopted. The Mariner

10 mission was launched in November of 1973 and utilized a close encounter with Venus in February of 1974 to fly by Mercury in March of that year. Subsequently, essentially all missions to the outer planets have utilized gravitational assists. Notably, the Voyager missions utilized a fortuitous alignment of the outer planets to make a grand tour of Jupiter, Saturn, Uranus and Neptune with only twelve years elapsing between launch and passing Neptune's orbit. This is significantly faster than would have been achieved with a Hohmann transfer orbit.

More recently, the Messenger mission to Mercury utilized a series of encounters with Earth, Venus (twice) and then Mercury itself four times before the onboard engines were fired to enter orbit around Mercury. This feat is even more impressive when one realizes that all of the planets have elliptical orbits, not the circular orbits that we have been studying and that the orbital planes of the Earth, Venus and Mercury are inclined with respect to one another. This means that calculating the real trajectory is a problem in three dimensions.

> EXERCISE 7.27.   Suppose that we add another planet at a radius $r_4 =$ 5, with $GM_4 = 0.1$. Use the results from the previous exercise as the starting point to design an encounter with the outer planet. This will involve adding another term to the force law to represent the gravitational force on M_3 due to M_4. We assume that the orbit of M_4, like M_2, is fixed and that the orbital period of this planet is related by Kepler's law. Turn off the mass initially and determine the phase of the orbit required so that the spacecraft M_3 crosses the orbit of M_4 at approximately the same time as M_4.
>
> Can you define an engine firing sequence that results in capture into an elliptical orbit around M_4?

The principle limitation to using gravity-assisted trajectories is time. For example, in order for a spacecraft to receive a gravitational assist from Mars to expedite a trajectory to Jupiter, it is necessary that the earth, Mars and Jupiter be in the appropriate alignment. That is, opportunities for gravitational assists depend on the orbital positions of the planets and these positions change over the time scale of years. Thus, if one has a particular launch date in mind, Mars may not be in the appropriate position to provide an assist to a Jupiter mission. Alternatively, if a gravitational assist is necessary to provide sufficient kinetic energy to reach a distant target, then the possible launch dates will be constrained by the planetary positions.

7.6. Orbital Resonance

The solar system is considerably more complex than just the sun and planets. There are a host of minor planets that also fall under the influence of the gravitational force of the sun. Where ancient philosophers tried to seek meaning from the ratios of orbital periods of the planets, we now understand that the orbital period is, as Kepler found, related to the orbital geometry. As we have also found, the orbital geometry depends upon the orbital energy. Changing the kinetic energy changes the orbit. The present orbital parameters that characterize the solar system constituents reflect the evolution of the solar system according to Newton's Universal Law of Gravitation from its initial configuration (and collisions amongst the early constituents) and do not reflect hidden meanings.

As it happens, though, that the orbital periods of some solar system bodies are related. The Jovian moons Io, Europa and Ganymede have orbital periods of 1.769, 3.551 and 7.154 days, respectively. This is a ratio of 1:2:4 and is not coincidental. The synchronous orbits reflect the phenomenon of **resonance**.[9] Resonances arise as a result of periodic forces. In the case of the Galilean moons of Jupiter, the three innermost moons align on the same side of the planet at integral multiples of the Io orbital period. This produces forces that couple the orbital periods through tidal interactions amongst the moons that we have not considered in this text. Resonant phenomena occur throughout physics and we shall see further examples as we proceed.

Another example within the solar system are the Kirkwood gaps in the asteroid belt. In 1866, the American astronomer Daniel Kirkwood noted that, of the 87 asteroids then known, there were significant gaps in the distance between some of the asteroids that occurred at positions that corresponded to the orbital radii associated with specific ratios of the Jupiter orbital period (3/1, 5/2, 7/2).[10] As of this writing, there are nearly one hundred million cataloged minor planets and comets and the gaps originally observed by Kirkwood are even more evident.

As a simple model of the problem, we can return to our standard model of a body in orbit at a circular radius of $r_2 = 1$ and period $T_2 = 2\pi$, with masses $GM_1 = 1$ and $GM_2 = 0.001$, which is close to the mass ratio for the sun and Jupiter. From Kepler's third law, we know that the orbital radius and period are related. Given our unit system, where $GM_1 = 1$, if a body has an orbital radius of $r_3 = a^{2/3}$, then we can show that the orbital velocity will be $v_3 = a^{-1/3}$.

---

[9]The word derives from the Latin *resonantia* meaning echo.
[10]Kirkwood's "On the theory of meteors" was published in the Proceedings of the American Association for the Advancement of Science in 1866.

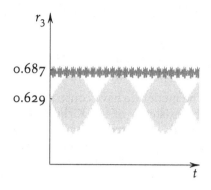

FIGURE 7.13. A mass M_3 that starts at an orbital radius of approximately $r_3 = 0.62996$ has an orbital period that is half that of the mass M_2 ($T_3/T_2 = 0.5$). The resulting trajectory (*light gray curve*) displays large oscillations about the nominal value. By comparison, a mass that starts at a radius $r_3 = 0.687$ ($T_3/T_2 = 0.57$) does not display such large oscillations

The results illustrated in figure 7.13 demonstrate resonant behavior in the vicinity of the ($T_3/T_2 = 1/2$) resonance. When the orbital period of M_3 is half that of the larger mass M_2, there is a point on the orbit (on the positive x axis) where the two masses are close. Here, the force on M_3 due to M_2 is greatest. This occurs periodically and leads to large oscillations in the radius r_3. When the orbital periods are not synchronous, the perturbations occur at various points along the orbits and do not add in a constructive fashion. As a result, the variations in the orbital radius off resonance are much smaller.

EXERCISE 7.28. Use the model defined in Exercise 7.3 as a starting point. Set the initial position to be $(x_3, y_3) = (a^{2/3}, 0)$ and the initial velocity to be $(v_{3x}, v_{3y}) = (0, a^{-1/3})$, where $a = 0.5$. Initially, set $GM_2 = 0$. You may need to set MaxSteps->20000 in the call to NDSolve. Do you obtain a circular orbit?

If so, now set $GM_2 = 0.001$ and compute the solution for 100 orbits $t_{max} = 200\pi$. What happens to the orbit? Plot the radius $r_3 = (x_3^2 + y_3^2)^{1/2}$ as a function of time. How does the orbital radius change? What is the average radius?

Now change a to 0.43 and 0.57. What happens to the orbital radius when the orbits are no longer synchronized?

If we now try to extrapolate our results to the asteroid belt, we can conclude that any object with an orbit at one of the special orbital periods actually occupies an orbit that oscillates with large amplitude around the average orbital radius. An added complexity in the real asteroid belt is that the asteroids can interact amongst themselves. These interactions will tend to depopulate the orbits near resonance, reinforcing the effect. A full explanation of the Kirkwood gaps is more complicated than we have presented here but the driving force is the periodic interactions with Jupiter.

> EXERCISE 7.29.   Consider resonant orbits for $T_3/T_2 = 1/3$ and 2/3. Do you find results like those in the previous exercise?

The model we have constructed seems to give a plausible explanation for the existence of the Kirkwood gaps in the asteroid belt. Periodic interactions reinforce to produce large oscillations in the orbital radius. The model does not apply to the resonant behavior of the Jovian moons. This reflects a greater complexity in that system. Where asteroids have very small masses compared to the sun and Jupiter, the masses of the first three Jovian moons are comparable. Hence, the restricted problem that we have faced is not applicable. In addition, the real moons are subject to gravitationally induced tidal forces that provide a mechanism for energy loss. This energy loss helps to damp out the oscillatory behavior that we observed in our model system but adds a level of complexity that we shall avoid at this time.

7.7. Chaos

We have but scratched the surface of the celestial mechanics problem, limiting ourselves to the restricted problem. This approach enables us to gain insights into the behavior of the systems without becoming overwhelmed with the complexity of real systems. This is the standard approach in dealing with physical systems. Start simple and then add complexity as necessary.

One simple thing that we can do is to allow the mass M_3 to move beyond the orbital plane defined by M_1 and M_2. By simple, we mean that we can add terms to the equations defined in Exercise 7.3 that provide for nonzero values of z_3 and v_{3z}. This will provide the opportunity to examine the behavior of systems in which the orbital planes do not coincide. One can easily revisit the previous exercises and add three-dimensional behavior to the models. The resulting behavior is, of course, generally not simple and forms the basis of significant research efforts to characterize the behavior of systems.

> EXERCISE 7.30.   Take the model from Exercise 7.3 and add equations for z3'[t] and vz3'[t]. Choose an initial value of v_{3z} that is not zero. You will have to use the `ParametricPlot3D` function to visualize the dynamics. What happens as you increase the initial value of v_{3z}?

EXERCISE 7.31. Consider the problem of resonant orbits where the orbit of M_3 is inclined slightly to the M_1-M_2 orbital plane. Do you still observe large oscillations for resonant orbits?

A more complex thing that we could attempt would be to try to compute the non-restricted problem, in which r_2 and v_2 are not assumed to be given but are computed. This adds significantly to the computational complexity of the problem. Indeed, others have followed precisely this path, with the most adventurous devising means for solving the problem for the complete solar system. We shall not attempt such a project here, as one needs to worry about numerical efficiency in order to make any significant progress.

One surprising result that arose from simulations of the solar system for hundreds of million years into the future was that the orbits of the planets are **chaotic**.[11] By chaotic, we really mean that if we conduct two simulations with starting conditions that are very close then after a long time, the two solutions will not be close. Imagine starting our model simulations with M_2 at a position $r_2 = (\cos(\phi), \sin(\phi))$ with velocity $v_2 = (-\sin(\phi), \cos(\phi))$, where ϕ is a small angle. We would expect that the solutions we obtain would not be hugely different than those conducted with $\phi = 0$. That is, if $\phi = 0.01$ we expect that the position of mass M_3 would be very close to the position it has when $\phi = 0$, and that it would stay that way forever. An example, is illustrated in figure 7.14, where the distance between two solutions for r_3 is shown as a function of time. Initially, the two solutions are close but move away from one another as time progresses. This example is somewhat contrived, as the starting difference $\phi = 0.1$ was relatively large, but it illustrates the point.

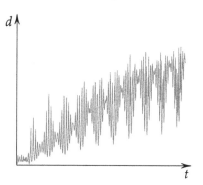

FIGURE 7.14. The distance $d = |r_3 - r_3'|$ is illustrated for two simulations r_3 and r_3' that differed in the initial position of M_2. As the simulations evolve, the two solutions diverge

EXERCISE 7.32. Use the basic model equations to add a phase shift ϕ to the definition of x2 and y2. Use the Animate function

---

[11]Gerald Sussmann and Jack Wisdom demonstrated that Pluto's orbit was chaotic in 1988.

to draw the positions of solutions for $\phi = 0$ and $\phi = 0.01$. (Use the Graphics[Point[]] function to draw the positions.) Plot the distance between the two solutions as a function of time. Do the solutions remain close? What happens if $\phi = 0.1$?

In simulations of the solar system, where the bodies gravitationally interact, the positions of the planets from two close starting points do not wind up close to one another. This remarkable result means that we cannot, with any certainty, predict where earth will be in a hundred million years. It is not knowable. Conversely, one cannot integrate backwards in time to determine how the solar system came to be in the position that it occupies now. This is, perhaps, deflating for those who want to achieve total predictive success. There are no practical reasons why one would need to predict the state of the solar system millions of years into the future. Certainly, such times are well beyond the scale of human lives and we have ruled out time travel in previous chapters. Hence, no one will be building machines to revisit the age of the dinosaurs and, thus, somehow have to fly back seventy or eighty million years. There simply are limitations to our predictive capabilities.

Constituents of the Atom

Following Rutherford's pathbreaking discovery that the atom had a nuclear center, physicists found themselves with a host of new problems to solve. As we have mentioned, a large number of physicists set to work on the problem of how atoms work. At a coarse level of scrutiny, it seems that the Coulombic force between the positively charged nucleus and negatively charged electron(s) is responsible for the attractive force that holds the electrons to the nucleus. Yet, when energy is pumped into a cell containing hydrogen gas, spectroscopists discovered that there were a specific set of wavelengths of light emitted from the cell. Apparently, the electrons cannot orbit randomly; they are required to occupy specific orbits. Actually, the use of the word *orbit* here is entirely incorrect in light of modern, quantum theories of the atom but does reflect the classical view of the universe held by physicists in the early 1900s.

In addition to the physicists working on building mathematical models of the atom, a number also tried develop a more detailed understanding of the properties of the constituents of the atom. What exactly is an electron? What is the nature of the nuclear matter? As one can well surmise, experimentation posed a number of practical difficulties. The "size" of the atom is of the order of 10^{-10} m and the nucleus is even smaller: of the order of 10^{-14} m. The visible light spectrum roughly spans the range from 400–700 nm, or 4–7×10^{-7} m. This is several hundred times the atomic size and several million times the nuclear size. It is simply not possible to see atoms or nuclei with visible light. Instead, physicists rely on other means of discerning the nature of matter. As it happens, static magnetic fields play a significant rôle in our understanding of matter.

Lodestones were known to the ancient Greeks as possessing the remarkable capability of attracting one another and objects made from iron.[1]

---

[1] In Plato's dialog $I\omega\nu$, Socrates tells Ion: "… it is a divine influence which moves you, like that which resides in the stone called Magnet by Euripides, and Heraclea by the people. For not only does this stone possess the power of attracting iron rings but it can communicate to them the power of attracting other rings; so that you may see sometimes a long chain of rings and other iron substances attached and suspended one to the other by this influence."

© Mark A. Cunningham 2015
M.A. Cunningham, *Neoclassical Physics*, Undergraduate Lecture Notes in Physics, DOI 10.1007/978-3-319-10647-2_8

In addition, if suspended by a thin string, a lodestone would seek a specific direction that we have come to call North. This property was eventually refined into devices that we now call compasses. Until the recent invention of the Global Positioning System and satellite navigation, compasses were essential equipment for navigators. From our previous discussions, we can infer that there must be some force acting on the compass needles that cause them to align in a specific direction and that some force acts upon the iron rings to hold them to the lodestone. Historically, this force has been termed the magnetic force but, in one of the great advances of physics, the Scottish physicist James Clerk Maxwell demonstrated that the separate electric and magnetic forces known to the ancients were actually just two different aspects of a common force that we today call electromagnetism.[2]

Maxwell's efforts unified the various known descriptions of the behavior of electric and magnetic phenomena into a single coherent framework. His theory represented the culmination of significant progress in the early 1800s in the scientific understanding of electric and magnetic phenomena. This progress was due, in large part, to the discovery of the electrochemical cell (battery) by Alessandro Volta in 1799 that provided a means for conducting systematic studies of electrical phenomena. There followed a series of insightful discoveries that captivated not only the scientific community but the populace as a whole. Volta was invited to Paris in 1801 by Napoleon and gave three public lectures on his research to the French National Academy, all of which were attended by Napoleon himself. Napoleon was so impressed by the Italian's charm and intelligence that he awarded Volta a gold medal for his achievements, established an annual prize for research into electrical phenomena and, in 1810, named Volta a count.[3] Legend has it that, upon leaving the library of the National Institute where Volta delivered his lectures, Napoleon came across a bronze plaque with the inscription "Au grand Voltaire" and personally erased the final three letters.

8.1. Lorentz Force

One of the consequences of Maxwell's system of equations is that the force on a charged particle must be expanded to include the influence of magnetic fields as well as the electric field, as we have previously discussed.

---

[2]Maxwell's four part paper "On physical lines of force" was published in the *Philosophical Magazine* in 1861 and 1862. His subsequent "A dynamical theory of the electromagnetic field," published in 1864 established that light was an electromagnetic wave.

[3]Volta communicated his results in a letter to the Royal Society of London on March 20, 1800. *Le Prix du Galvanisme* was awarded annually in the years 1802–1815.

A series of physicists contributed to the problem but, today, we associate the result to the Dutch physicist Hendrik Lorentz:

$$(8.1) \qquad\qquad \mathbf{F} = q(\mathbf{E} + \mathbf{v} \times \mathbf{B}).$$

Here, the force on the charge q due to the magnetic field $\mathbf{B}$ is seen to depend also on the velocity $\mathbf{v}$ of the charge. Indeed, the vector cross product in Equation 8.1 results in the force being directed perpendicular to the motion of the charge. Hence, the magnetic force is on a moving charged particle is a centripetal force and does no work.

If we want to consider the motion of a charged particle in a magnetic field, with no electric field present, we note that it is astute to use the direction of the magnetic field as one of the coordinate axes. We can then decompose the velocity vector $\mathbf{v}$ in terms of a component that is parallel to the magnetic field and two components that are perpendicular:

$$\mathbf{v}_{\parallel} = \frac{\mathbf{v} \cdot \mathbf{B}}{|\mathbf{B}|^2} \mathbf{B} \quad \text{and} \quad \mathbf{v}_{\perp} = \mathbf{v} - \mathbf{v}_{\parallel}.$$

We see immediately that $\mathbf{v}_{\parallel} \times \mathbf{B} = 0$. As a consequence, motion along the direction of the magnetic field is unaccelerated. Only the motion perpendicular to the magnetic field is subject to a force.

EXERCISE 8.1. Consider the vector $\mathbf{v} = (v_x, v_y, v_z)$. The vector $\mathbf{B}/|\mathbf{B}|$ is a unit vector in the direction of the magnetic field. Suppose that $\mathbf{B} = B_y \hat{\mathbf{y}}$. Compute the parallel and perpendicular components of $\mathbf{v}$. Suppose now that $\mathbf{B} = (0, B_y, B_z)$. What are the parallel and perpendicular components of $\mathbf{v}$?

Recall that the angular momentum is conserved for a centripetal force. Hence, we can write the following relation:

$$(8.2) \qquad\qquad \frac{d}{dt} M\mathbf{r} \times \mathbf{v} = M\mathbf{r} \times \frac{d\mathbf{v}}{dt} = 0,$$

where $\mathbf{r}$ is the position of the charge in some coordinate system. Note that for a particle of mass M, we have $M\, d\mathbf{v}/dt = \mathbf{F}$. So, if we substitute the magnetic component of the Lorentz force into Equation 8.2 we find the following result:

$$(8.3) \qquad\qquad q\mathbf{r} \times [\mathbf{v}_{\perp} \times \mathbf{B}] = 0.$$

Now for Equation 8.3 to hold, we must have that $\mathbf{r}$ is parallel (or antiparallel) to the vector $\mathbf{v}_{\perp} \times \mathbf{B}$.

EXERCISE 8.2. Use the vector identity $\mathbf{a} \times (\mathbf{b} \times \mathbf{c}) = (\mathbf{a} \cdot \mathbf{c})\mathbf{b} - (\mathbf{a} \cdot \mathbf{b})\mathbf{c}$ to prove that $\mathbf{r}$ is perpendicular to both $\mathbf{v}_{\perp}$ and $\mathbf{B}$.

EXERCISE 8.3. If no work is done on the charged particle, then the kinetic energy $T = \frac{1}{2}M\mathbf{v} \cdot \mathbf{v}$ is also conserved. Show that the magnetic component of the Lorentz force does not alter the kinetic energy.

Let us simplify the calculations and align the magnetic field with the z axis. Then $\mathbf{B} = B\hat{z}$ and $\mathbf{v}_\perp = (v_x, v_y, 0)$., We can now observe that $\mathbf{v}_\perp \times \mathbf{B} = (v_y B, -v_x B, 0)$ and, from this result, that the following relation holds:

$$\mathbf{r} \times (\mathbf{v}_\perp \times \mathbf{B}) = (zv_x B, -zv_y B, -xv_x B - yv_y B) = 0.$$

For the first two terms to vanish, we require that $z = 0$. This means that motion will take place in the plane perpendicular to $\mathbf{B}$. For the last component of the vector to vanish, it is necessary that

$$0 = xv_x + yv_y = x\frac{dx}{dt} + y\frac{dy}{dt}$$

(8.4)
$$= \frac{d}{dt}(x^2 + y^2).$$

Note that the term $x^2 + y^2$ is the (squared) distance to the origin and it is constant in time. Hence, Equation 8.4 represents circular motion in the plane perpendicular to $\mathbf{B}$. We must have that $x = r\cos\varphi$ and $y = r\sin\varphi$, where r is the radius of revolution and φ is the angle from the x-axis, as indicated in figure 8.1.

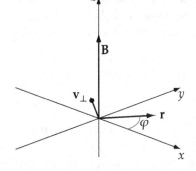

FIGURE 8.1. The vector $\mathbf{r}$, $\mathbf{v}_\perp$ and $\mathbf{B}$ are mutually perpendicular

We note further that the particle possesses a constant angular velocity. We can see this from the fact that the magnitude of velocity is also constant; we have

(8.5)
$$\mathbf{v}_\perp = \frac{d\mathbf{r}}{dt} = \left(-r\sin\varphi\frac{d\varphi}{dt}, r\cos\varphi\frac{d\varphi}{dt}, 0\right).$$

From this result, we can show that $v_\perp = r\,d\varphi/dt = r\omega$, where ω is the angular velocity. Thus, if $v_\perp$ is constant, then so is ω.

The derivative of velocity is, of course, acceleration. From Equation 8.5, we can write that

$$(8.6) \qquad \frac{d\mathbf{v}_\perp}{dt} = \left[-r\cos\varphi \left(\frac{d\varphi}{dt} \right)^2, -r\sin\varphi \left(\frac{d\varphi}{dt} \right)^2, 0 \right] = -\mathbf{r}\,\omega^2.$$

The magnitude of the acceleration is then given by the following:

$$(8.7) \qquad\qquad\qquad |\mathbf{a}| = \frac{v_\perp^2}{r}.$$

This result arises from the centripetal nature of the force.

From the definition of the Lorentz force, we also have that the magnitude of the acceleration is given by $|\mathbf{a}| = qv_\perp B/M$. We can combine this fact with results in Equation 8.7 to solve for the radius r:

$$(8.8) \qquad\qquad\qquad r = \frac{Mv_\perp}{qB}.$$

Thus, the radius of gyration is proportional to the component of momentum perpendicular to the magnetic field. If the particle also has a component of the velocity parallel to the magnetic field, as illustrated in figure 8.2, then the vertical distance d the particle travels during one revolution is given by $d = v_\parallel T$, where T is the period of revolution. In this instance, the period is given by $T = 2\pi r/v_\perp = 2\pi M/qB$.

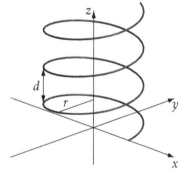

FIGURE 8.2. If the magnetic field is oriented in the z direction, then a charged particle will follow a helical trajectory. The axis of the helix is also parallel to the z axis

EXERCISE 8.4.   Define the vector $\mathbf{r} = (a\cos(t), a\sin(t), bt)$. Use the Manipulate function to examine the resulting trajectory over the range $1 \le a \le 10$ and $0 \le b \le 1$. Use the ParametricPlot3D function to display the trajectory. How does the trajectory change as a and b change?

Suppose now that a charged particle were to enter into a finite region where the magnetic field is nonzero. Such a situation is illustrated in figure 8.3, where the magnetic field is confined to the area inside the dashed circle. The particle is deflected through an angle θ, where $\sin\theta = L/r$ and L represents the horizontal distance covered by the particle while it is still in the magnetic field.

EXERCISE 8.5. Suppose the magnetic field is nonzero in a rectangular region of area $A = LW$ and that the radius of curvature r is large compared to L. Draw an arc of radius r that extends for a horizontal distance L, as in figure 8.3. Use geometry to prove the assertion that $\sin\theta = L/r$.

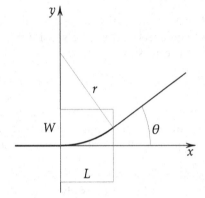

FIGURE 8.3. A charged particle entering a region (box) where the magnetic field is nonzero will be deflected through an angle θ. The radius of curvature r is defined in Equation 8.8

8.2. Cathode Rays

One of the most intriguing devices constructed in the latter part of the 1800s was the cathode ray tube. The device itself is quite simple: two electrodes were fitted inside of a glass bulb in which rarefied gases were present. When a source of high voltage was connected across the electrodes, the bulb glowed. This phenomenon is, of course, not particularly spectacular to modern students, for whom neon lights are commonplace. For people in the late 1800s, who derived their light from gas lamps or candles, the observation of light emanating from a glass bulb in which no combustion was occurring was quite spectacular indeed.

Physicists, of course, were interested in the nature of these cathode rays and the definitive experiments were conducted by the British physicist J. J. Thomson. In Thomson's initial experiment, using the apparatus sketched in figure 8.4, application of voltages V_a and V_c to the anode and cathode, respectively, gave rise to a cathode ray beam, provided that the cathode voltage was relatively negative to the anode voltage and provided that the voltage difference $V_a - V_c$ was sufficiently large. A narrow beam was obtained by drilling a small hole through the center of the anode. The cathode ray beam could be visualized through its interactions with

FIGURE 8.4. Thomson's first experiment showed that a beam of cathode rays would travel in a straight line unless deflected by an external magnetic field. In that case, the rays would pass through the grounded shield (earth) and strike the detector. Accumulated charge was measured by an electrometer

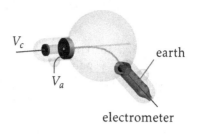

gas molecules in the glass container, generating a glowing trail along the trajectory of the beam. When a transverse magnetic field was applied, the beam was deflected downward where it entered the slot cut in the grounded shield. The beam then struck the detector, which was simply another slotted cylinder attached to a wire that passed outside the glass cell. Attaching the wire to an external electrometer enabled Thomson to prove that the cathode rays were negatively charged corpuscles. This was consistent with the predictions of the Lorentz force law for the interaction of negatively charged particles with a magnetic field.

Other researchers had failed to deflect the cathode ray beam with applied electric fields, giving rise to a puzzle. Apparently, cathode rays were affected by magnetic fields but not by electric fields. These early experiments indicated that cathode rays do not obey the Lorentz force law and, possibly, represented some new form of matter. Thomson recognized that the passage of cathode rays through a gas-filled tube could potentially generate ionic currents that would mask the effect of an applied electric field. So, he constructed a tube that contained electrodes but was evacuated of the residual gas. This approach would eliminate the ionic currents but also eliminated the ability to directly visualize the beam. To solve this problem, Thomson painted the end of the tube opposite the cathode with phosphorescent material that emitted light when struck by the cathode rays. The results were quite striking.[4] Attaching a battery across the electrodes produced a measurable deflection of the electron beam. Reversing the battery connections led to an equivalent deflection in the opposite direction. Clearly, cathode rays were charged particles and behaved in the manner predicted by the Lorentz force law.

---

[4]Thomson published his quantitative results in the *Philosophical Magazine* in 1897. Thomson was awarded the Nobel Prize in Physics in 1906 "in recognition of the great merits of his theoretical and experimental investigations on the conduction of electricity by gases."

Thomson made a series of systematic studies of the cathode rays, replacing the anode and cathodes with different metals and filling the tubes with

FIGURE 8.5. In Thomson's second experiment, a pair of electrodes were placed inside a tube that was evacuated of (nearly) all gas. Thomson painted the inside of the large sphere at the end of the tube with a phosphorescent material that emitted light when struck by the cathode ray beam

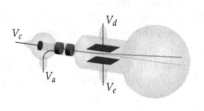

different gases before evacuating them (to study residual gas effects). All cathode rays appeared to be the same, independent of materials used.

To quantify Thomson's results, let us begin by approximating the electric field in the region between the plates to be given by $E = (V_d - V_e)/d$, where d is the distance between the two electrode plates. The electric field is assumed to be zero outside of the plates. This result is exact for infinitely large plates but is not a bad approximation if the distance d is relatively small compared to the length L of the plates.

The electron enters the region between the plates with a kinetic energy that is given by $T_1 = e(V_a - V_c)$, where e is the magnitude of the charge on the electron. Hence, the electron will have a velocity given by $T_1 = \frac{1}{2}m_e v_1^2$, that is directed along the long axis of the tube shown in figure 8.5 and where m_e is the mass of the electron. The electron will therefore spend a time $t = L/v_1$ traversing the plates while subject to an acceleration given by $a = eE/m_e$. This results in a component of velocity perpendicular to the original velocity, where the direction depends on the sign of the bias voltage $V_d - V_e$. Putting all of these results, together, we can determine that the electron will be deflected by an angle θ, where

(8.9) $$\tan\theta = \frac{at}{v_1} = \frac{[e(V_d - V_e)/d][L/v_1]}{v_1} = \frac{L(V_d - V_e)}{2d(V_a - V_c)}.$$

EXERCISE 8.6. Fill in the details of the derivation of Equation 8.9.

Recall now that we can also produce a deflection with an applied magnetic field. Thomson found that, with perpendicular electric and magnetic fields, he could arrange for the deflections to cancel. If the electric and magnetic fields extended over the same region L, then the Lorentz force vanishes if $vB = -E$. We have already found a value for v in terms of the anode and cathode potentials and the ratio of the electron charge to its mass can be determined to be given by the following:

$$(8.10) \qquad \frac{e}{m_e} = \frac{(V_d - V_e)^2}{2(V_a - V_c)B^2 d^2}.$$

After extensive studies, Thomson found that the ratio of the electron's charge to mass was more than a thousand times larger than the comparable ratio for the hydrogen ion. This could be due either to the electron possessing a charge vastly greater than the charge of a hydrogen ion or to the electron possessing a mass that was vastly smaller. Over the next several years, the charge of the electron was measured independently and shown to be the same as that of the hydrogen ion (although with the opposite sign). This means that the electron mass is over a thousand times smaller than the mass of the hydrogen nucleus (the proton). (The currently accepted value (2010) is $m_e/m_p = 5.446\ 170\ 2178(22) \times 10^{-4}$ or $m_p/m_e = 1836.15267245(75)$, where the numbers in parentheses represent the uncertainty in the last two digits.)

As there seemed to be an inexhaustible supply of cathode rays from any set of electrodes, Thomson proposed that the atoms must be somehow studded with these bits of negative charge. As we have discussed previously, Thomson's model came to be known as the plum pudding model and was shown to be incorrect by Rutherford's experiments a decade later. Nevertheless, Thomson's proposal fit the data available at that time.

EXERCISE 8.7. Define a coordinate system in which particles travel initially in the z-direction. If we utilize an electric field with only an x-component $\mathbf{E} = E\,\hat{x}$, what would the direction of a magnetic field need to be to produce a Lorentz force in the opposite direction of the force due to the electric field?

Fill in the details of the derivation of Equation 8.10.

EXERCISE 8.8. We have treated the kinematics in this section using the non-relativistic equations developed by Newton. Suppose that the accelerating potential $(V_a - V_c)$ was 500 V. Should we have used Einstein's relativistic kinematics? Would your answer change if the accelerating potential was 5 kV?

8.3. Canal Rays

In 1886, the German scientist Eugen Goldstein noted that, if he filled a tube with a dilute gas and inserted a perforated cathode, not only did he observe a glowing discharge between the anode and cathode but a series of faint lines emanated from the cathode in the opposite direction, as sketched in figure 8.6. Goldstein named these emanations Kanalstrahlen ("channel rays" or "canal rays") because they only occurred when the cathode had holes in it.

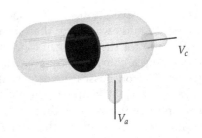

FIGURE 8.6. When a large voltage difference $(V_a - V_c)$ was applied across the anode and cathode, a luminous discharge on the right side of the tube (between the anode and cathode) could be observed. If the cathode had holes in it, rays extending in the opposite direction could also be observed

Following his work on electrons, Thomson conducted a series of detailed experiments on these canal rays.[5] A sketch of his apparatus is illustrated in figure 8.7. A beam of canal rays was produced by placing a cathode with a narrow orifice in the neck between two glass chambers. The beam then entered a region where an electric field was created by a pair of electrodes and a parallel magnetic field was generated by a large electromagnet. The end of the flared bottle was painted with a phosphorescent material that glowed when struck by the canal rays. Photographic plates were then affixed to the end of the flared tube and exposed for several hours.

FIGURE 8.7. A glass bottle containing rarefied gases was connected by a narrow neck to a second bottle. Voltages applied to the anode (V_a) and cathode (V_c) resulted in canal rays that were deflected by an electric field created by the plates $(V_e - V_d)$ and a parallel magnetic field generated between the poles $(P_1$ and $P_2)$ of a large electromagnet

EXERCISE 8.9. Define the following equations in the *Mathematica* program:

---

[5]Thomson's experiments were detailed in his report "Rays of positive electricity," made on May 22, 1913 to the Royal Society of London.

```
eqs={x'[t]==vx[t],y'[t]==vy[t],z'[t]==vz[t].
  vx'[t]==QM(Ex+vz[t] By-vy[t] Bz),
  vy'[t]==QM(Ey+vx[t] Bz-vz[t] Bx),
  vz'[t]==QM(Ez+vy[t] Bx-vx[t] By)}
ics={x[0]==x0,y[0]==y0,z[0]==0,
  vx[0]==vx0,vy[0]==vy0,vz[0]==vz0}
soln=NDSolve[Join[eqs,ics]/.{QM->1,Ex->0,Ey->0,Ez->0,
  Bx->1,By->0,Bz->0,x0->0,y0->0.0,z0->0,
  vx0->0,vy0->0,vz0->1},{x,y,z,vx,vy,vz},{t,0,5}]
ParametricPlot3D[Evaluate[{x[t],y[t],z[t]}/.soln],
  {t,0,5}]
```

Describe the resulting trajectory. What happens if there are non-zero components of the electric field? What happens if the sign of Q/M is negative?

EXERCISE 8.10.   Use the equations from Exercise 8.9 but define E_x and B_x to be Piecewise[{{1,0<z[t]<=1},{0,True}}]. This defines the fields to be non-zero only in the interval $0 < z \leq 1$. Plot the trajectory over the range $-2 \leq x \leq 2$, $-2 \leq y \leq 2$ and $0 \leq z \leq 6$. Describe the resulting trajectory. What value of the initial velocity in the z-direction is required in order for the trajectory to hit the screen at $z = 6$?

EXERCISE 8.11.   The fields are not exactly constant over the interval $0 \leq z \leq 1$. The function $f(z) = 1 - 16(z - 1/2)^4$ is one that is approximately constant in the center and drops to zero on either end. Plot $f(x)$. Now, using the equations from Exercise 8.9, define E_x and B_x to be Piecewise[{{1-16(z[t]-1/2)^4,0<z[t]<=1},{0,True}}]. Does this significantly alter the behavior of the beam? Can you alter the length of the piecewise constant region and obtain comparable results?

In this experiment, Thomson utilized aligned electric and magnetic fields that were perpendicular to the motion of the canal rays. If we define a coordinate system in which the velocity of the rays is in the z-direction and the fields are in the x-direction, then the action of the electric field is to produce an acceleration in the x-direction. This will generate a component of the velocity in the x-direction but, as this is parallel to the magnetic field, there is no additional magnetic acceleration. The magnetic field will initially produce an acceleration in the $\hat{z} \times \hat{x} = \hat{y}$ direction but, as we have seen, this will lead to a circular path in the y-z plane. If the initial velocity v_z is too small, the beam will not make it to the screen at the end of the flared bottle.

For relatively large velocities, then, the motions in the x- and y-directions nearly decouple (v_z is approximately constant.) If we assume that the electric field is constant for some length L and zero elsewhere, then the x-component of the velocity at the end of the acceleration interval will be given by the following:

$$(8.11) \qquad v_x = \frac{q}{M} E_x t_1 = \frac{q}{M v_z} E_x L,$$

where t_1 is the length of time the beam takes to traverse the distance L and we have assumed that the velocity in the z-direction is constant. In this case, the beam has moved a distance x_1 along the x-direction, where we have:

$$(8.12) \qquad x_1 = \frac{q}{2M} E_x t_1^2 = \frac{q}{M v_z^2} \frac{E_x L^2}{2}.$$

At the end of the acceleration interval, we have that the component of the velocity in the y-direction will be approximately given by the following:

$$(8.13) \qquad v_y = \frac{q}{M} v_z B_x t_1 = \frac{q}{M} B_x L$$

The beam will have moved a distance y_1 along the y-direction, where we have:

$$(8.14) \qquad y_1 = \frac{q}{2M} v_z B_x t_1^2 = \frac{q}{M v_z} \frac{B_x L^2}{2}$$

The canal ray beam then traveled a further distance d along the z-direction before striking the screen. This will take a time $t_2 = d/v_z$. This motion is not accelerated, so the beam will strike the screen at a position (x_2, y_2) given by the following:

$$(x_2, y_2) = (x_1 + v_x t_2, y_1 + v_y t_2)$$

$$(8.15) \qquad = \left(\frac{q}{M v_z^2} E_x (L^2/2 + Ld), \frac{q}{M v_z} B_x (L^2/2 + Ld) \right).$$

We see that the position on the screen depends on the charge to mass ratio q/M and on the initial velocity v_z. The remaining dependencies on the fields and the geometry of the apparatus can be lumped into additional constant terms.

EXERCISE 8.12. Use the ParametricPlot function to plot $(1/v^2, 1/v)$ over the range $2 \leq v \leq 20$. Describe the results. Now add plots for $(1/3v^2, 1/3v)$ and $(1/5v^2, 1/5v)$. Describe the results.

What Thomson discovered in his experiments was a series of parabolic traces on his photographic films like those depicted in figure 8.8. A bright spot was produced by uncharged components of the canal ray beam that were not deflected by the electric and magnetic fields. The photographs also contained a series of parabolic arcs. Thomson deduced that the arcs were generated because the constituents of the beam did not all have the same velocity. Presumably, the positively-charged constituents were generated throughout the volume of the bottle and so were not accelerated to the maximum potential $V_c - V_a$. The maximum kinetic energies would be expected to be $T_{max} = q(V_c - V_a)$ and hence the maximum velocities would be given by $v_{max} = [q(V_c - V_a)/2M]^{1/2}$. As a result, the arcs generally began at the same value of x.

EXERCISE 8.13. Use the ParametricPlot function to plot the function $f = (1/Mv^2, 1/Mv)$. The maximum velocity will be mass dependent. For mass $M = 1$, assume that the velocity is in the range $3 \leq v \leq 10$. Scale the ranges appropriately for masses $M = 2, 16, 20$ and 28. What is the minimum value of x?

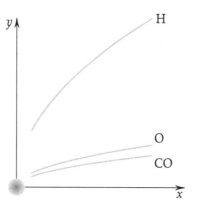

FIGURE 8.8. Thomson's apparatus produced images similar to the one sketched at right. A bright spot marked the point struck by uncharged components of the canal rays that were not deflected by the fields. A series of parabolic lines could be used to identify the elements contained in the gas

EXERCISE 8.14. Thomson observed that hydrogen was present in nearly every gas sample he tested. How can the measured y-values of the parabolas (at the minimum x-distance) of different lines like those in figure 8.8 be used to determine the masses? (Hint: Use Equation 8.15 and consider the ratio with respect to hydrogen.)

EXERCISE 8.15. It is not necessary to utilize the approximations that led to Equation 8.15. Use the equations from Exercise 8.9 but define E_x and B_x again to be piecewise constant over the range $0 \leq z \leq 1$. Compute the trajectories for a range of velocities. Now evaluate the values of x and y at the distance $z = 6$. Plot the resulting values and show that the points fall along a parabola.

Thomson also observed a number of different effects in his photographs that deviated from the nominal behavior displayed in figure 8.8. In some cases, he observed that the traces extended to smaller values of x than would be achievable with the known accelerating potential. Thomson attributed these results to the production of ions with charge $+2q$. In those films, Thomson would generally observe traces that corresponded to both q/M and $2q/M$. The extension of the q/M trace to smaller x values could be explained if the ions were initially doubly charged while being accelerated by the potential $V_c - V_a$ but then lost a charge before entering the deflection region of the apparatus.

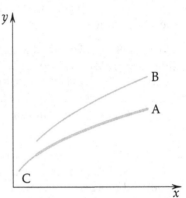

FIGURE 8.9. The trace A is produced by the charge to mass ratio q/M. The trace B is produced by the charge to mass ratio $2q/M$. The extension C of trace A is generated by ions that were doubly charged when accelerated but then lost a charge when being sorted

In one of his photographs that contained neon gas, Thomson observed parabolas that corresponded to masses of $M = 20$ and 22. He also found parabolas that could be attributed to masses $M = 10$ and 11. These values are precisely half of the values of the heavier masses. Because the traces for the heavier masses were extended to low values of x like trace A from figure 8.9, Thomson concluded that the traces corresponding to $M = 10$ and 11 were due to multiple ionization. Nevertheless, neon gas was composed of two fractions of different masses. The mass of neon was then known to be about 20.2 from other measurements, so Thomson concluded that about ten percent of neon has mass 22 and the remainder a mass of 20. Today, we refer to the two different fractions as the **isotopes** of neon.

Thomson's assistant in many of these experiments was Francis Aston, who recognized that several improvements to Thomson's apparatus could lead to improved resolution. Aston's developments led to precision mass spectrometry.[6] His work was interrupted by the first world war but when

---

[6]Aston was awarded the 1922 Nobel Prize in Chemistry "for his discovery, by means of his mass spectrograph, of isotopes, in a large number of non-radioactive elements, and for his enunciation of the whole-number rule."

he returned to the problem in 1919, Aston's improved devices rapidly produced evidence for over two hundred isotopes of non-radioactive elements. This solved a long-standing problem with the masses of the chemical elements. It was known that the masses are approximately integral multiples of the mass of the hydrogen atom but there were significant discrepancies. For example, magnesium (A=12) has an atomic mass of 24.31, which is not particularly close to 24. Intriguingly, cobalt (A=27) has an atomic mass of 58.93, which is close to 59, but nickel (A=28) has an atomic mass of 58.69, which is surprisingly lighter than cobalt. The non-integral values could now be explained by simply adding different fractions of the isotopes. This, of course, leads to the subsequent problem of why we see the particular isotopic ratios and whether or not those are universal ratios.

> EXERCISE 8.16.   Chlorine has an atomic mass of 35.45. Aston determined that chlorine has two isotopes ^{35}Cl and ^{37}Cl. What fraction of isotopes leads to the observed atomic mass?

Aston's major improvement to Thomson's apparatus was involved separating the electric and magnetic field regions and rotating the magnetic field perpendicular to the electric field. This had the consequence that the charged particles in the beam were deflected in a plane along the beam axis, rather than being deflected into three dimensions. Moreover, with judicious choices for field intensities, all of the particles with a common charge to mass ratio (q/M) could be focussed to a single point, independent of the particle velocity. This allowed Aston to simplify his data collection. All beam constituents with the same q/M struck the same point on the phosphorescent material, reducing the required exposure time and increasing the signal to noise ratio. By 1922, Aston had identified 212 isotopes.

> EXERCISE 8.17.   Use the equations of motion from Exercise 8.9 but define E_x to be Piecewise[{{1,0<z[t]<=1/2},{0,True}}] and B_y to be Piecewise[{{Bval,2<z[t]<=3.5},{0,True}}]. Use the ParametricPlot function to plot $x(t)$ versus $z(t)$. Group the NDSolve and plot functions inside a Module so that you can use Manipulate to vary the value of Bval from zero to four. Choose an initial charge to mass ratio of $q/M = 1/20$. Compute separate solutions for velocities ranging from $v = 3/\sqrt{20}$ to $v = 7/\sqrt{20}$ in steps of $\delta v = 0.5/\sqrt{20}$.
>
> What happens as the applied magnetic field is increased? How could you improve the focus? What happens when you decrease the charge to mass ratio to $q/M = 1/25$ and $q/M = 1/30$?

A depiction of the electromagnetic focussing achieved by Aston is illustrated in figure 8.10, where the trajectories of particles with the same

FIGURE 8.10. Charged particles are first deflected by an electric field and then a magnetic field in the indicated zones. Those with large velocities (*black*) are deflected relatively less than those with smaller velocities (*gray*). The trajectories converge after passing through the magnetic field

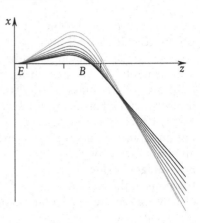

charge to mass ratio but with different velocities are shown. As can be seen from the figure, the trajectories diverge after passing through the electric field but they again converge after passing through the magnetic field. Aston's spectrometer was constructed to place the phosphorescent screen at the focus of the beam. Thus, all particles with a mass ratio of q/M would strike at the same point, independent of velocity. Particles with different charge to mass ratios would strike the screen at different points.

FIGURE 8.11. Aston's mass spectrometer produced results like those depicted above. Spots occurred for integral values of the charge to mass ratio. The broad peak at a relative mass of 28 is due to carbon monoxide (CO)

Aston observed that it was difficult to operate his apparatus with pure gases, so most of the spectra that he produced included contributions from several molecular species. This turned out to be useful in interpreting the results of his experiments. For example, in figure 8.11, we depict a representation of Aston's investigations utilizing chlorine. The broad peak at the mass of 28 can be interpreted as due to the carbon monoxide (CO) gas that was frequently present. The peaks at masses 13–16 were the "carbon ladder" and are interpreted as due to the molecules CH, CH_2, CH_3 and CH_4, where carbon is assumed to have mass 12.

EXERCISE 8.18.   From these data, Aston identified two isotopes of chlorine: ^{35}Cl and ^{37}Cl. To which molecules do the unidentified peaks in the spectrum correspond?

8.4. Neutrons

After conducting numerous experiments, Aston was able to quantify his whole number rule. If the mass of oxygen is taken as 16 exactly, then all atoms and molecules have integral masses.[7] In his spectrographs, Aston sometimes encountered lines at half-integral values of mass but these corresponded to doubly charged ions. For example, chlorine would provide faint peaks at apparent mass values of 17.5 and 18.5. Occasionally, weak lines at third-integral values of mass, corresponding to triply charged ions, could be observed. As Aston refined his instruments and the results became more precise, the whole number rule was repeatedly validated. Of course, why some elements had different isotopes was not understood.

The resolution to the isotope problem was produced in 1932 by the English physicist James Chadwick, who was investigating the curious phenomenon of a new penetrating radiation observed when α particles emitted from a polonium source struck a beryllium target. Before we reveal Chadwick's solution, we should first revisit the principle of momentum conservation.

As discussed in previous chapters, linear momentum is conserved. This means that the total momentum of a system is independent of time. We can express this mathematically as follows:

$$(8.16) \qquad\qquad \frac{d}{dt} \sum \mathbf{p}_i = 0,$$

where the summation extends over all the constituents in the system. So, at any time—before and after a scattering event, for example—the total momentum is constant. In three dimensions and if there are N constituents in the system, then Equation 8.16 provides three equations relating the $3N$ momentum components. This does not provide enough information to specify the final state of the system uniquely. As a result, we must obtain further information about the state of the constituents before we can provide a complete analysis.

At relatively low impact velocities, the kinetic energy may also be conserved; such collisions are called elastic collisions. This is, of course, not a common result in the collisions of macroscopic objects. The familiar click of two billiard balls striking one another is a sound wave that represents energy being radiated away from the collision. Infrared imaging technology employed in cricket matches clearly identifies the "hot spot" where the ball struck the bat. This hot spot is, indeed, hotter than the surrounding portions of the bat and represents the conversion of a

---

[7]The atomic mass unit utilized Aston's definition: $M(^{16}O) = 16$. In the SI system of units, the Dalton (Da) is defined as $1/12$ of the mass of ^{12}C in its ground state.

portion of the kinetic energy in the ball/bat system to thermal energy. For anyone who has played baseball or softball, the stinging sensation produced by striking the ball away from the "sweet spot" of the bat is produced by the vibrational modes of the bat excited during the collision. Additionally, the collision of two automobiles results in remodelling of their aerodynamic forms, representing the work done in reshaping fenders and breaking glass, in addition to the sound of the impact that announces the calamity to those in close proximity. In microscopic systems, however, it is possible for collisions to approximate the elastic ideal, where no kinetic energy present initially is converted into some other form of energy as a result of the collision.

EXERCISE 8.19. Consider the case of a one-dimensional collision. Initially, mass M_1 has a velocity v_0 and mass M_2 is at rest. After the collision, mass M_1 has velocity v_1 and mass M_2 has velocity v_2. If the kinetic energy is also conserved, solve for v_1 and v_2 in terms of v_0 and the masses.

What final velocities v_1 are possible when $M_1 < M_2$? What final velocities v_1 are possible when $M_1 > M_2$?

If we consider the elastic scattering of one object from a second, then it suffices to consider motion in a plane to describe the final state. (This is the result of angular momentum conservation.) In this case, the first particle can be considered to have an initial momentum $\mathbf{p}_0 = p_0 \hat{\mathbf{z}}$. After the scattering event, the first particle has a momentum $\mathbf{p}_1$ and the trajectory asymptotically makes an angle θ_1 from the initial direction, as is illustrated in figure 8.12. The second particle, initially at rest, travels along a direction that is at an angle θ_2 with respect to the initial direction of the projectile.

EXERCISE 8.20. In figure 8.12, we defined the x-z plane to be the plane that contains the trajectory of the first particle. Use momentum conservation to prove that the second particle cannot have a nonzero y-component of its momentum.

Chadwick's problem was that the penetrating radiation emitted from the beryllium target was electrically neutral. This meant that it was not deflected by electric or magnetic fields and, consequently, was exceptionally difficult to analyze. His ingenious solution to the problem was to use collisions between the penetrating radiation and a proton source to convert the hard-to-analyze penetrating radiation into a beam of protons that would be easier to analyze. There was a suggestion at the time that the penetrating radiation could be γ rays. At first blush, it would seem that

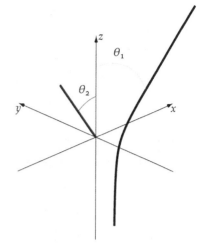

FIGURE 8.12. One particle travels initially along the z-direction before scattering, making an angle θ_1 from its initial direction after the collision. The second particle is scattered at an angle θ_2

Equations 8.16 are not applicable to γ radiation, as light has no mass and, thus, the product mv would vanish. Recall, though, that the relativistic energy of a particle was defined by the following relation:

$$\mathcal{E}^2 = p^2 c^2 + m^2 c^4.$$

Recall also that the energy of a photon was related to its wavelength by Planck's constant: $\mathcal{E} = hc/\lambda$. Combining these two results provides us with the momentum of the photon: $p = \mathcal{E}/c = h/\lambda$. So, we can apply the principles of momentum and energy conservation but, when photons are involved, we shall have to use the relativistic formulas.[8]

We can systematize the mathematics of relativistic kinematics if we use a matrix representation. Consider that the four-momentum is defined as the following column vector:

(8.17)
$$\mathbf{p} = \begin{bmatrix} \mathcal{E}/c \\ p_x \\ p_y \\ p_z \end{bmatrix} \equiv \begin{bmatrix} \mathcal{E}/c & p_x & p_y & p_z \end{bmatrix}^T,$$

where by the superscript T in the last part of the equation we mean the transpose (row vector). The dual vector to the vector $\mathbf{p}$ is obtained by taking the transpose and flipping the signs of the spatial components:

(8.18)
$$\tilde{\mathbf{p}} = \begin{bmatrix} \mathcal{E}/c & -p_x & -p_y & -p_z \end{bmatrix}.$$

In this matrix representation, the dual vector is a row vector. Similar to the situation we encountered when dealing with the lattice vectors and

---

[8]Equation 8.16 is equally valid if we consider the vectors $\mathbf{p}_i$ to be the four-dimensional momentum vectors.

the reciprocal lattice vectors, the magnitude of the four-momentum is obtained by taking the dot product of the dual vector with the vector:

$$\tilde{\mathbf{p}} \cdot \mathbf{p} \equiv \begin{bmatrix} \mathcal{E}/c & -p_x & -p_y & -p_z \end{bmatrix} \begin{bmatrix} \mathcal{E}/c \\ p_x \\ p_y \\ p_z \end{bmatrix}$$

$$= (\mathcal{E}/c)^2 - p_x^2 - p_y^2 - p_z^2,$$

which is just the invariant quantity $m^2 c^2$ defined in Chapter 4.

From figure 8.12, the initial photon has the following four-momentum:

$$\mathbf{p}_\gamma^0 = \begin{bmatrix} h/\lambda_0 & 0 & 0 & h/\lambda_0 \end{bmatrix}^T.$$

Note that $\tilde{\mathbf{p}}_\gamma^0 \cdot \mathbf{p}_\gamma^0 = 0$, reflecting that the mass of the photon is zero.

From figure 8.12, the scattered photon has the following four-momentum:

$$\mathbf{p}_\gamma^1 = \begin{bmatrix} h/\lambda_1 & h/\lambda_1 \sin\theta_1 & 0 & h/\lambda_1 \cos\theta_1 \end{bmatrix}^T.$$

The second particle has mass M_p and is initially considered to be at rest, so its initial four-momentum is just

$$\mathbf{p}_p^0 = \begin{bmatrix} M_p c & 0 & 0 & 0 \end{bmatrix}^T$$

and, in the final state, the four-momentum is given by the following relation:

$$\mathbf{p}_p^1 = \begin{bmatrix} \sqrt{p_p^2 + M_p^2 c^2} & -p_p \sin\theta_2 & 0 & p_p \cos\theta_2 \end{bmatrix}^T,$$

where the magnitude of the three-momentum is p_p. The momenta are conserved, so we must have additional relations amongst them [9]:

(8.19) $$\mathbf{p}_\gamma^0 + \mathbf{p}_p^0 = \mathbf{p}_\gamma^1 + \mathbf{p}_p^1.$$

The first component of the four-momentum conservation equation (Equation 8.19) provides us with the following relation:

$$\frac{h}{\lambda_0} + M_p c = \frac{h}{\lambda_1} + \sqrt{p_p^2 + M_p^2 c^2}.$$

We can solve this for the momentum p_p of the second particle:

(8.20) $$p_p^2 = \left[\frac{h}{\lambda_0} - \frac{h}{\lambda_1} + M_p c \right]^2 - M_p^2 c^2.$$

---

[9]This analysis was applied by the American physicist Arthur Holly Compton to describe the scattering of electrons by γ rays. Chadwick argued that the same basic equations should also apply to other charged masses like protons. Compton was awarded the 1927 Nobel Prize in Physics "for his discovery of the effect named after him." Compton shared the prize with the Scottish physicist Charles Thomson Rees Wilson, who earned the award "for his method of making the paths of electrically charged particles visible by condensation of vapour."

The spatial components of Equation 8.19 provide us with the following relations:

(8.21) $\qquad 0 = \dfrac{h}{\lambda_1}\sin\theta_1 - p_p\sin\theta_2 \quad$ and $\quad \dfrac{h}{\lambda_0} = \dfrac{h}{\lambda_1}\cos\theta_1 + p_p\cos\theta_2.$

If we square both equations and add the result, we can demonstrate that the momentum of the scattered particle can also be written as follows:

(8.22) $$p_p^2 = \frac{h^2}{\lambda_0^2} + \frac{h^2}{\lambda_1^2} - 2\frac{h^2}{\lambda_0\lambda_1}\cos\theta_1.$$

We can equate the results of Equations 8.20 and 8.22 to eliminate the unknown particle momentum. We find then that the wavelength of the scattered photon can be written as follows:

(8.23) $$\lambda_1 = \lambda_0 + \frac{h}{M_p c}(1 - \cos\theta_1).$$

This value is largest (and the energy of the scattered photon smallest) when $\theta_1 = \pi$, i.e., when the photon is scattered back along its initial direction. The factor $h/M_p c$ is known as the Compton wavelength of the particle. The maximum kinetic energy attained by the recoiling particle is then given by the following formula:

$$T_{max} = \frac{hc}{\lambda_0} - \frac{hc}{\lambda_0 + 2h/M_p c}$$

(8.24) $$= \frac{2(hc/\lambda_0)^2}{M_p c^2 + 2hc/\lambda_0}.$$

EXERCISE 8.21. Write the four-momentum vectors $\mathbf{p}_\gamma^i$ and $\mathbf{p}_p^i$ explicitly. Use these to fill in the missing details in the derivation of Equation 8.24. Plot the function $f(x) = 1 - \cos x$ and show that it is maximal at $x = \pi$.

EXERCISE 8.22. The mass of the proton is about $938\,\text{MeV}/c^2$. Plot the maximum kinetic energy of the scattered proton as a function of photon energy over the domain $0 \leq hc/\lambda_0 \leq 100\,\text{MeV}$.

In Chadwick's apparatus, a polonium source was placed in a vacuum chamber near a beryllium disk, as illustrated in figure 8.13. The penetrating radiation generated in the beryllium passed into an ionization detector: a gas-filled cell in which ions produced in the gas by the passage of an ionizing particle generated a measurable current. Chadwick observed that placing a thin sheet of paraffin between the beryllium and detector greatly increased the event rate. He reasoned that this was due to collisions between the penetrating radiation and protons in the paraffin that resulted in the protons being projected forward.

FIGURE 8.13. Radioactive polonium was precipitated onto a thin disk (Po) that was mounted in close proximity to a beryllium disk (Be) and the pair enclosed in a vacuum cell. Energetic particles were detected in the gas cell (D)

By placing a series of aluminum sheets between the paraffin and the detector and observing the subsequent reduction in event rates, Chadwick was able to determine the range of the protons and, thereby, their energy. Chadwick deduced that the most energetic protons had kinetic energies of about 5.7 MeV. From Equation 8.24, an initial photon energy of nearly 55 MeV was required to produce protons with that energy. Such an energy is completely inconsistent with the known energies of nuclear γ rays, where typical energies ranged from about 100 keV to a few MeV. If the protons were being accelerated by photons, it would require photons ten times as energetic as had been observed previously. As a result, Chadwick concluded that the penetrating radiation was not γ rays.

EXERCISE 8.23. Use Equation 8.24 to solve for the initial photon energy hc/λ_0 in terms of the recoil energy $p_2^2 c^2$. Compute the photon energy required to accelerate a proton ($M_p = 938\,\text{MeV}/c^2$) to an energy of 5.7 MeV.

Chadwick recognized that if the penetrating radiation had mass, then the requisite initial energy would be lower. Utilizing Aston's whole number rule as a constraint, Chadwick then suggested that the penetrating radiation was an *electrically neutral* body with the same mass (approximately) as the proton.[10] Today we refer to these particles as neutrons. If we look back to the results of Exercise 8.19, we find that if the two colliding bodies have the same mass, then the kinetic energy of the initial particle can be entirely transferred to the second particle. This solves the problem of the observed proton energies in Chadwick's experiments. The radiation from the beryllium disk consisted of neutrons, which have approximately the same mass as the proton. It is also possible that these neutrons could possess kinetic energies of 5–6 MeV, typical of the observed energies in nuclear systems. When such neutrons encounter protons in the paraffin, some will exchange (nearly) all of their energy with the protons.

---

[10]Chadwick was awarded the Nobel Prize in Physics in 1935 "for the discovery of the neutron."

As a result, we have arrived at the modern model of the nucleus. It is composed of some number of positively-charged protons that determine the element and an additional number of neutrons. It is possible to have different numbers of neutrons associated with each number of protons: these are the observed isotopes. We shall refrain here from producing a drawing of a cluster of two-differently colored grapes to represent the nucleus. Such drawings are ubiquitous but are hugely misleading. As we have mentioned previously, neutrons are waves. An imaginary super microscope capable of resolving the nucleus would not observe a defined surface any more than zooming in on the edges of a cloud in the sky will delineate the cloud surface.

At the time of Chadwick's experiments, the scientific consensus was that the nucleus was composed of some number of protons equal to the atomic mass and a number of additional protons bound somehow to electrons that produced the particular isotope. In his initial papers, Chadwick's displays his conception that the neutron was, indeed, was some sort of bound state of a proton and an electron. He reasoned that the mass of the neutron therefore should be roughly the sum of the proton and electron masses. In fact, we know today (2010 CO-DATA tables) that the neutron mass is $M_n = 939.565379(21)\,\text{MeV/c}^2$, the proton mass is $M_p = 938.272046(21)\,\text{MeV/c}^2$ and the electron mass is $M_e = 0.510998928(11)\,\text{MeV/c}^2$. So, the neutron has a mass that is the same as the proton's to within about a part per thousand and it is somewhat larger than the sum of the proton and electron masses.

In his original experiments with radioactive decay, Rutherford identified two different components of the radiation emanating from uranium salts that he called α and β. We have seen that the α particles are just the nuclei of ^{4}He atoms and β particles are, in fact, electrons. Chadwick's vision of the neutron as a bound proton and electron provides us with a simple explanation for β radiation: it is the result of the neutron decaying into its constituents. This decay is the hallmark of what physicists term the fourth fundamental force: the weak nuclear force.

EXERCISE 8.24. If a neutron at rest decays into an electron and proton, what is the kinetic energy of the electron in the final state? What is the kinetic energy of the proton in the final state? Hint: momentum must be conserved.

8.5. The World is not so Simple

In some sense, we have now reached the end of the trail that began with our first studies of gravitating masses. We have encountered, at least, some aspects of the electromagnetic force that binds the nuclei and electrons into atoms. We have recounted Rutherford's discovery that the nucleus was an amazingly smaller object than anyone anticipated and interpreted this result as evidence for the existence of a strong nuclear force that binds the positively-charged protons into such a small volume. The discovery of the weak nuclear force (that accounts for neutron decay) completes the modern picture of the fundamental forces of nature. There are just four.

Chadwick's discovery of the neutron and the subsequent discoveries that neutrons can, indeed, decay into protons and electrons seems, at first glance, to provide further evidence of the rather tidy picture we have just described. In truth, experimenters investigating β decay were almost immediately confronted with a conundrum. When measuring the energies of the emitted β particles, they observed a continuous distribution of electron energies like that observed in figure 8.14. If we are to think of the neutron as some sort of bound proton and electron, then the neutron decay process in the nucleus should be approximately described by the formulas derived in Exercise 8.19. In fact, a first approximation would consider the neutron to be initially at rest. In this case, the momentum of the electron would be opposite to that of the recoiling proton (or nucleus) and the kinetic energy of the electron would be vastly greater than that of the proton. Effectively, all of the kinetic energy would be carried away by the electron.

One might suppose then, that the electron spectrum would be a large peak at the mass difference between the parent and daughter nucleus. One might further suppose that structure in the spectrum would provide some sort of insights into how neutrons were actually bound in the nucleus. No one expected the results of figure 8.14 because it represents a clear case of momentum non-conservation!

FIGURE 8.14. The electron energy spectrum from the β-decay of ^{210}Bi is continuous. The number of electrons counted in a specific interval (in arbitrary units) is plotted versus the energy of the emitted electrons

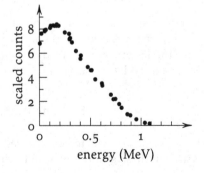

EXERCISE 8.25. Use the results derived in Exercise 8.19 and use the masses $M_n = 939.565$ MeV, $M_p = 938.272$ MeV and $M_e = 0.511$ MeV. What are the expected kinetic energies T_p and T_e?

A potential solution to the dilemma was proposed by the Austrian physicist Wolfgang Pauli in 1930. Pauli suggested that there existed a very light, neutral particle that was also emitted during the β decay process. This new particle, which Pauli originally called the neutron, would also carry away some of the kinetic energy released during the decay. The particle was eventually renamed the **neutrino** by the Italian physicist Enrico Fermi after Chadwick's discovery of the neutral particle that Chadwick also called the neutron. Because Pauli's particle was very light (possibly massless), Fermi decided that the Italian diminutive *ino* should be added to the root neutron to distinguish the very different entities.

In some sense, Pauli's solution is quite elegant. In converting the two-body decay of the neutron into a three-body decay, he could explain the continuous spectrum of β decay electrons. On the other hand, Pauli had just proposed a massless, chargeless, colorless, flavorless particle whose sole reason for existence was to preserve the principle of momentum conservation. In this light, one might also characterize Pauli's neutrino as a desperate, nonsensical attempt to bolster a principle of physics that was in direct conflict with experiment. In fact, Pauli's original suggestion in 1930 came in the form of an open letter to the "Dear radioactive ladies and gentlemen" of the Physical Institute at the ETH in Zürich. Pauli himself was concerned that his suggestion would not survive the review process that was required to publish his idea in a scientific journal.

Indeed, other physicists were not terribly enthusiastic about Pauli's neutrino initially. Given the implausibility of such an entity, notable physicists of the day entertained the idea that momentum conservation might not hold at microscopic scales. Further experimentation, though, revealed that the nature of nuclear matter was significantly more complex than anyone had envisioned in 1930. A key instrument used in enabling physicists to visualize the complexity was the bubble chamber, invented by the American physicist Donald Glaser in 1952.[11] This apparatus provided physicists with the ability to visualize particle trajectories directly and provides another example of the use of the magnetic force to study the spectrum of particle masses.

The bubble chamber consists of a large volume of fluid, usually liquid hydrogen, under high pressure. Just before a beam of charged particles enters the chamber, a large piston is cycled and the pressure in the

---

[11]Glaser was awarded the Nobel Prize in Physics in 1960 "for his invention of the bubble chamber."

chamber reduced dramatically. This leaves the fluid in a supercritical state. When charged particles then enter the chamber, they scatter electrons in the fluid, depositing energy into the fluid and causing the liquid to boil. The trajectory of a charged particle is thereby marked by a trail of bubbles. Photographs of the bubble trails taken at different angles permit researchers to perform three-dimensional reconstructions of the trajectories. Recycling the piston and repressurizing the chamber erases the bubbles and readies the system for the next batch of beam particles. Glaser's original prototype was a few centimeters in diameter; later versions expanded to several meters in diameter.

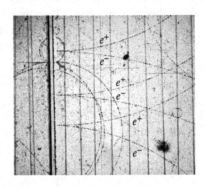

FIGURE 8.15. In this negative image, γ rays enter the bubble chamber from the left. Upon striking a lead sheet (about 1/4 of the way from the left edge of the image), three of the γ rays produce V-shaped tracks that indicate the production of positron-electron (e^+e^-) pairs (Image provided courtesy of Lawrence Berkeley Laboratory)

An example of a bubble chamber photograph is shown in figure 8.15, where γ rays enter from the left side of the image.[12] The γ rays are uncharged and do not leave tracks in the chamber. Upon striking a thin lead sheet, three of the γ rays are converted into charged particle pairs that leave V-shaped tracks.[13] The magnetic field in this case is directed into the page, so the deflections up and down represent two different charges for the created particles. The radii of curvature are similar in each case, so each particle carries the same momentum. Knowing the energies of the initial γ rays, it is possible to infer that the negative particles are electrons (e^-). That means that there also exist positively charged particles that possess the same mass as the electron. These are called positrons (e^+) and represent a new state of matter that has been given the name antimatter. Apparently, all particles possess antiparticle analogs and one of

---

[12]This reproduction is actually a negative image that has better contrast when printed. The original images have bright trails of reflected light from the bubbles against a black background.

[13]Note that we use here the common word *particle* to describe the various states of matter. As we have indicated previously, these particles are inherently waves, so one should not attempt to produce a mental image of these fundamental states of matter as small spheres or beads, despite the very tempting inclination to interpret the bubble trails as having substance.

the most significant open questions in physics is why the universe is filled
with matter and not antimatter.

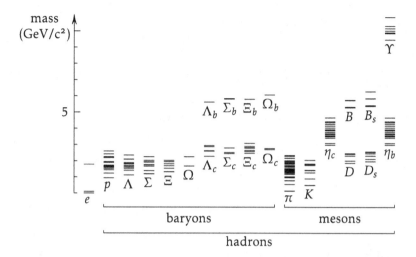

FIGURE 8.16. The list of known particles includes the electron and
other leptons such as the neutrinos and the muon. Particles that interact through the strong nuclear force are called hadrons. These are
subdivided into two main categories: the baryons and the mesons

In addition to the protons and neutrons that are the constituents of atomic
nuclei, there are a host of other particles that also interact via the strong
nuclear force. None of these particles is stable; all decay quite rapidly.
So, particle is perhaps not a particularly good word to describe these
excitations of nuclear matter. Nevertheless, it is the word most commonly
utilized. A sketch of the particle mass spectrum is provided in figure 8.16.
Recall that in our previous discussions (like those in figures 5.14 and 8.11)
we used the word spectrum to mean a plot of intensity versus frequency
or, equivalently, wavelength. Here, the vertical axis represents the masses
of the different particles and, in Einstein's relativistic approach to dynamics, mass is energy. We have seen that the frequency and wavelength are
also related to the energy through Planck's constant, so our terminology
of "mass spectrum" is reasonable.

In figure 8.16, particles with similar properties have been grouped along
the horizontal axis. The first column (labelled e) contains the leptons,
those are particles that do not interact via the strong nuclear force. This
column includes the electron and Pauli's neutrino, along with heavier particles known as the muon and tau lepton. (On this scale, where the mass
of the proton is approximately 1 GeV/c², the 0.511 MeV/c² electron is

indistinguishable from the (nearly) massless neutrino(s).) The remaining columns of particles are those that interact via the strong nuclear force and are known collectively as hadrons.[14] These have been classified into lighter particles like the pions and kaons, known as mesons, and heavier particles, known as baryons, that include protons and neutrons and a host of others. The mesons and baryons have been further grouped into categories that are described by what is now called the Standard Model of elementary particle physics. Without going into further detail at this time, suffice to say that physicists quickly ran out of Greek letters to describe all of the different excitations of nuclear matter.

Bubble chambers provided a number of key images that helped refine our ideas about nuclear matter. For example, the image depicted in figure 8.17 illustrates the production of the Ξ^- hyperon ($M_\Xi = 1321.71(07)\,\text{MeV}/c^2$) by a collision between an incident kaon K^- and a proton. In examining bubble chamber photographs, the first order of business is to establish the direction of the magnetic field. This is easily accomplished by finding the spiral tracks of (relatively) low-energy electrons. These tracks are inevitably created by interactions between beam particles and electrons in the bubble chamber. One such track in the image is found near the center, between the registration marks 1 and 2. As the electron traverses the bubble chamber, the Lorentz force bends the path into a circle, as we have described previously. A significant portion of the electron's energy is lost to bubble formation, so the electron loses energy as it travels and the radius of curvature decreases. Hence, the electrons leave characteristic spiral tracks. Knowing that the electron has a negative charge, it is possible to determine that the magnetic field in figure 8.17 must be directed out of the page.

> EXERCISE 8.26. Electrons traversing the fluid in the bubble chamber experience a resistive force that scales like the square of the velocity. Construct a numerical solution of the electron trajectory by modifying the equations from Exercise 8.9 to include a velocity-dependent, resistive force. Demonstrate that the helical trajectories that arise without loss become spiral trajectories when loss is included.

Having now established the direction of the magnetic field, we can further identify the charges of the various particles by the curvature of their trajectories. Particles with negative charges will bend in the same direction as the electrons and particles with positive charges will bend in the

---

[14]Originally, the classifications utilized the Greek roots λεπτός (thin or slender), μέσος (middle), βαρύς (heavy) and ἁδρός (thick) to indicate the masses of the particles. As physicists continued their experiments, mesons were discovered that were heavier than the proton, so the term meson no longer refers to the mass of the particle.

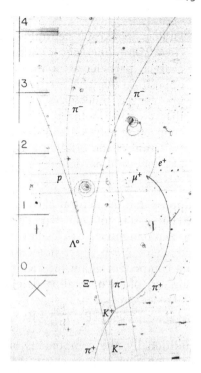

FIGURE 8.17. In this image, a kaon K^- entered the bubble chamber from the bottom. Upon striking a proton in the chamber, the K^- was converted into a positive kaon K^+ and the proton into a negative hyperon Ξ^-. The hyperon then decays into a neutral Λ° and a π^-. The Λ° does not leave a track but decays into a proton p and a π^-. In the bottom of the image, the scattered K^+ decays into three pions: $\pi^+\pi^+\pi^-$. One of the π^+ mesons then decays into a muon (μ^+) that subsequently decays into a positron (e^+) (Image courtesy of Lawrence Berkeley Laboratory)

opposite direction. While the assignment of particular particles to the various tracks in the figure has not yet been explained, it is obvious how the charge assignments were made.

EXERCISE 8.27. There is an unidentified track just to the right of the K^- that traverses the entire image without interacting. From the curvature, what is the charge of this particle?

Figure 8.17 represents one of the earliest records of the existence of the Ξ^- particle (historically called the cascade hyperon). It is created in the reaction $K^- p \rightarrow K^+ \Xi^-$ that occurs near the bottom of the image. The track identified as the Ξ^- trajectory terminates with a kink that represents the decay of the hyperon into daughter particles $\Xi^- \rightarrow \Lambda^\circ \pi^-$. The Λ° baryon is a neutral particle and does not leave a track in the chamber. Its existence can be inferred from the tracks left by its decay products: the proton and pion. At higher resolution, these daughter particle tracks can be seen to cross, like the tracks depicted in figure 8.18.

Figure 8.18. A neutral particle (*dashed line*) decays into two
charged particles at point A. The
oppositely-charged particles follow curved trajectories (1 and 2)
that cross at point B. The line
connecting points A and B is a continuation of the original trajectory
of the neutral particle

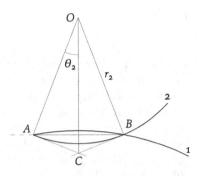

Suppose that the neutral particle has three-momentum $\mathbf{p}_0$. Because it has
no charge, it is not deflected in a magnetic field. When the particle decays,
the two daughter particles have three-momenta $\mathbf{p}_1$ and $\mathbf{p}_2$, respectively,
and where $\mathbf{p}_0 = \mathbf{p}_1 + \mathbf{p}_2$. Momentum conservation means that the momentum component along the original direction of motion (the unobserved
trajectory of the neutral particle) must be unchanged after the decay:

$$(8.25) \qquad |\mathbf{p}_0| = \mathbf{p}_1 \cdot \frac{\mathbf{p}_0}{|\mathbf{p}_0|} + \mathbf{p}_2 \cdot \frac{\mathbf{p}_0}{|\mathbf{p}_0|} = |\mathbf{p}_1| \cos\theta_1 + |\mathbf{p}_2| \cos\theta_2.$$

Additionally, the components of momentum transverse to the original
direction of motion must vanish:

$$(8.26) \qquad 0 = \left[\mathbf{p}_1 - \frac{\mathbf{p}_1 \cdot \mathbf{p}_0}{|\mathbf{p}_0|^2} \mathbf{p}_0 \right] + \left[\mathbf{p}_2 - \frac{\mathbf{p}_2 \cdot \mathbf{p}_0}{|\mathbf{p}_0|^2} \mathbf{p}_0 \right] = |\mathbf{p}_1| \sin\theta_1 + |\mathbf{p}_2| \sin\theta_2.$$

Consider the track made by particle 2 in figure 8.18. If we ignore losses
and assume a constant radius of curvature r_2, then the initial momentum
vector of daughter particle 2 lies along the line connecting the points A
and C. (This line is the tangent to the trajectory at point A.) If the distance
between A and B is L, then $\sin\theta_2 = L/2r_2$. The radius of curvature is
defined by the following:

$$(8.27) \qquad\qquad\qquad r_2 = \frac{|\mathbf{p}_2|}{q|\mathbf{B}|},$$

where q is the particle charge and $|\mathbf{B}|$ is the magnitude of the magnetic
field. We then find that

$$(8.28) \qquad\qquad |\mathbf{p}_2 \sin\theta_2| = |\mathbf{p}_2| \frac{qL|\mathbf{B}|}{2|\mathbf{p}_2|} = qL|\mathbf{B}|/2.$$

EXERCISE 8.28. The line segment AC in figure 8.18 is tangent to the
trajectory at point A. Show that the angle BAC is equal to the angle
$AOC \equiv \theta_2$. Show that the angle θ_2 is half of the angle AOB.

EXERCISE 8.29. Show that, if particle 1 has an initial angle θ_1 and radius of curvature r_1, that $|\mathbf{p}_1 \sin \theta_1| = qL|\mathbf{B}|/2$. Show that this implies that the line A–B lies along the vector $\mathbf{p}_0/|\mathbf{p}_0|$.

The bubble chamber image represents a projection of the three-dimensional trajectories onto a plane. The above argument is still valid in this case. If we take the original momentum direction to be the z-direction, for example, then the transverse momentum would lie in the x-y plane. The requirement that the sum of the transverse momenta vanish in the final state must hold for all components: it must be true in both the x- and y-directions separately. As a result, in the particular view of the trajectory in the image (call it the x-z plane) the line connecting the points A and B, will indeed provide a continuation of the (invisible) trajectory of the neutral particle, as projected onto the x-z plane.

EXERCISE 8.30. Use the numerical model created in Exercise 8.26 to study what happens when the losses are not negligible. Create two trajectories of oppositely charged particles, subject to the same resistive force. Is it still the case that the particle crossing points can be used to infer the direction of the initial momentum?

The problem of assigning particles to the individual trajectories is not trivial. In a keynote speech at the Rochester Conference (a scientific meeting devoted to particle physics), the theoretical physicist Victor Weisskopf once famously displayed a completely blank cloud chamber[15] photograph that he claimed provided evidence for a previously undiscovered neutral particle that decayed into two neutral particles, thus explaining the lack of any tracks whatsoever. Reportedly, Weisskopf's joke elicited hearty laughter from the audience. We can assume that this was due, in part, to Weisskopf's backhanded admission that analysis of bubble chamber images is a challenging enterprise.

To make assignments, researchers would try several different scenarios and then select the most likely explanation. In a sizable fraction of cases, the assignments cannot be performed unambiguously, particularly in situations where the final state included a number of neutral (and thereby invisible) particles. Consider, for example, analysis of the initial decay event in figure 8.17. The original beam of K^- mesons was produced by collisions of a high-energy proton beam that impacted a copper target. The K^- particles were selected by bending them through a magnetic field before sending them through the bubble chamber.[16] As a result, we know

---

[15]The cloud chamber was the forerunner of the bubble chamber. It was filled with gas not fluid and produced far fewer events than the bubble chambers.

[16]There is a recurrent theme here.

the initial state consisted of a K^- with a particular kinetic energy and a proton that we can assume to be essentially at rest.

Suppose now that we want to consider that the reaction actually produced a final state in which there were three particles A, B and C, and C was a neutral particle that did not leave a track. Conservation of the four-momentum of the system requires the following relation:

$$\mathbf{p}_K - \mathbf{p}_p = \mathbf{p}_A + \mathbf{p}_B + \mathbf{p}_C,$$

where the $\mathbf{p}_i$ are the four-momenta of the particles. The momentum of the track that is labelled Ξ^- in figure 8.17, can be determined by measuring the radius of curvature. Similarly, we can also obtain the momentum of the track labelled K^+. For this particular example, the measured momenta of the Ξ^- and K^+ tracks in figure 8.17 are equal (to within the experimental precision) to the momentum of the initial K^-, thereby excluding a possible neutral third particle.

The experimenters then tried various different combinations of choices for A and B, before settling on Ξ^- and K^+. Other possible choices could be excluded on the basis of kinematic grounds. For example, some of the much heavier particles identified in figure 8.16 could not have been produced, given the initial K^- energy. Further constraints on the identification of the tracks can be inferred from the subsequent behavior of the particles. For example, the K^+ track ends in a three-particle spray that is identified as three pions. One decay mode of the charged kaons that occurs about six percent of the time is three pions. Few other particles decay into three charged daughter particles so this is something of a characteristic feature that helps to identify the track as that of a charged kaon.

> EXERCISE 8.31. The mass of the charged kaon is now (2010) known to be $M_{K^\pm} = 493.677(16)\,\mathrm{MeV}/c^2$ and the masses of the charged pions are $M_{\pi^\pm} = 139.57018(35)\,\mathrm{MeV}/c^2$. If we consider a decay from a frame in which the K^+ is initially at rest, is it possible for the kaon to decay into three pions? If so, what kinetic energy is available to the pions in the final state?

> EXERCISE 8.32. In the decay shown in figure 8.17, if the initial momentum of the K^- is $p_{K^-} = 1.2$ GeV/c, is it possible to produce an Ω^- particle and a K^+ in the final state? The mass of the Ω^- is $M_{\Omega^-} = 1672.45(29)\,\mathrm{MeV}/c^2$.

Bubble chambers have been eclipsed by newer detection methods that are more amenable to automatic computer analysis. Analysis of the photographs, measuring tracks and merging the results from multiple camera images, was a very labor-intensive task. Experimental groups often employed hundreds of part-time, undergraduate students to assist in the monumental task of sorting through the millions of images that were generated in a single experiment. Professional physicists were not used much in this part of the data analysis because they tended to find "interesting" things to study in every image. Consequently, physicists were much less efficient than non-physics-major students, who were given specific instructions to look for characteristic features like Vs or the tridents of three-pion decays. After these initial screens, interesting events were measured and then analyzed to try to make the track assignments.

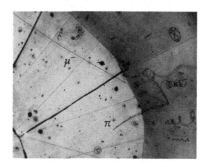

FIGURE 8.19. In this image, a neutrino entered the bubble chamber from the right. Upon striking a proton in the chamber, the neutrino was converted into a spray of charged particles, revealing its existence (Image provided courtesy of Argonne National Laboratory)

Despite their limitations, bubble chambers did play an important rôle in helping physicists construct the current picture of the nature of nuclear matter and, in fact, confirmed Pauli's speculation about the existence of a new, neutral particle. Figure 8.19 represents the first observation of a neutrino event in a bubble chamber, which was obtained in November of 1970 in the 12-foot bubble chamber at Argonne National Laboratory. The reaction here is $\nu p \rightarrow \mu^- p \pi^+$. The spiralling electron trajectories throughout the volume provide the direction of the magnetic field. The neutrino entered from the right hand side of the image but, of course, did not leave a track. We can infer its presence from the recoil proton and the production of the μ^- and π^+. The chamber was shielded externally, so that other neutral particles, such as γ rays can be excluded as the source of the event.

EXERCISE 8.33. What is the direction of the magnetic field in figure 8.19? Can you explain the charge assignments in the final state?

TABLE 8.1. Particle lifetimes

Leptons	τ (s)	Mesons	τ (s)	Baryons	τ (s)
$\mu^\pm$	2.2×10^{-6}	$\pi^\pm$	2.6×10^{-8}	Ξ	1.6×10^{-10}
$\tau^\pm$	2.9×10^{-13}	$K^\pm$	1.2×10^{-8}	Λ°	2.6×10^{-10}
		π°	8.5×10^{-17}	Ω^-	8.2×10^{-11}

Exercise 8.34. It is not a bad approximation to assume that most of the particles in bubble chamber interactions are traveling near the velocity of light. From the lifetime data in Table 8.1, how far would each travel in a lifetime? (What is $c\tau$?) Very short-lived particles would not leave visible tracks in a bubble chamber; this is another reason why bubble chambers fell out of favor.

IX

The Classical Electron

In 1820, the Danish physicist Hans Christian Ørsted was performing a demonstration of the temperature rise in a wire connected to a voltaic cell. It was his intent to also subsequently demonstrate magnetic phenomena and, so, had a compass in close proximity to the wire. In the midst of his demonstration, Ørsted observed that connection of the wire to the voltaic cell caused the compass to deflect from its normal position. After several months of further experimentation, Ørsted was able to confirm that electrical currents generated magnetic fields and published his results.[1] In September of 1820, the French physicist Dominique François Jean Arago presented Ørsted's paper to the French Academy. Among those in attendance were the French physicists André-Marie Ampère and Jean-Baptiste Biot. Within weeks of Arago's presentation, Ampère had constructed a mathematical relation between the magnetic field and currents and Biot, with his assistant Félix Savart, had derived an approach to compute the magnetic field from a current distribution.

This episode from the history of physics should serve to reinforce our previous assertions that physics is an experimental science. Ørsted and others had sought connections between electrical and magnetic phenomena for some time before Ørsted conducted his fateful demonstration that finally solidified the connection. What Ørsted observed was that the magnetic field created by the current in the wire was not directed along the wire or even away from the wire. Instead, the field was directed perpendicular to both of those directions!

9.1. Currents

In developing their theories of electromagnetism, researchers of the day drew on their previous experiences, such as from heat conduction and

---

[1]Ørsted wrote a short note in Latin "Experimenta circa Efficaciam Conflictus Electrici in acum Magneticam" that he sent to the leading academic societies in Europe.

© Mark A. Cunningham 2015
M.A. Cunningham, *Neoclassical Physics*, Undergraduate Lecture Notes in Physics, DOI 10.1007/978-3-319-10647-2_9

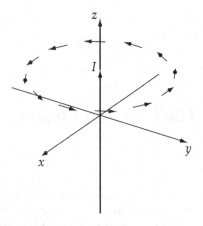

Figure 9.1. For a current I aligned with the z-axis, an array of compasses arranged in the x-y plane will align in a circular pattern. At each compass, the magnetic field of the wire points in the direction perpendicular to the current and the direction vector from the wire to the compass

fluid dynamics. As a result, many of the concepts applied to electromagnetic phenomena have analogous formulations in other fields. For example, the British physicist Michael Faraday is generally credited with popularizing the idea of utilizing field lines as a means of visualizing the abstract electric and magnetic fields. An electric field line represents the path that an infinitesimal charge would follow if released from rest in a region of space where an electric field is present.[2] The use of field lines should, of course, be familiar to anyone who has studied fluid mechanics, where *stream lines* are utilized to visualize fluid flow. These are, again, the paths that small particles would follow if immersed in the fluid.

Similarly, magnetic field lines represent the paths followed by infinitesimal magnetic charges (if such things actually existed). Note that the magnetic field lines *do not* represent the path followed by an electric charge subject to the Lorentz force. As we have observed, electrically-charged particles will spiral around the direction of the magnetic field; using the Lorentz force to define magnetic field lines leads to complex visual representations when the fields have a simple spatial structure. Hence, Faraday simplified matters by defining the magnetic field lines in terms of fictitious magnetic charges. From figure 9.1, we can observe that the magnetic field line at the radius at which the compasses are located would form a closed circle, tangent to the compass directions. In this instance, the magnitude of the magnetic field is constant along the field line but this is not true in general.

Another concept adapted from fluid dynamics is the electrical current. The name alone conjures the image of charged particles coursing through

---

[2]Strictly speaking, the field lines are constructed from the tangent vectors to the fields at each point in space. The two constructs are essentially the same for a particle with infinitesimal mass and charge.

a wire just as water flows through pipes and river channels. One might be led to fear that rapid disconnection of an electrical circuit could give rise to a stream of electrons spilling out onto the floor, as one might expect if a water hose is disconnected before the shut-off valve is completely closed. Such an expectation is completely out of place, as is the visual picture of a current as a stream of electrons flowing through a wire like water through a pipe. Nevertheless, current is the word used to describe the phenomenon, so we shall have to discover what we mean by the word.

As we have mentioned previously, Alessandro Volta invented what we today call the battery (voltaic pile). His stable electrical source provided the means for systematic studies of the behavior of electrical phenomena. The German physicist Georg Simon Ohm conducted just such a series of experiments on electrical circuits and formulated a theory of their behavior.[3] Ohm concluded from his studies that the battery exerted a force on charged particles that would cause them to circulate but that materials resisted the flow. In modern terminology, Ohm's law can be stated mathematically as follows[4]:

$$(9.1) \qquad \mathbf{J} = \sigma \mathbf{E},$$

where $\mathbf{J}$ is the current density and $\mathbf{E}$ is the electric field. The quantity σ is a material-dependent proportionality factor that is called the *conductivity*.

If we consider a volume of space V, containing a distribution of charges $\rho(\mathbf{r}, t)$, then the total charge in the volume is given by the following:

$$(9.2) \qquad q(t) = \int_V d^3\mathbf{r}\, \rho(\mathbf{r}, t).$$

We should pause to consider the meaning of this expression. First, we recognize that matter is composed of atoms, which are, in turn, composed of nuclei and electrons. So, by the charge distribution ρ, we really mean the sum of all of the charged entities—nuclei and electrons—within the volume, as sketched in figure 9.2. As such, the distribution is not a continuous function in the strict mathematical sense. Formally, the mathematical limiting process by which we define derivatives and integrals does not exist when charge is discrete. Yet, we use the integral sign in the equation instead of a summation because macroscopic quantities of matter possess something like Avogadro's number of atoms. In this case, the addition or subtraction of a few nuclei or electrons amounts to an infinitesimal

---

[3]In 1827, Ohm published his pamphlet *Die galvanische Kette mathematisch bearbeit* while on a leave of absence from his high school teaching duties in Berlin. While it can be considered a landmark in physics history, the publication did not lead to any immediate job offers from German Universities. Ohm's dream of a professorship did not materialize until 1849.

[4]Note that we are reusing the symbol $\mathbf{J}$ but its meaning is something completely different from the original definition of angular momentum in Chapter 2.

change in the total charge. Thus, as a practical matter, treating the charge distribution as a continuous function is a reasonable approximation.

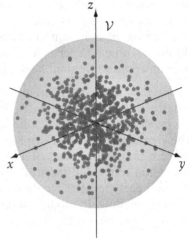

FIGURE 9.2. The distribution of charges $\rho(\mathbf{r}, t)$ within a volume $\mathcal{V}$ can be approximated by a continuous function

If we consider the volume $\mathcal{V}$ to be constant in time, then the total charge will be constant as well, unless charge leaves the volume. We can state this mathematically as follows:

$$(9.3) \qquad \frac{dq(t)}{dt} = \int_{\mathcal{V}} d^3\mathbf{r} \, \frac{\partial \rho(\mathbf{r}, t)}{\partial t} + \int_{\partial \mathcal{V}} d\mathbf{a} \cdot \mathbf{J}(\mathbf{r}, t) = 0,$$

where the current density $\mathbf{J}$ represents the charge flux through the boundary of the volume $\partial \mathcal{V}$. Equation 9.3 serves as the defining equation for the current density $\mathbf{J}$. It also serves as a statement of the conservation of charge. As we shall see shortly, this relation is a consequence of Noether's theorem applied to electromagnetic phenomena. The Maxwell equations that define the nature of the fields possess a symmetry beyond that of just translational and rotational invariance; the result of this so-called gauge invariance is charge conservation. This connection between conserved quantities and symmetries of the equations of motion lies at the heart of modern physical theories.

As a practical matter, there are some instances in which the electrical current is precisely a number of charged particles moving with a nominal velocity in a particular direction. This happens, for example, in particle accelerators, where protons and/or electrons are formed into beams and accelerated to high energies as a prelude to conducting experiments on the nature of matter. It happens also in outer space, where charged particles stream in all directions, accelerated by the fields that arise in the vicinity of stars and heavier objects. These currents undoubtedly align more closely to the other uses of the word current and with one's initial prejudices.

In a wire, though, the physical picture is much more complicated. Firstly, the atoms that make up the wire are generally electrically neutral. The number of electrons balances the charges of the nuclei. Electrons feel a strong electrical force due to nuclear charge and also from the charges of neighboring electrons. An applied electrical field $\mathbf{E}_{app}$, it would seem, must be somehow comparable to that of the fields to which the electrons are already subjected in the atoms, in order for there to be any net transport of charge. Indeed, we can make the further observation that not all materials will support current flows (at least with the sort of voltages obtained by batteries). We usually designate the materials that do not conduct currents as **insulators** and the materials that will support current flows as **conductors**.

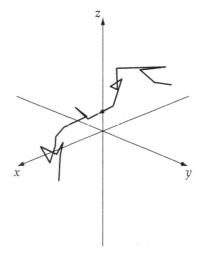

FIGURE 9.3. The path of a charged particle subject to an electric field in the x-direction will not follow streamlines due to interactions with scattering sites within the material. Instead, the path will be a rather tortuous one

In 1900, the German physicist Paul Karl Ludwig Drude published a theory of electrons in metals that provides an explanation for Ohm's earlier results. If we think of applying an electric field in, say, the x-direction, then an electron would nominally experience an acceleration in that direction. On its path, however, the electron will frequently encounter the nuclear centers that make up the material (and other electrons) and scatter from those sites, as depicted in figure 9.3. As a result, the net motion along the x-direction is reduced from that we would expect with simple accelerated motion in a vacuum. Drude recognized that the equations of motion of such a particle could be represented as follows:

(9.4)
$$m\frac{d\mathbf{v}}{dt} = q\mathbf{E}_{app} - \alpha\mathbf{v},$$

where α is a parameter that represents a *resistive* force acting on the electron. As we have already seen, such a force leads to an exponential evolution of the velocity to a constant final value $\mathbf{v}_f$ at which point the force due to the applied electric field is balanced by the resistive force.

> EXERCISE 9.1. Solve the one-dimensional version of Equation 9.4 for the case where the initial velocity is $v(0) = 0$. Plot the solution as a function of time. What is the asymptotic velocity in terms of m, q, α and the magnitude of the field $|\mathbf{E}_{app}|$? What are the dimensional units of α?

Further, what Drude recognized is that the motion of a number of electrons through the metal constituted a current. If there are n electrons per unit volume of material, then the current density is given by the following expression:

$$\mathbf{J} = nq\mathbf{v}_f.$$

Using the result from the previous Exercise, we can rewrite this as follows:

(9.5)
$$\mathbf{J} = \frac{nq^2}{\alpha}\mathbf{E}_{app}.$$

This is just Ohm's Law, with the identification that the conductivity σ is related to the physical properties: the density of charge carriers n, the electron charge q and the parameter α.

> EXERCISE 9.2. The parameter α in Equation 9.4 must have units of M/T in order for the equation to be dimensionally correct. A sensible choice for the mass would, of course, be the mass of the electron. A reasonable interpretation of the time associated with α would be the average time τ_{avg} between scattering events in the electron trajectory. So let us assume $\alpha = m_e/\tau_{avg}$. A mole of copper has a mass of 63.546 g, a density of 8.96 g/cm³ and a conductivity of 4×10^7 S/m at room temperature. Use this information to compute the number of charge carriers per unit volume n. Assume that there is one conduction electron per atom. What is the characteristic time τ_{avg}? If a typical current density in copper is of the order of $J = 10^6$ C/s·m², what is the magnitude of the characteristic velocity $|\mathbf{v}_f|$?

As one can see from the previous exercise, the current density does lead to the net transport of charge but, in conducting materials like metals, the picture is not that of electrons travelling through the wire at high velocities. Instead, the transport velocity $\mathbf{v}_f$, often called the drift velocity, of any individual electron is quite small. Appreciable currents arise because large numbers of electrons move in response to the applied field.

9.2. Magnetic Fields

Shortly after Ørsted's experimental results became public knowledge, Biot and Savart produced the magnetic equivalent of Coulomb's Law. They recognized that the form of the equation had to be modified to include the observation that the magnetic fields were perpendicular to both the source current and the direction from the source. This they achieved by utilizing the properties of the vector cross product. In our modern terminology, we express the Biot-Savart equation as follows:

$$(9.6) \qquad \mathbf{B}(\mathbf{r}_2) = \frac{\mu_0}{4\pi} \int d^3\mathbf{r}_1 \, \frac{\mathbf{J}(\mathbf{r}_1) \times (\mathbf{r}_2 - \mathbf{r}_1)}{|\mathbf{r}_2 - \mathbf{r}_1|^3},$$

where μ_0 is a proportionality factor called the *magnetic permeability* and the integral extends over the support of the current density $\mathbf{J}$, i.e., where the current is non-vanishing.

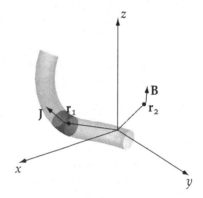

FIGURE 9.4. A current density $\mathbf{J}$, depicted as the *gray tube*, will generate a magnetic field. The contribution of a small segment (*dark gray ring*) of the current density produces a magnetic field that is perpendicular to both $\mathbf{J}$ and the vector $\mathbf{r}_2 - \mathbf{r}_1$

An illustration of the magnetic field produced by a small current element is provided in figure 9.4. At the point $\mathbf{r}_2$, the total magnetic field would be obtained by integrating over all of the current elements. In general, such integrations are quite involved but for a few simple cases we can produce analytical results.

As Georg Ohm noticed, electrical currents are largely restricted to conducting bodies. In the circuits he studied, Ohm utilized wires of various lengths and diameters and considered the entire current to be restricted to the volume of the wire. Strictly speaking, this is not quite correct as air also possesses a conductivity, albeit one that is twenty orders of magnitude less than that of most metals.[5] As a consequence, Ohm was quite justified in neglecting currents in the air in his experiments.

---

[5]The conductivity of air varies, depending upon humidity, temperature and pressure but is generally in the range of $\sigma = 3\text{--}8 \times 10^{-15}$ S/m. Of course, with very high voltages, the air molecules ionize and the conductivity increases dramatically. Witness lightning.

Restricting the current density to a one-dimensional path simplifies matters considerably. Instead of integrating over a volume, as is indicated in Equation 9.6, we need only integrate along a (directed) path. Let us first examine the simplest path: a straight line segment that extends along the z-axis, as depicted in figure 9.1. This is, of course, not a practical current source. Charge conservation requires that the current entering or exiting the wire at the endpoints be continuous. Mathematically, however, there is no reason why we cannot compute the field of a physical circuit as a sum over different current segments: the fields of different segments simply add. Let us now also make a further simplifying assumption that the wire is very thin and we are only interested in computing the field at some reasonable distance away from the wire. We shall relax this latter requirement later but will proceed with a very simple model first.

In order to utilize Equation 9.6, we need to define the current density $\mathbf{J}$. If we make the assumption that the wire has no physical extent, we can represent this with the Dirac delta function[6]:

$$(9.7) \qquad \mathbf{J}(\mathbf{r}_1) = \hat{\mathbf{z}} I \, \delta(x_1)\delta(y_1),$$

where I is a constant current, with dimensions of charge per unit time (Q/T). The delta function has the following property:

$$\int_a^b dx\, F(x)\delta(x - x_0) = \begin{cases} F(x_0) & a \le x_0 \le b \\ 0 & \text{otherwise.} \end{cases}$$

As long as the point x_0 lies within the bounds of the integral, the result of the integration is simply the integrand evaluated at x_0. Note that the delta functions have dimensions of the inverse of their arguments. This is reflected in the scaling rule: $\delta(ax) = \delta(x)/|a|$. In our particular case, the current density $\mathbf{J}$ has dimension $(Q \cdot T^{-1} \cdot L^{-2})$, so for Equation 9.7 to make sense dimensionally, the delta functions must each have dimension of (L^{-1}).

For a line segment that extends from say z_a to z_b, the support of the integral in Equation 9.6 will be only for points $\mathbf{r}_1 = (0,0,z_1)$, where z_1 is restricted to the interval $z_a \le z_1 \le z_b$, because the delta functions collapse the integrals in the x- and y-directions. Hence, for an arbitrary location $\mathbf{r}_2 = (x_2, y_2, z_2)$, we then have that $\mathbf{r}_2 - \mathbf{r}_1 = (x_2, y_2, z_2 - z_1)$ and $|\mathbf{r}_2 - \mathbf{r}_1| = [x_2^2 + y_2^2 + (z_2 - z_1)^2]^{1/2}$. The magnetic field is therefore obtained

---

[6]Strictly speaking, the delta function is a *functional* not a function in the usual mathematical sense. Dirac's simple idea required a significant amount of mathematical work to justify the existence of such an object and led to development of the mathematical theory of distributions.

from the following integral:

$$B(r_2) = \frac{\mu_0 I}{4\pi} \int_{z_a}^{z_b} dz_1 \frac{\hat{z} \times (x_2, y_2, z_2 - z_1)}{[x_2^2 + y_2^2 + (z_2 - z_1)^2]^{3/2}},$$

which will be defined as long as x_2 and y_2 are not both simultaneously zero. In that case, the denominator would vanish when $z_2 = z_1$ and the integrand would become singular.

We can evaluate the numerator readily:

$$\hat{z} \times (x_2, y_2, z_2 - z_1) = (-y_2, x_2, 0).$$

It is also possible to perform the resulting integral, with the following result:

(9.8) $$B(r_2) = \frac{\mu_0 I}{4\pi} \frac{(-y_2, x_2, 0)}{x_2^2 + y_2^2} \left[\frac{z_1 - z_2}{[x_2^2 + y_2^2 + (z_2 - z_1)^2]^{1/2}} \right]_{z_1 = z_a}^{z_b},$$

where we have grouped the terms in a particular order for a reason. First, we note that the term $x_2^2 + y_2^2$ is the squared distance ζ_2^2 from the z-axis (in a cylindrical coordinate system). Second, we can recall that $\cos\varphi = x/\zeta$ and $\sin\varphi = y/\zeta$ and that the unit vector in the azimuthal direction is defined as $\hat{\varphi} = (-y/\zeta, x/\zeta, 0)$. So, we can rewrite Equation 9.8 as follows:

(9.9) $$B(r_2) = \hat{\varphi} \frac{\mu_0 I}{4\pi\zeta_2} \left[\frac{z_b - z_2}{[\zeta_2^2 + (z_2 - z_b)^2]^{1/2}} - \frac{z_a - z_2}{[\zeta_2^2 + (z_2 - z_a)^2]^{1/2}} \right].$$

This defines a magnetic field that is directed azimuthally around the wire segment, as is seen in figure 9.1.

> EXERCISE 9.3.   Fill in the missing steps in the derivation of Equation 9.9. Use the Plot3D function to plot the magnitude of the magnetic field for $0 \le \zeta_2 \le 4$ and for $-2 \le z_2 \le 2$. Start with $z_a = -20$ and $z_b = 0$ and then vary the endpoints.

> EXERCISE 9.4.   Use the StreamPlot function to plot magnetic field lines in the x-y plane at $z = 0$. Consider the cases where $z_a = -20$ and $z_b = 0$ and 20.

We can utilize Equation 9.9 to examine the limiting case of an infinite wire. If we take the limit where $z_a \to -\infty$ and $z_b \to \infty$, then the term in brackets becomes 2. Hence, the magnetic field of an infinite wire would be given by the following:

(9.10) $$B(r) = \hat{\varphi} \frac{\mu_0 I}{2\pi\zeta_2}.$$

In practice, there are no infinite wires. The expression given in Equation 9.10 will serve as a reasonable approximation for suitably long wires.

EXERCISE 9.5. Take the limit of Equation 9.9 for $z_a \to -\infty$ and $z_b \to \infty$. Show that you obtain the result in Equation 9.10. (Hint: Factor out z_b from the numerator and denominator of the first term and z_a from the second term.)

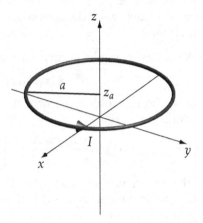

FIGURE 9.5. A thin loop carrying a constant current I will generate a magnetic field with components in the radial and vertical directions

In the case of a thin wire loop, as depicted in figure 9.5, we can repeat the above analysis but the symmetry of a circular loop suggests that we utilize cylindrical coordinates to perform the integration, where the z-axis is aligned with the axis of the loop. Heretofore, we have utilized Cartesian coordinates for representing vectors. At this juncture, we shall retain that strategy, wherein we will compute the x-, y- and z-components of the magnetic field but will utilize cylindrical coordinate systems to form the integral representations. This is something of a mixed metaphor, as it were, but defers the headlong plunge into non-Cartesian systems for the time being. In a cylindrical coordinate system, then, we can approximate the current density as follows:

$$(9.11) \qquad \mathbf{J}(\mathbf{r}_1) = \hat{\boldsymbol{\varphi}} I \, \delta(\zeta_1 - a)\delta(z_1 - z_a),$$

where the loop has a radius a and is located at the position $z_1 = z_a$. The differential volume element is given by $d^3\mathbf{r}_1 = \zeta_1 \, d\zeta_1 \, d\varphi_1 \, dz_1$, whereupon the Biot-Savart equation provides that the magnetic field is determined by the following expression:

$$(9.12) \qquad \mathbf{B}(\mathbf{r}_2) = \frac{\mu_0 I a}{4\pi} \int_0^{2\pi} d\varphi_1 \, \frac{\hat{\boldsymbol{\varphi}} \times (\mathbf{r}_2 - \mathbf{r}_1)}{|\mathbf{r}_2 - \mathbf{r}_1|^3},$$

where the ζ_1 and z_1 integrals have been collapsed by the delta functions.

The first thing that we should note from Equation 9.12 is that the magnetic field will, by virtue of the cross product, be perpendicular to the azimuthal direction $\hat{\varphi}$.[7] This means that the magnetic field of a loop will only have radial ($\hat{\zeta}$) and vertical ($\hat{z}$) components. We can use this observation to help us simplify the computation of the field.

In principle, we would like to evaluate the field at some arbitrary point $\mathbf{r}_2 = (x_2, y_2, z_2)$ but the field will only be a function of $\zeta_2 = (x_2^2 + y_2^2)^{1/2}$ and not x_2 and y_2 independently. (The field does not depend upon the azimuthal coordinate φ. In cylindrical coordinates, it must be only a function of the radial ζ and vertical z coordinates.) So, without loss of generality, we can consider just the point $\mathbf{r}_2 = (\zeta_2, 0, z_2)$.[8] Using the result that, due to the delta functions, $\mathbf{r}_1 = (a\cos\varphi_1, a\sin\varphi_1, z_a)$. we have

$$\mathbf{r}_2 - \mathbf{r}_1 = (\zeta_2 - a\cos\varphi_1, -a\sin\varphi_1, z_2 - z_a)$$

and

$$|\mathbf{r}_2 - \mathbf{r}_1| = [\zeta_2^2 + a^2 - 2a\zeta_2\cos\varphi_1 + (z_2 - z_a)^2]^{1/2}.$$

Note that the unit vector $\hat{\varphi}$ is not a constant vector. In Cartesian coordinates, we have $\hat{\varphi} = (-\sin\varphi_1, \cos\varphi_1, 0)$, and, consequently,

$$\hat{\varphi} \times (\mathbf{r}_2 - \mathbf{r}_1) = \left((z_2 - z_a)\cos\varphi_1, (z_2 - z_a)\sin\varphi_1, a - \zeta_2\cos\varphi_1\right).$$

We also note that the unit vector $\hat{\zeta}$ in Cartesian coordinates is given by $\hat{\zeta} = (\cos\varphi_1, \sin\varphi_1, 0)$, so that we can write the following;

$$\hat{\varphi} \times (\mathbf{r}_2 - \mathbf{r}_1) = \hat{\zeta}(z_2 - z_a) + \hat{z}(a - \zeta_2\cos\varphi_1).$$

As we expected, the magnetic field will only have components in the $\hat{\zeta}$ and $\hat{z}$ directions.

Substituting back into Equation 9.12, we obtain the following result:

$$\mathbf{B}(\mathbf{r}_2) = \frac{\mu_0 I a}{4\pi}\left[\int_0^{2\pi} d\varphi_1 \frac{\hat{\zeta}(z_2 - z_a)}{[\zeta_2^2 + a^2 - 2a\zeta_2\cos\varphi_1 + (z_2 - z_a)^2]^{3/2}}\right.$$

(9.13)

$$\left. +\hat{z}\int_0^{2\pi} d\varphi_1 \frac{a - \zeta_2\cos\varphi_1}{[\zeta_2^2 + a^2 - 2a\zeta_2\cos\varphi_1 + (z_2 - z_a)^2]^{3/2}}\right].$$

Note that we cannot slide the unit vector $\hat{\zeta}$ outside the φ_1 integral because $\hat{\zeta}$ is a function of φ_1, whereas $\hat{z} = (0, 0, 1)$ is a constant vector.

EXERCISE 9.6. Repeat the derivation of Equation 9.13, filling in the steps missing from the text. Show that when $\zeta_2 = 0$ that the magnetic field only has a z-component.

---

[7]This will not be true in general but follows from the fact that we have chosen a *constant* current I.

[8]We can recover the field at some point in the x-y plane by rotating about the z-axis by some angle φ_2: $\mathbf{r}_2 \rightarrow (\zeta_2\cos\varphi_2, \zeta_2\sin\varphi_2, z_2)$.

EXERCISE 9.7. The dimensional scale of the integral representation is set by the loop radius a. Rewrite Equation 9.13 in a dimensionless form by dividing numerator and denominator by the appropriate powers of a, producing terms like (ζ_2/a) and (z_2/a), etc. How does $|\mathbf{B}|$ scale as a function of a? Use the Plot3D function to plot the integrands. Set $\zeta_2/a = 1.1$ and plot the integrands over the range $0 \le \varphi_1 \le 2\pi$ and $-2 \le z_2/a \le 2$ (assuming $z_a = 0$). Set $z_2/a = 0.1$ and plot the integrands over the range $0 \le \varphi_1 \le 2\pi$ and $0 \le \zeta_2/a \le 3$ (again assuming $z_a = 0$).

Somewhat to the consternation of nineteenth Century mathematicians (and, of course, modern-day students), while it is a straightforward exercise to derive the integral representation of the magnetic field (Equation 9.13), it is not at all straightforward to compute the integrals. As long as we do not try to evaluate the integral at the location of the current source, where the denominators vanish and the integrals become singular, we saw in the previous exercise that the integrands are smooth, well-behaved functions. Nevertheless, the solutions of Equation 9.13 are not expressible in terms of simple functions.

As it happens, though, the solutions are related to the mathematical problem of determining the path length along an ellipse. We saw in the early chapters that planets move along elliptical trajectories. It is well known that the arc length ds along a circle of radius r is just $ds = r\,d\psi$ for some small angular interval $d\psi$. This is also true for elliptical paths but, as we have seen, the radius r is not a constant along an elliptical path. The problem was initially investigated by the Italian mathematician Guilio Fagnano and the Swiss Leonhard Euler but the form presented below is due to the French mathematician Adrien-Marie Legendre, who demonstrated that all elliptic integrals can be reduced to one of three fundamental forms.[9] The elliptic integrals of the first, second and third kinds are defined as follows:

$$(9.14) \qquad F(\varphi|m) = \int_0^\varphi d\psi\, \frac{1}{[1 - m\sin^2\psi]^{1/2}},$$

$$(9.15) \qquad E(\varphi|m) = \int_0^\varphi d\psi\,[1 - m\sin^2\psi]^{1/2} \quad \text{and}$$

$$(9.16) \qquad \Pi(n;\varphi|m) = \int_0^\varphi d\psi\, \frac{1}{1 - n\sin^2\psi}\, \frac{1}{[1 - m\sin^2\psi]^{1/2}},$$

---

[9]Different authors use somewhat different notation for the elliptic integrals. We shall utilize the notation provided by the *Mathematica* software, for reasons of ease of use. Be wary, though, if using other sources.

respectively. We shall need to make use of the complete elliptic integrals of the first, second and third kind, which are defined as $K(m) = F(\pi/2|m)$, $E(m) = E(\pi/2|m)$ and $\Pi(n|m) = \Pi(n; \pi/2|m)$, respectively.

> EXERCISE 9.8.   Plot the complete elliptic functions of the first and second kind, using the *Mathematica* functions EllipticK and EllipticE over the range $-1 \leq m \leq 1$. Plot the complete elliptic function of the third kind over the range $-1 \leq m \leq 1$ and use the Manipulate function to study the behavior as n varies over the range $0 \leq n \leq 1$.

If we now return to the problem of the magnetic field defined in Equation 9.13, we note that it is not usually a productive strategy to simply call the *Mathematica* Integrate function for complicated arguments. One runs the risk of locking up whatever computer you have for extended periods. A better strategy is to first understand the properties of the integrand: is it finite and well-behaved or is it singular? One can provide guidance about the known properties of the integrand through the use of the Assumptions options and it is often possible to simplify the formulas provided to the Integrate function.[10] In particular, while we are utilizing the variables a, r_2 and z_2 to be the real numbers that specify the radius of the loop and a point in space, the Integrate function makes no particular assumptions about the values of unspecified parameters and will, in general, assume that unspecified parameters are complex numbers.

Suppose that we define the new variable ξ as follows:

$$\xi = \frac{2a\zeta_2}{\zeta_2^2 + a^2 + (z_2 - z_a)^2}.$$

Then we can rewrite Equation 9.13 as follows:

$$\mathbf{B}(\mathbf{r}) = \frac{\mu_0 I a}{4\pi[\zeta_2^2 + a^2 + (z_2 - z_a)^2]^{3/2}} \left[\int_0^{2\pi} d\varphi_1 \frac{\hat{\zeta}(z_2 - z_a)}{[1 - \xi \cos\varphi_1]^{3/2}} \right.$$
$$\left. + \hat{z} \int_0^{2\pi} d\varphi_1 \frac{a - \zeta_2 \cos\varphi_1}{[1 - \xi \cos\varphi_1]^{3/2}} \right].$$

We observe now that providing a solution to Equation 9.13 will require us to perform three basic integrals, with integrands of 1, $\cos\varphi_1$ and $\sin\varphi_1$, each divided by $[1 - \xi \cos\varphi_1]^{3/2}$.

---

[10]We can state that the *Mathematica* package is an exceptionally useful tool but it is not capable of doing one's homework autonomously. Thinking remains the province of the student.

EXERCISE 9.9.   Define three functions that represent the three integrands. Plot the integrands over the range $0 \leq \varphi_1 \leq 2\pi$. What symmetries exist?

From the previous exercise, we note that the integral involving $\sin\varphi_1$ will vanish because the integrand is an odd function on the domain $0 \leq \theta_1 \leq 2\pi$. The remaining two integrands are even functions, so we can use that symmetry to reduce the domain to the interval $0 \leq \varphi_1 \leq \pi$ and double that result. If we now utilize the *Mathematica* Integrate function, we can obtain the following results:

$$\int_0^\pi d\varphi_1 \frac{1}{[1 - \xi \cos\varphi_1]^{3/2}} = \frac{2E\left(\frac{2\xi}{1+\xi}\right)}{(1-\xi)[1+\xi]^{1/2}}$$

(9.17)
$$\int_0^\pi d\varphi_1 \frac{\cos\varphi_1}{[1 - \xi \cos\varphi_1]^{3/2}} = 2\frac{E\left(\frac{2\xi}{1+\xi}\right) - (1-\xi)K\left(\frac{2\xi}{1+\xi}\right)}{(1-\xi)\xi[1+\xi]^{1/2}},$$

provided that the variable ξ is within the range $-1 \leq \xi \leq 1$.

EXERCISE 9.10.   If we rewrite the variable ξ in dimensionless terms, we have $\xi = 2(\zeta_2/a)/[1 + (\zeta_2/a)^2 + (z_2/a)^2]$, where we set $z_a = 0$ for the moment. Plot the function ξ for $0 \leq (\zeta_2/a) \leq 4$ and $-2 \leq (z_2/a) \leq 2$. Is the magnitude of ξ less than one?

We can now utilize the results of Equations 9.17 to solve for the magnetic field. After some extensive algebra, we find the following:

$$(9.18) \quad \mathbf{B}(\mathbf{r}) = \frac{\mu_0 I}{2\pi[(\zeta_2 + a)^2 + (z_2 - z_a)^2]^{1/2}}$$

$$\left\{ \hat{\zeta} \frac{(z_2 - z_a)}{\zeta_2} \left[\frac{\zeta_2^2 + a^2 + (z_2 - z_a)^2}{(\zeta_2 - a)^2 + (z_2 - z_a)^2} E(\eta) - K(\eta) \right] \right.$$

$$\left. + \hat{z} \left[K(\eta) - \frac{\zeta_2^2 - a^2 + (z_2 - z_a)^2}{(\zeta_2 - a)^2 + (z_2 - z_a)^2} E(\eta) \right] \right\},$$

where $\eta = 2\xi/(1 + \xi) = 4a\zeta_2/[(\zeta_2 + a)^2 + (z_2 - z_a)^2]$.

After a lengthy derivation like that we encountered in producing Equation 9.18, it is useful to perform a dimensional analysis to ensure that we haven't gone astray.[11] While we haven't as yet discussed the units of magnetic fields, what we observed in computing the field of a straight wire is that the field was proportional to the term $\mu_0 I$ divided by a length. In Equation 9.18, we observe that the prefactor has precisely that dimension and the terms within the curly brackets are all dimensionless, including

---

[11]In fact, it is always a good idea to perform dimensional analysis *during* derivations as a check on the algebraic manipulations.

the argument η of the complete elliptic integrals. As a result, we can have some confidence that the results are correct.

EXERCISE 9.11. Repeat the derivation of Equation 9.18, filling in the missing steps.

FIGURE 9.6. The magnitude of the magnetic field is singular at the position of the wire and falls rapidly away from the wire

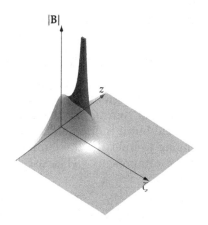

Of course, the best way to see whether or not Equation 9.18 is correct is to plot the results, whereupon we encounter the non-trivial problem of visualizing three-dimensional vector fields. Plotting the results is always a good idea; we are, after all, visual creatures. We expect that the fields will be smoothly varying, at least away from the source location. If the plotted results include kinks or random spikes, there is a good chance that we have erred somewhere in deriving the defining equations.

Plotting a vector field in three dimensions is not a trivial exercise. As a first step, consider plotting just the magnitude of the field. This is a scalar function of space and, for the current loop, is only a function of the radial and vertical components. In figure 9.6, we illustrate the magnitude of the magnetic field of a loop. At the position of the current, the field is singular, which should not be particularly surprising. Away from the source location, the field falls relatively rapidly but is smoothly-varying everywhere.

Another possible representation of the field can be achieved by drawing a vector aligned with the local direction of the field and with the length of the vector scaled by the local magnitude of the field. As it happens, the *Mathematica* package has a VectorPlot function that will perform precisely that action. In practice, this approach does not lead to particularly satisfactory results, due to the fact that the field decays rapidly away from the current source. Often, this leads to a single large vector with array of dots around it; the sense of local direction is lost due to the poor scaling of the fields.

An alternative approach uses the magnetic field lines defined originally by Faraday. The magnetic field lines of a current loop are illustrated in figure 9.7. As we have mentioned, the field lines depict the trajectories that infinitesimal magnetic charges would follow in the magnetic field. As such, we should remember that the magnetic field is not a constant along the magnetic field lines: these do not represent constant contours of the field. Nevertheless, plotting the magnetic field lines is one of the most common approaches used in visualizing the magnetic field. The field lines, even in three dimensions, can be readily traced and drawn in perspective. One can also, of course, plot the magnitudes of the individual components of the field but it is often quite difficult to interpret those results; it is challenging to infer the three-dimensional structure of the field from plots of the components. This last option is usually reserved for debugging activities.

EXERCISE 9.12.　Define two functions that represent the ζ- and z-components of the magnetic field (scaled by the factor $\mu_0 I/2\pi$), using the functions EllipticE and EllipticK. Use the VectorPlot function to visualize the field and then use the StreamPlot function to plot magnetic field lines for the case where $a = 1$ and $z = 0$. Plot over the range $-3 \le \zeta \le 3$ and $-3 \le z \le 3$.

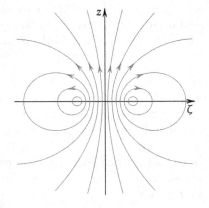

FIGURE 9.7. Magnetic field lines of a current loop. The current direction is in the azimuthal direction (With the thumb of your right hand oriented along the z-axis, your fingers will curl in the direction of the current)

EXERCISE 9.13.　Use the functions defined in the previous exercise to examine the results of adding the fields of several loops. (Use the Sum function.) Plot the magnetic field lines and field intensity (with the StreamDensityPlot function) for an Helmholtz coil. The Helmholtz coil is composed of two loops of radius 1, separated by a distance $z = 1$.

Plot the magnetic field lines and field intensity for a solenoid, which we can approximate as the sum of several loops spaced closely together. Try summing over the range $0 \leq z_1 \leq 2$ in steps of $\delta z_1 = 0.1$. Describe the magnetic field in the interior of the solenoid. How does this compare to the field of the Helmholtz coil?

As one might surmise, more complex current paths and time-varying currents will lead to significant complications in determining the fields. The Biot-Savart equation provides a general framework for computing magnetic fields due to current sources but, in practice, one must generally resort to numerical methods to obtain field distributions. As we have seen, even the field of a simple, circular loop is not expressible in terms of simple functions. Nevertheless, except for the problem of the field being singular at the source, the Biot-Savart equation does provide a means for systematically determining an integral representation of the field.

As we mentioned earlier, the French physicist Ampère was also present when Ørsted's experiments were first announced in Paris. Shortly thereafter, Ampère derived a connection between the magnetic field and currents that held in the limit of constant currents. Ampère's law, as it has become known, was subsequently modified by the Scottish physicist James Clerk Maxwell.[12] In our modern notation, we write the Ampère-Maxwell equation as follows:

$$(9.19) \qquad \oint_{\partial S} d\mathbf{s} \cdot \mathbf{B} = \mu_0 \int_S d\mathbf{A} \cdot \mathbf{J} + \mu_0 \epsilon_0 \frac{\partial}{\partial t} \int_S d\mathbf{A} \cdot \mathbf{E},$$

where $\mathbf{B}$ is the magnetic field, $\mathbf{E}$ is the electric field and $\mathbf{J}$ is the current density. The area integrals on the right-hand side of Equation 9.19 extend over some (finite) surface S and the path integral on the left-hand side extends over the closed boundary ∂S of S. The factor ϵ_0 is known as the dielectric permittivity and is related to the electric field proportionality factor by $\kappa = 1/4\pi\epsilon_0$.

Consider now the fact that physical wires have finite dimensions. We can use Ampère's law to determine the magnetic field inside the wire. For a constant current, we can ignore the last term in Equation 9.19 as the time derivative of the electric field will vanish. If the current density, for simplicity, is constant, then for a total current I in a wire of radius R, we would have $\mathbf{J} = \hat{z}I/(\pi R^2)$. Consider the surface S_1 illustrated in figure 9.8. The current flux through the surface with differential element $d\mathbf{A} = \hat{z}\,\zeta\,d\zeta\,d\varphi_1$ is given by the following:

$$\int_{S_1} d\mathbf{A} \cdot \mathbf{J} = \int_0^{\zeta_1} d\zeta \int_0^{2\pi} d\varphi\,\hat{z} \cdot \left[\hat{z}\,\frac{I}{\pi R^2}\right] = I\,\frac{\zeta_1^2}{R^2}.$$

---

[12]Maxwell's contribution to Equation 9.19 is the last term on the right-hand side. It is an essential element in defining the wave nature of the electromagnetic field.

By symmetry, the magnetic field at the radius ζ_1 will only have an azimuthal component, $\mathbf{B} = \hat{\varphi} B_\varphi$, and will be a constant along the path along the boundary of S_1. The path integral is written as follows:

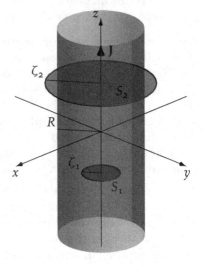

FIGURE 9.8. A current density $\mathbf{J}$ flows in the wire of radius R in the positive z-direction. Two Ampèrian loops are illustrated. One has a radius $\zeta_1 < R$ and the second has a radius $\zeta_2 > R$

$$\oint_0^{2\pi} d\varphi\, \zeta_1 \hat{\varphi} \cdot \mathbf{B} = 2\pi \zeta_1 B_\varphi.$$

Equating these two results, we obtain the following:

(9.20)
$$B_\varphi = \frac{\mu_0 I \zeta_1}{2\pi R^2}.$$

Consider now the second surface, in which the current is entirely enclosed within the loop. Here, the radial integral would nominally extend to ζ_2 but the current density vanishes beyond the radius R. As a result, the radial integral only extends to $\zeta_1 = R$. We can show that the field at the radius ζ_2 is given by the following:

(9.21)
$$B_\varphi = \frac{\mu_0 I}{2\pi \zeta_2},$$

which is precisely the result we obtained previously for the field outside a long wire.

For a constant current the magnetic field in the interior of the wire increases linearly with the radius ζ and then decreases inversely with ζ for radii larger than the wire radius R. For more complex current distributions, this simple result will not hold but the principle result of Ampère's law is that the magnetic field will remain finite, even inside conductors carrying current, as long as the current density remains finite.

The singular behavior that we found previously was simply the result of using a singular current density.

As a practical matter, we will often use singular source currents because such a choice leads to simplified mathematical results. Integrations with delta functions are trivial to perform. As long as we do not try to evaluate the field on top of such a source, we will not encounter any difficulties, and as a practical matter, it is difficult to imagine how one could get a magnetic field probe inside a wire in the first place. Physicists may seem nonchalant in their treatment of singularities but it is important to understand the nature of how the singularities arise and whether or not singularities have physical significance. Mathematically, our choices of different representations of the fields can be generally justified by careful study. In the current state of mathematical sophistication of typical students, we shall not always be able to justify every choice.

EXERCISE 9.14.   Fill in the missing details in the derivations of Equations 9.20 and 9.21. Plot the magnetic field as a function of ζ.

EXERCISE 9.15.   Consider a current density that is a function of ζ:

$$J = \begin{cases} \hat{z} J_0 (1 - e^{-a\zeta}) & \zeta \leq R \\ 0 & \zeta > R. \end{cases}$$

What is the total current I through the circular surface of radius R? What is the magnetic field at a radius $\zeta_1 < R$? What is the magnetic field at a radius $\zeta_2 > R$? Write this in terms of the total current I. How does your result compare to Equation 9.21?

9.3. Magnetic Materials

Of course, the study of magnetic fields did not begin with current sources, as we have just been investigating. It was well known that certain materials possess static magnetic fields. Iron, in particular, can be magnetized and this was certainly known to the ancient Greeks. So, how is this possible if we are to believe that magnetic fields arise from currents? No doubt each of us has held a bar magnet at one time or another or affixed some important document to the face of a refrigerator with a "refrigerator magnet." There are no batteries embedded within these magnets; there is no current flowing. Yet there is a magnetic field.

In fact, it is the same kind of magnetic field as that generated by currents. There are not two different kinds of magnetic fields. We can perform an experiment to verify this assertion. Consider constructing a C-shaped permanent magnet, as is depicted in figure 9.9. Between the poles of the magnet, the field is quite uniform. If we now send an electron beam through

FIGURE 9.9. A permanent magnet (*dark C-shaped object*) will have a magnetic field that is approximately constant and vertical in the region between the poles and zero elsewhere. An electron beam passing through the magnet will be deflected ($I = 0$ curve). If a current I_c is passed through the Helmholtz coils (*light gray rings*), the resulting magnetic field will cancel that of the permanent magnets and the electron beam will emerge undeflected

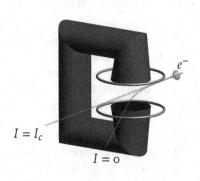

the region between the poles, we will find it is deflected according to the Lorentz force law. Now consider building a Helmholtz coil. As we have observed in the previous section, the field of the Helmholtz coil is quite uniform near the center. If we again send a beam of electrons through the field of the Helmholtz coil, we will again observe that the beam is deflected in accordance with the Lorentz force law. Indeed, if we align the permanent magnet and Helmholtz coil, we can arrange for the magnetic field to cancel in the central region. The electron beam will emerge from the space undeflected! Insofar as we know, the magnetic fields of permanent magnets have precisely the same electromagnetic properties as those created by currents.

Prior to Rutherford's discovery of the nuclear structure of the atom, early physicists postulated that the static magnetic fields of permanent magnets must somehow be due to microscopic currents embedded within the material, although they could not find a plausible explanation as to why those currents did not dissipate. As Ohm discovered, any current in a circuit experiences a drop in voltage proportional to the product of the current and the resistance: $V = IR$. There is a loss of energy per unit time (power) due to this resistive force that is proportional to I^2R. As all known materials have resistance, it was unclear as to how currents in permanent magnets could be maintained indefinitely.

With the discovery that atoms are composed of electrons and nuclei, an alternative explanation for the static magnetic fields of permanent magnets became available. Suppose that the electrons possess a static magnetic field. That is, let us suppose that electrons are little blobs (spherical is the obvious choice for shape) of matter that possess mass, electric charge and are little magnets. We can explain the static magnetic fields of permanent magnets as arising due to the (fortuitous) alignment of the magnetic fields of many electrons.[13] This explains the experimental observation that iron magnets lose their magnetic fields when heated. Presumably, the energetic electrons produced when a material is heated will begin bouncing around, scattering from one another and, thereby, destroying their alignment.

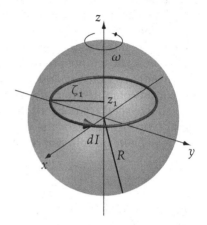

FIGURE 9.10. A rotating, charged sphere will generate an external magnetic field due to the creation of infinitesimal current loops within the sphere

In fact, we don't even need to require our electron to be made of some magnetic material. Suppose that the electron is made of a material that has charge. For simplicity, let us further suppose that the charge is evenly distributed throughout the volume. In this case, we will have a charge density $\rho = q/(4/3 \pi R^3)$, where q is the total charge in the sphere and R is the radius. Now let us spin the electron around the z-axis, as is depicted in figure 9.10. The rotation gives rise to a current density that can be written as $\mathbf{J} = \hat{\phi}\zeta_1\rho\omega$. As we have seen, current densities generate magnetic fields.

EXERCISE 9.16.   Use dimensional analysis to show that $\mathbf{J}(\mathbf{r}_1) = \hat{\phi}\zeta_1\rho\omega$ has the correct dimensionality of a current density. Show that, if we consider an infinitesimal area $dA = d\zeta_1\,dz_1$, that the amount of charge dq flowing through the area in a time dt will be given by the following:

$$dq = \zeta_1\rho\omega\,d\zeta_1\,dz_1\,dt.$$

(Hint: recall that the angular velocity is defined as $\omega = d\varphi/dt$ and the volume element in cylindrical coordinates is $dV = \zeta_1\,d\zeta_1\,d\varphi_1\,dz_1$.)

---

[13]The astute student might also be asking why we don't think that the nucleus might also be a magnet. As it happens, nuclei do possess magnetic moments but these are dramatically smaller than the magnetic moment of the electron.

EXERCISE 9.17. For a fixed charge distribution ρ, the current density is $J = \rho v$, where v is the velocity of the distribution. For a *rotating*, fixed charge distribution, show that the tangential velocity at the radius r is given by $v = \omega \times r$, where ω is the angular velocity vector. (Hint: work in cylindrical coordinates and align ω with the z-axis.)

We now want to obtain the magnetic field due to a rotating, spherical charge density. We note that the differential current dI flowing through an infinitesimal area $d\mathbf{A} = \hat{\varphi}\, d\zeta_1\, dz_1$ is just $dI = \mathbf{J}(\mathbf{r}_1) \cdot d\mathbf{A} = \zeta_1 \rho \omega\, d\zeta_1\, dz_1$. Hence, the total magnetic field produced by the rotating sphere can be obtained by integration over all of the infinitesimal loops dI. One might, at this point, be tempted to plug the previous results (Equation 9.18) into the **Integrate** function and see what happens.[14] This approach will require us to compute integrals of elliptic integrals, which sounds dreadful. Instead, we shall try to avoid that difficulty and begin by writing the integral representation for the field with the source current provided by a rotating sphere of uniform, constant charge density ρ and constant angular velocity ω:

(9.22)

$$\mathbf{B}(\mathbf{r}_2) = \frac{\mu_0 \rho \omega}{4\pi} \int\limits_{-R}^{R} dz_1 \int\limits_{0}^{\sqrt{R^2 - z_1^2}} d\zeta_1 \int\limits_{0}^{2\pi} d\varphi_1 \frac{\zeta_1 [\hat{\zeta}(z_2 - z_1) + \hat{z}(\zeta_1 - \zeta_2 \cos\varphi_1)]}{[\zeta_2^2 + \zeta_1^2 - 2\zeta_1\zeta_2 \cos\varphi_1 + (z_2 - z_1)^2]^{3/2}}.$$

Note that, if the charge density ρ is not a constant, then we cannot remove it from the integration.

EXERCISE 9.18. Return to the Biot-Savart law (Equation 9.6). Use the definition of $J = \hat{\varphi}\zeta_1 \rho \omega$ and fill in the missing details of the derivation of Equation 9.22. Note that the order of integration matters.

In deriving Equation 9.22, we selected a particular order of integration, which amounts to building up the spherical volume by stacking thin (dz_1) circular plates. One might also reverse the order of integration to construct the volume by the stacking of thin ($d\zeta_1$) cylinders but doing so will require a revision to the limits of integration. Now we have previously computed the φ_1 integrals: this approach led to elliptic integrals. As we know little enough about elliptic integrals, much less integrals of elliptic integrals, let us consider reversing the order of the φ_1 and ζ_1 integrals. This turns out to be a reasonable thing to try but, unfortunately, does not lead to useful results. As it turns out, no reordering of integrations leads to a simple result. So, as practical people, we might be tempted to just

---

[14]Unfortunately, computers do not *think*, they merely perform calculations. To reiterate: thinking remains the province of the student.

integrate the formula numerically and plot the results; we have certainly taken this path previously. As it happens, though, a change of coordinates will prove fruitful.

Consider recasting the problem into spherical coordinates, where points in the plane are determined by $\mathbf{r} = (r, \theta, \varphi)$, where r is the radius vector, θ is the polar angle and φ is the azimuthal angle, as was discussed previously in Chapter 6. In spherical coordinates, the current density becomes $\mathbf{J}(\mathbf{r}_1) = \hat{\boldsymbol{\varphi}} r_1 \sin \theta_1 \rho \omega$, where r_1 here is the radial distance to the origin, not the z-axis. With the restriction that we wish to compute the field at a point in the x-z plane, then $\mathbf{r}_2 = (x_2, 0, z_2) = (r_2 \sin \theta_2, 0, r_2 \cos \theta_2)$, where $r_2 = (x_2^2 + z_2^2)^{1/2}$. An arbitrary point $\mathbf{r}_1$ in the sphere is given by $\mathbf{r}_1 = (r_1 \sin \theta_1 \cos \varphi_1, r_1 \sin \theta_1 \sin \varphi_1, r_1 \cos \theta_1)$. The magnetic field due to a rotating charged sphere can thus be shown to be given by the following expression:

$$(9.23) \quad \mathbf{B}(\mathbf{r}_2) = \frac{\mu_0 \rho \omega}{4\pi} \int_0^R dr_1 \int_0^\pi d\theta_1 \int_0^{2\pi} d\varphi_1 \, r_1^3 \sin^2 \theta_1$$

$$\left\{ \hat{\mathbf{x}} \cos \varphi_1 (r_2 \cos \theta - r_1 \cos \theta_1) + \hat{\mathbf{y}} \sin \varphi_1 (r_2 \cos \theta_2 - r_1 \cos \theta_1) + \right.$$

$$\left. \hat{\mathbf{z}} (r_1 \sin \theta_1 - r_2 \sin \theta_2 \cos \varphi_1) \right\}$$

$$[r_1^2 + r_2^2 - 2 r_1 r_2 (\cos \theta_1 \cos \theta_2 + \sin \theta_1 \sin \theta_2 \cos \varphi_1)]^{-3/2}.$$

This is another integral representation of the magnetic field. It represents the same magnetic field as the one defined in Equation 9.22.

EXERCISE 9.19. Fill in the details of the derivation of Equation 9.23.

Unfortunately, performing the integrals in Equation 9.23 turns out to be just as challenging as those in Equation 9.22. Mathematically, the problem arises from the occurrence of the factor $\cos \varphi_1$ in the denominator. The φ_1 integration results in elliptic integrals and we are back at square one. As it happens, though, there is a trick that we can invoke that will render the problem tractable.

Let us consider rotating the sphere in figure 9.10 around the y-axis such that the field will be evaluated on the z-axis: $\mathbf{r}_2' = (0, 0, r_2)$. This will require a rotation of $-\theta_2$ as described by the rotation matrix $\mathbf{R}_2$ in Equation 4.26. In the rotated (primed) frame, we have that the following relation holds[15]:

$$\mathbf{r}_2' - \mathbf{r}_1 = (-r_1 \sin \theta_1 \cos \varphi_1, -r_1 \sin \theta_1 \sin \varphi_1, r_2 - r_1 \cos \theta_1).$$

---

[15]Note that here we have not rotated the $\mathbf{r}_1$ vector. We justify this by recognizing that we intend to integrate over *all* of the vectors $\mathbf{r}_1$ within the sphere and, thus, do not need to also rotate a particular point $\mathbf{r}_1$.

From this result, we can compute $|\mathbf{r}'_2 - \mathbf{r}_1| = [r_1^2 + r_2^2 - 2r_1 r_2 \cos\theta_1]^{1/2}$. Magically, there is no φ_1 dependence in $|\mathbf{r}'_2 - \mathbf{r}_1|$!

EXERCISE 9.20. Define the vector $\mathbf{r}_2 = (r_2 \sin\theta_2, 0, r_2 \cos\theta_2)$. Define the matrix R_2 as follows:

$$\mathbf{R}_2(\theta_2) = \begin{bmatrix} \cos\theta_2 & 0 & \sin\theta_2 \\ 0 & 1 & 0 \\ -\sin\theta_2 & 0 & \cos\theta_2 \end{bmatrix}.$$

(This definition of $\mathbf{R}_2$ is the spatial part of the matrix $\mathbf{R}_2$ defined in Equation 4.26.) Show that $\mathbf{R}_2(-\theta_2)\mathbf{r}_2 = (0, 0, r_2)$.

Of course, we now have to adjust the current to conform to the new coordinate system. We note that the current can be written as $\mathbf{J}(\mathbf{r}_1) = \rho\boldsymbol{\omega}\times\mathbf{r}_1$, where initially $\boldsymbol{\omega} = \hat{\mathbf{z}}\omega$. Rotating around the y-axis by the angle $-\theta_2$, changes the angular velocity vector to $\boldsymbol{\omega}' = (-\omega\sin\theta_2, 0, \omega\cos\theta_2)$. As a result, the current in the rotated frame becomes:

$$(9.24) \quad \mathbf{J}'(\mathbf{r}_1) = r_1\rho\omega\Big\{-\hat{\mathbf{x}}\sin\varphi_1 \sin\theta_1 \cos\theta_2$$
$$+ \hat{\mathbf{y}}[\cos\varphi_1 \sin\theta_1 \cos\theta_2 + \cos\theta_1 \sin\theta_2] - \hat{\mathbf{z}}\sin\varphi_1 \sin\theta_1 \sin\theta_2\Big\}.$$

Whereupon we can now show that $\mathbf{J}'(\mathbf{r}_1)\times(\mathbf{r}'_2-\mathbf{r}_1)$ is given by the following:

$$(9.25) \quad \mathbf{J}'(\mathbf{r}_1)\times(\mathbf{r}'_2 - \mathbf{r}_1) = r_1\rho\omega\Big\{\hat{\mathbf{x}}\big[\cos\varphi_1(r_2 - r_1\cos\theta_1)\sin\theta_1 \cos\theta_2$$
$$+ (r_2\cos\theta_1 - r_1\cos^2\theta_1 - r_1\sin^2\varphi_1 \sin^2\theta_1)\sin\theta_2\big]$$
$$+ \hat{\mathbf{y}}\sin\varphi_1 \sin\theta_1\big[(r_2 - r_1\cos\theta_1)\cos\theta_2 + r_1\cos\varphi_1 \sin\theta_1 \sin\theta_2\big]$$
$$+ \hat{\mathbf{z}}r_1\sin\theta_1\big[\sin\theta_1 \cos\theta_2 + \cos\varphi_1 \cos\theta_1 \sin\theta_2\big]\Big\}.$$

While this form of the integral representation of **B** will have a much more complicated numerator, the only φ_1 dependence is in the numerator, as expressed by Equation 9.25.

EXERCISE 9.21. Define the vectors $\boldsymbol{\omega}$, $\mathbf{r}_1$ and $\mathbf{r}_2$ and the rotation matrix $\mathbf{R}_2$ (as lists within the *Mathematica* program). Check the validity of Equation 9.25. Integrate the results over the interval $0 \leq \varphi_1 \leq 2\pi$.

The φ_1 integrals can be performed readily: in the cases where there is no φ_1 dependence, we obtain a factor of 2π and for the cases that depend upon $\sin^2\varphi_1$ or $\cos^2\varphi_1$, we obtain a factor of π. The remaining integrals involving $\sin\varphi_1$ and $\cos\varphi_1$ vanish, eliminating all of the $\hat{\mathbf{y}}$ terms and several of the $\hat{\mathbf{x}}$ and $\hat{\mathbf{z}}$ terms.

Somewhat surprisingly, given the difficulties that we've encountered to this point, it is also possible to perform all of the θ_1 integrals. We owe this bit of mathematical good fortune to the factor of $\sin\theta_1$ that comes from the volume element in spherical coordinates. Presumably, students have, by this time in their academic careers, been exposed to the strategy of changing variables to make integration simpler. No doubt this current example is one of the most convoluted to date but it illustrates that one can utilize physical insights to assist in the mathematical process. Rotation around axes will not affect the physical interpretation of our results but will affect the mathematical representation.

We do have one further wrinkle to consider: the denominator $|\mathbf{r}_2 - \mathbf{r}_1|$ will vanish when $r_1 = r_2$. We shall have to treat the cases where (i) $r_1 > r_2$ and (ii) $r_1 < r_2$ separately. Using the Assumptions option of the Integrate function to provide this information, we find that, after performing the θ_1 integrations, the magnetic field now has the following representation:

$$(9.26) \quad \mathbf{B}'(\mathbf{r}_2') = \frac{\mu_0 \rho \omega}{3} \int_0^R dr_1\, r_1^2 \begin{cases} \left(-\dfrac{2\sin\theta_2}{r_1}, 0, \dfrac{2\cos\theta_2}{r_1} \right) & r_1 > r_2 \\[2ex] \left(\dfrac{r_1^2 \sin\theta_2}{r_2^3}, 0, \dfrac{2r_1^2 \cos\theta_2}{r_2^3} \right) & r_1 < r_2. \end{cases}$$

EXERCISE 9.22. Fill in the missing steps of the derivation of Equation 9.26.

EXERCISE 9.23. Plot the radial dependence of the integrands from Equation 9.26 over the range $0 \le r_1 \le 1$ and $0 \le r_2 \le 3$. What is the radial dependence of the integrands of the x- and z-components of the field? In particular, what happens at the point $r_1 = r_2$?

We now perform the integrations over r_1, taking care when $r_2 < R$, to find that the magnetic field in the rotated frame can be written as follows:

$$(9.27) \quad \mathbf{B}'(\mathbf{r}_2') = \frac{\mu_0 \rho \omega}{15} \begin{cases} \left([6r_2^2 - 5R^2]\sin\theta_2, 0, [5R^2 - 3r_2^2]\cos\theta_2 \right) & r_2 < R \\[2ex] \left(\dfrac{R^5 \sin\theta_2}{r_2^3}, 0, \dfrac{2R^5 \cos\theta_2}{r_2^3} \right) & r_2 > R. \end{cases}$$

EXERCISE 9.24. Plot the radial part of the field components from Equation 9.27 over the range $0 \le r_2 \le 3$, with $R = 1$. What is the radial dependence of the x- and z-components of the field? In particular, what happens at the point $r_2 = R$?

We, of course, now need to rotate back into the original orientation. The rotation can be accomplished by multiplying the result in Equation 9.27 by the rotation matrix $\mathbf{R}_2(\theta_2)$. The result is written below:

$$B(r_2) = \frac{\mu_0 \rho \omega}{15} \begin{cases} \left(3r_2^2 \sin \theta_2 \cos \theta_2, 0, 5R^2 - r_2^2 \dfrac{9 - 3[\cos^2 \theta_2 - \sin^2 \theta_2]}{2} \right) & r_2 < R \\ \left(\dfrac{3R^5 \sin \theta_2 \cos \theta_2}{r_2^3}, 0, R^5 \dfrac{1 + 3[\cos^2 \theta_2 - \sin^2 \theta_2]}{r_2^3} \right) & r_2 > R. \end{cases}$$

This is the representation of the magnetic field using spherical coordinates, albeit resolved into Cartesian coordinates. To facilitate plotting the field, let us recognize that $r_2 = (x_2^2 + z_2^2)^{1/2}$ and that $\cos \theta_2 = z_2/r_2$ and $\sin \theta_2 = x_2/r_2$. Using these results, we can write that the field is given by the following:

$$(9.28) \qquad B(r_2) = \frac{\mu_0 \rho \omega}{15} \begin{cases} \left(3x_2 z_2, 0, 5R^2 - 6x_2^2 - 3z_2^2 \right) & x_2^2 + z_2^2 < R^2 \\ \left(\dfrac{3R^5 x_2 z_2}{[x_2^2 + z_2^2]^{5/2}}, 0, \dfrac{R^5(2z_2^2 - x_2^2)}{[x_2^2 + z_2^2]^{5/2}} \right) & x_2^2 + z_2^2 > R^2. \end{cases}$$

The field of a spinning sphere is depicted in figure 9.11.

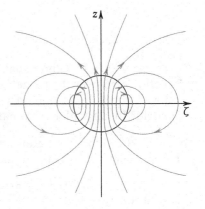

FIGURE 9.11. The magnetic field lines of a rotating, charged sphere resemble those of a simple current loop. The field is finite everywhere

EXERCISE 9.25. Fill in the details of the derivation of Equation 9.28.

EXERCISE 9.26. Use the StreamPlot and StreamDensityPlot functions to visualize the magnetic field of a spinning, charged sphere, using Equation 9.28.

From our derivation, Equation 9.28 also serves as the representation of the magnetic field in cylindrical coordinates, if we make the identifications that $\hat{x} \to \hat{\zeta}$ and $x_2 \to \zeta_2$. It will be useful to also provide a representation of the field in spherical coordinates. We can utilize the fact that

$\hat{\zeta} = \hat{r} \sin\theta_2 + \hat{\theta} \cos\theta_2$ and $\hat{z} = \hat{r}\cos\theta_2 - \hat{\theta}\sin\theta_2$ to show that the magnetic field of a rotating sphere is written as follows:

$$(9.29) \quad \mathbf{B}(\mathbf{r}_2) = \frac{\mu_0 q \omega}{20\pi R3} \begin{cases} \dfrac{\hat{r}\cos\theta_2[5R^2 - 3r_2^2] + \hat{\theta}\sin\theta_2[6r_2^2 - 5R^2]}{R5} & r_2 < R \\[2mm] \dfrac{R5}{r_2^3}\left(\hat{r}\,2\cos\theta_2 + \hat{\theta}\sin\theta_2\right) & r_2 > R, \end{cases}$$

where we have utilized a constant charge density $\rho = q/(4\pi R^3/3)$. We can recognize that this result is the same as that of Equation 9.26, with the identification that $\hat{z} \to \hat{r}$ and $\hat{x} \to \hat{\theta}$.

Remarkably, the fields of a current loop and a spinning sphere are quite similar, at least at distances that are relatively large compared to the size of the loop or sphere. As can be seen from Equation 9.29, the magnetic field decays like r^{-3} at large distances from the center of the sphere. This behavior is characteristic of a dipole field. Recall that the electric field of a spherical charge decayed like r^{-2}, which we term a monopole field. There is, in fact, a systematic process by which one can expand the fields in powers of r^{-n}.[16] This process is useful at large distances because there the field will be dominated by terms that involve the smallest power of n.

EXERCISE 9.27. Plot the functions r^{-n} for the range $1 \leq r \leq 10$ and $n = 2, 3, 4$, and 5.

9.4. Magnetic Field of the Electron

So, if we cling to this mental picture of an electron as a small, charged sphere then, if the sphere is spinning about some axis, the electron will also possess a magnetic field. This has an important consequence. Electromagnetic fields can possess momentum and angular momentum. This may seem odd, owing to our initial definition of momentum as the product of mass and velocity $\mathbf{p} = m\mathbf{v}$. Nevertheless, the total momentum in the fields can be defined as follows:

$$(9.30) \qquad \mathbf{p}_{em} = \epsilon_0 \int d^3\mathbf{r}_2 \left[\mathbf{E}(\mathbf{r}_2) \times \mathbf{B}(\mathbf{r}_2)\right]$$

and the total angular momentum in the fields is then:

$$(9.31) \qquad \mathbf{L}_{em} = \epsilon_0 \int d^3\mathbf{r}_2 \left\{\mathbf{r}_2 \times \left[\mathbf{E}(\mathbf{r}_2) \times \mathbf{B}(\mathbf{r}_2)\right]\right\}.$$

---

[16]For historical reasons, terms in this series have somewhat arcane names: monopole, dipole, quadrupole, octupole, hexadecapole, etc., owing to the fact that one can generate such fields with one, two, four, eight and sixteen point sources in suitable locations, respectively. Predictably, the prefixes are the Latin roots for one, two, four, etc.

We solved for the field of a charged sphere in Chapter 3. Recall that it can be written as follows:

$$E(r_2) = \frac{q}{4\pi\epsilon_0}\hat{r}\begin{cases} \dfrac{r_2}{R^3} & r_2 < R \\ \dfrac{1}{r_2^2} & r_2 > R. \end{cases}$$

With our previous result for the magnetic field, we see that the momentum density will only have a φ-component:

(9.32) $$\epsilon_0 E \times B = \hat{\varphi}\frac{\mu_0 q^2 \omega}{80\pi^2 R3}\begin{cases} \dfrac{r_2 \sin\theta_2[6r_2^2 - 5R^2]}{R3} & r_2 < R \\ \dfrac{R^5 \sin\theta_2}{r_2^5} & r_2 > R. \end{cases}$$

When we integrate this result over all space, the integrals all vanish. Recall $\hat{\varphi} = (-\sin\varphi, \cos\varphi, 0)$. As a result, the spinning spherical charge has no linear momentum.

In spherical coordinates, we have $r_2 = \hat{r}r_2$, and $\hat{r} \times \hat{\varphi} = -\hat{\theta}$. Using these results, the angular momentum of the spinning sphere is given by the following:

$$L_{em} = -\frac{\mu_0 q^2 \omega}{80\pi^2 R3}\int d^3 r_2 \, \hat{\theta}\begin{cases} \dfrac{r_2^2 \sin\theta_2[6r_2^2 - 5R^2]}{R3} & r_2 < R \\ \dfrac{R^5 \sin\theta_2}{r_2^4} & r_2 > R. \end{cases}$$

We note that $\hat{\theta} = (\cos\theta\cos\varphi, \cos\theta\sin\varphi, -\sin\theta)$, so the φ integrations will leave only a $\hat{z}$ component. The θ integrals produce a factor of $4/3$. We are left then with the following:

$$L_{em} = \hat{z}\frac{3\mu_0 q^2 \omega}{10\pi R3}\left[\int_0^R dr_2 \frac{r_2^4[6r_2^2 - 5R^2]}{R3} + \int_R^\infty dr \frac{R^5}{r_2^2}\right]$$

(9.33) $$= \hat{z}\frac{\mu_0 q^2 \omega R}{35\pi}.$$

Hence, a spinning, charged sphere has angular momentum due to the fields.

EXERCISE 9.28.   Fill in the missing details of the derivation of Equation 9.33.

EXERCISE 9.29.   One possible interpretation of the angular momentum of the fields is that mass arises solely from the fields and is not a separate entity. A spinning sphere with mass M also has mechanical angular momentum given by $\mathbf{L}_{mech} = \hat{z}\,^2/_5 MR^2 \omega$. Suppose that this is equal to the angular momentum due to the fields in Equation 9.33. Given the measured mass and charge of the electron, what would be its radius R?

The question now is to ascertain whether or not electrons have such intrinsic magnetic fields. If we think about conducting an experiment to study this phenomenon, we can draw on our previous experiences and suggest that if we utilize a beam of electrons, then the presence of an electron's magnetic field should somehow lead to the deflection of the beam. Mapping trajectories, after all, has been our stock in trade throughout most of the text.

The Lorentz force on a charge distribution ρ is given by the following:

$$(9.34) \qquad \mathbf{F} = \int d^3\mathbf{r}_1 \left[\rho \mathbf{E}(\mathbf{r}_1) + \mathbf{J}(\mathbf{r}_1) \times \mathbf{B}(\mathbf{r}_1) \right],$$

where $\mathbf{J}$ is the current density. For a rigid charge distribution, like the spherical model of the electron that we are considering, the current density will consist of two parts. First, there is translation of the center of mass of the distribution: $\mathbf{J}_1 = \rho \mathbf{v}$. This is the component that we have studied previously and that causes charge particles to deflect when passing through a magnetic field. If the charge density is also rotating about its center of mass, then there will be a second component of the current: $\mathbf{J}_2 = \rho \boldsymbol{\omega} \times \mathbf{r}$. We have not yet studied the effect of rotation on the motion of the charge.

So, consider the force arising from an infinitesimal element of charge in our spinning sphere in an external magnetic field $\mathbf{B}$. In spherical coordinates, we have that $dq = \rho r_1^2 \sin\theta_1\, dr_1\, d\theta_1\, d\varphi_1$, so the total force acting on the sphere is the integral over the volume of the sphere:

$$(9.35) \qquad \mathbf{F} = \rho \int_0^R dr_1 \int_0^\pi d\theta_1 \int_0^{2\pi} d\varphi_1\, r_1^2 \sin\theta_1\, (\boldsymbol{\omega} \times \mathbf{r}_1) \times \mathbf{B}(\mathbf{r}_1).$$

Let us use the orientation of the sphere indicated in figure 9.10, where $\boldsymbol{\omega} = \hat{z}\,\omega$. We can then show the following relation is true:

$$(\boldsymbol{\omega} \times \mathbf{r}_1) \times \mathbf{B}(\mathbf{r}_1) = \omega r_1 \sin\theta_1 (B_z \cos\varphi_1, B_z \sin\varphi_1, -B_x \cos\varphi_1 - B_y \sin\varphi_1).$$

If the magnetic field $\mathbf{B}$ is a constant field, then the φ_1 integrals in Equation 9.35 all vanish. Hence, there will be no net force on a spinning sphere in a uniform magnetic field. This is a an interesting conclusion. It means that the trajectory of a spinning, charged sphere is identical to that of a non-spinning, charged sphere, at least in uniform magnetic fields. So, we can exclude using a uniform magnetic field experiment to determine whether or not the electron has an intrinsic magnetic field.

EXERCISE 9.30.   Fill in the details of the derivation of Equation 9.35. Demonstrate that all of the φ_1 integrals indeed vanish.

While there is no net force on a spinning sphere, there will be a torque acting on the sphere. Recall that the torque on an object is defined as $\boldsymbol{\tau} = \mathbf{r} \times \mathbf{F}$. So, the torque acting on the spinning sphere will be given by the following:

$$\boldsymbol{\tau} = \rho \int_0^R dr_1 \int_0^\pi d\theta_1 \int_0^{2\pi} d\varphi_1 \, r_1^2 \sin\theta_1 \mathbf{r}_1 \times \left[(\boldsymbol{\omega} \times \mathbf{r}_1) \times \mathbf{B}(\mathbf{r}_1) \right].$$

$$(9.36) \qquad = \frac{qR^2\omega}{5}(-B_y, B_x, 0) = \frac{qR^2}{5}\boldsymbol{\omega} \times \mathbf{B},$$

where we have assumed that $\mathbf{B}$ is a constant vector with components $\mathbf{B} = (B_x, B_y, B_z)$ and that the charge density was also constant $\rho = q/(4\pi R^3/3)$. We have also used the condition that, initially, $\boldsymbol{\omega} = \hat{\mathbf{z}}\omega$ in the last step.

EXERCISE 9.31.   Equation 9.36 defines the time evolution of the vector $\boldsymbol{\omega}$ because $\boldsymbol{\tau} = I\, d\boldsymbol{\omega}/dt$, where I is the moment of intertia. Define the following equations:

```
eqs={wx'[t]==B wy[t],wy'[t]==-B wx[t],wz'[t]==0}

ics={wx[0]==0.3,wy[0]==0,wz[0]==0.7}

soln=NDSolve[Join[eqs,ics]/.{B->1.0},{wx,wy,wz},
    {t,0,10}]

ParametricPlot3D[Evaluate[{wx[t],wy[t],wz[t]}/.soln],
    {t,0,10}]
```

The plot will trace the time evolution of the angular velocity vector. What happens if you change the initial value of B to other values?

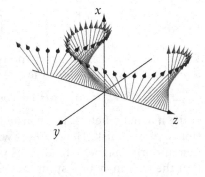

FIGURE 9.12. The angular velocity vector precesses as the spinning sphere moves along the z-axis through a constant magnetic field $\mathbf{B} = \hat{\mathbf{x}}B$

In a constant magnetic field, the torque on the sphere causes the angular velocity vector to precess around the magnetic field direction. This precession is depicted in figure 9.12, where the spinning sphere (and the vector ω) are moving in the z-direction in a region where the magnetic field has only an x-component. Here the x-component of ω will remain constant, the remaining components will vary in a sinusoidal fashion.

In 1922, the German physicists Otto Stern and Walther Gerlach reported the results of their exceptionally clever experiment to determine the intrinsic magnetic moment of the electron.[17] First, the two recognized that sending a beam of electrons through a magnetic field was impractical: the Lorentz force on the electrons would cause a large deflection that could mask the smaller force due to the field gradient. Instead, Stern and Gerlach used a beam of electrically neutral silver atoms. Silver has an atomic number of 47 and, rather fortuitously, 46 of those electrons form a relatively stable core, leaving the electronic properties of silver to be established by just one of the electrons. As a result, the magnetic moment of silver is essentially that of a single electron. Of course, silver had long been used in photographic emulsions, so Stern and Gerlach proposed to analyze their results by the standard methods of film development available to photographers.[18]

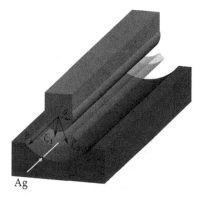

FIGURE 9.13. The magnetic field used by Stern and Gerlach was fashioned from two pole pieces with very different shapes. This produces a large gradient in the magnetic field in the vertical direction. The silver beam (Ag) passes through the region where the gradient is large

A simplified version of the apparatus utilized by Stern and Gerlach is depicted in figure 9.13. The upper pole piece (light gray) has a radius

---

[17]Their papers "Der experimentelle Nachweis der Richtungsquantelung im Magnetfeld," and "Das magnetische Moment des Silberatoms," were published in the German journal *Zeitschrift für Physik* in the spring of 1922.

[18]In Stern's recollection of the experiment, Gerlach handed him the plate onto which silver had been deposited but there was no evidence of silver until Stern's cigar smoke wafted onto the plate. Sulfur in the cigar smoke combined with the silver to make jet black silver sulfide, which was immediately apparent. Stern received the 1943 Nobel Prize in Physics "for his contribution to the development of the molecular ray method and his discovery of the magnetic moment of the proton."

of curvature a and the lower pole piece (dark gray) has a radius of curvature b. From the symmetry of the magnet design, we should find, in the region between the two pole pieces, that the magnetic field is radially directed ($\mathbf{B} = \hat{\boldsymbol{\zeta}}\, B_\zeta$). Consider now a Gaussian surface constructed from two arc segments of radius ζ_1 and ζ_2 that extend for some distance L along the magnet, two arc segments of radius ζ_1 and ζ_2 that extend from some angle $\pm\varphi_1$ in the azimuthal direction and two plane segments that close the surface. (Light gray surfaces in the figure.) From Gauss's law for magnetic fields, we know that the following is true:

$$\oint d\mathbf{A} \cdot \mathbf{B} = 0.$$

(There are no magnetic charges.) Because the field is radially directed, the integrals over the arc segments on the ends ($d\mathbf{A} = \pm\hat{\mathbf{z}}\,\zeta\, d\zeta\, d\varphi$) and the plane segments along the sides ($d\mathbf{A} = \pm\hat{\boldsymbol{\varphi}}\, d\zeta\, dz$) will vanish. The only non-zero contributions to the integral will therefore come from the arc segments that run parallel to the z-axis ($d\mathbf{A} = \pm\hat{\boldsymbol{\zeta}}\,\zeta\, d\varphi\, dz$). If we further assume that the symmetry of the system leads to the magnetic field being approximately constant on the remaining surfaces, then we find the following:

$$\int d\mathbf{A} \cdot \mathbf{B} = -2\zeta_1\varphi_1 L B_\zeta(\zeta_1) + 2\zeta_2\varphi_1 L B_\zeta(\zeta_2) = 0.$$

We can rearrange this result to show that the magnetic field has the following behavior:

$$\frac{B_\zeta(\zeta_2)}{B_\zeta(\zeta_1)} = \frac{\zeta_1}{\zeta_2}.$$

This implies that the magnetic field inside the Gaussian box has the form $\mathbf{B} = \hat{\boldsymbol{\zeta}} B_0/\zeta$, where B_0 is some proportionality constant with dimension of the magnetic field times a length.

Let us now think about what the results of the experiment should be. Let us choose a coordinate system in which x is the vertical direction, z extends along the beam axis and y is in the horizontal direction perpendicular to x and z. We are representing the silver atom as effectively a point charge of $+e$ and an electron of radius R and charge $-e$ centered on the point charge. (This is a dubious model to be sure but it will suffice for now. In truth, it is probably not a bad model for hydrogen, which was studied in 1927 by the American physicists Thomas Phipps and John Taylor. The hydrogen results confirmed that the electron has a magnetic moment.) By virtue of the point charge at the center of the electron, our silver atom will possess the intrinsic magnetic field associated with rotation of the negative charge but will not possess a total charge. As a result, silver atoms passing through the apparatus illustrated in figure 9.13 will not be deflected by the Lorentz force but should feel a force due to the inhomogeneous field.

The angular velocity vectors of our electrons will be randomly oriented $\omega = (\omega_x, \omega_y, \omega_z)$. The magnetic field is vertically oriented at the location of the beam but we want to use a coordinate system centered on the electron, which is displaced initially some distance x_0 from the center of curvature of the upper pole piece of the magnet. The magnetic field of the magnet then has the following form:

$$(9.37) \qquad \mathbf{B} = B_0 \frac{(r_1 \cos \varphi_1 - x_0, r_1 \sin \varphi_1, 0)}{r_1^2 + x_0^2 - 2 r_1 x_0 \cos \varphi_1}.$$

The force on the sphere can be computed from Equation 9.35, with the following result:

$$(9.38) \qquad \mathbf{F} = \frac{3\pi q R^2 B_0}{40 x_0^2} (-\omega_x, \omega_y, 0).$$

The sphere feels a force in both the x- and y-directions, proportional to the field and the components of ω perpendicular to the direction of motion.

> EXERCISE 9.32. To derive Equation 9.38, we assumed that the angular velocity vector had a random orientation $\omega = (\omega_x, \omega_y, \omega_z)$. Fill in the missing details in the derivation of Equation 9.38.

> EXERCISE 9.33. Plot the magnetic field lines of the field defined in Equation 9.37. The StreamPlot function utilizes a rectangular grid, so recall that $\cos \varphi_1 = x_1/r_1$ and $\sin \varphi_1 = y_1/r_1$. Consider the case where $x_0 = 3$ and plot the fields over the domains $-1 \leq x_1 \leq 1$ and $-1 \leq y_1 \leq 1$. Is the field radially directed? Use the StreamDensity-Plot to assess the change in magnitude of the field as a function of x_1.

The sphere also feels a torque, that we can compute from Equation 9.36. We find the following result:

$$(9.39) \qquad \boldsymbol{\tau} = \frac{q R^2 B_0}{5 x_0} (0, -\omega_z, \omega_y).$$

This torque will cause the sphere to precess around the local field direction. At the beam location, the magnetic field is oriented primarily in the x-direction. As a result, the ω_y component of the spin will change as the sphere passes through the magnetic field, as depicted in figure 9.12. The force acting in the y-direction will be alternately positive and negative, modulated in a sinusoidal fashion depending on the orientation of the angular momentum vector. Consequently, the force will tend to cancel over time and there will be no net displacement of the sphere in the y-direction as it traverses the apparatus.

EXERCISE 9.34.   Fill in the missing steps in the derivation of Equation 9.39.

The precession of the vector $\boldsymbol{\omega}$ during the transit does not affect the x-component and there will be a net motion in the x-direction, depending upon the magnitude (and sign) of ω_x. There will be a net acceleration $\mathbf{a}$ of the sphere in the x-direction which is just $a_x = F_x/M$, where M here is the mass of the silver atom not the electron. Motion of the silver atoms in the x-direction is complicated by the fact that the acceleration is not a constant. Nevertheless, there will be a displacement that is proportional to the unknown quantity $\omega_x R^2$.

EXERCISE 9.35.   Define the following equations:

$$\frac{dx(t)}{dt} = v_x(t) \qquad \frac{dv_x(t)}{dt} = \frac{\alpha}{x^2(t)} \qquad x(0) = x_0 \quad v_x(0) = 0.$$

Use NDSolve to solve the equations for $\alpha = 10$ and $x_0 = 3$ over the range $0 \leq t \leq 10$. Plot the solutions. How do these change as α changes?

As $\boldsymbol{\omega}$ is expected to be randomly oriented for silver atoms entering the beam, we should expect that silver will be deposited in a band that extends from some maximum value $x(\omega_{\max})$ to a minimum value $x(-\omega_{\max})$ that depends upon the maximal rotation $\omega_{\max}$. Measuring the width of the band will provide us with an estimate of the angular velocity ω and, hence, the size of the intrinsic magnetic field of the electron. Instead of observing a continuous band of silver, Stern and Gerlach found that the silver atoms that passed through their apparatus separated into just two components, as illustrated below.

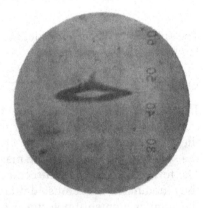

FIGURE 9.14. The first image obtained by Stern and Gerlach utilized a relatively wide beam. At the edges of the beam, there was no deflection of the silver atoms. In the center of the beam, the atoms split into two components (Image provided with kind permission of Springer Science+Business Media)

This image has enormous implications for our classical model of the electron. First, it appears that the angular velocity ω of every electron is the same. Second, the orientation of the angular velocity is restricted to be either aligned with the magnetic field or anti-aligned with the magnetic field: there are no spinning spheres where ω_x is anything but $\pm\omega$. These results are quite difficult to reconcile with our macroscopic experiences with spinning spheres. As billiard balls bang around the surface of a felt-covered tabletop, we can certainly observe them to spin about some axis or another but they do not spin at the same rates. Additionally, as they encounter other balls, whatever spin they had originally is inevitably altered by the collision process.

9.5. Electron Charge

Instead, the results of Stern and Gerlach echo the previous observations of **quantization** in microscopic systems. By the time Stern and Gerlach conducted their experiments, Thompson had already determined the charge to mass ratio of the electron. Subsequently in 1913, the American physicist Robert Millikan published the results of a series of his experiments on oil droplets in which he determined that the charges on the droplets were all integral multiples of some fundamental charge.[19]

Millikan used an atomizer to produce a fine mist of oil droplets between two metal plates. When a large (~10 kV) potential was applied to the plates, most of the droplets were rapidly swept away but a few drifted slowly through the apparatus. Millikan and Fletcher were able to track some of the droplets for tens of minutes, alternately turning off the electric field and allowing the droplets to fall under the influence of gravity and then turning on the electric field and watching the droplets rise under the influence of the electromagnetic force. In each mode, falling or rising, the drops achieved a terminal velocity that Millikan could obtain by measuring the time it took each droplet to travel a specific distance.

In Table 9.1, we display Millikan's data for what he called drop #6. The columns labelled t_g represent the amount of time that the drop took to fall a measured distance under the influence of gravity. The potential was then applied to the metal plates, inducing an electric field in the region where the drop was falling. The Coulomb force on the charged drop was sufficient to cause it to rise back through the measured distance. The columns labelled t_F are the (corrected) times for the drop to travel the

---

[19]Millikan and his student Harvey Fletcher began the experiments in 1908. Unfortunately, University of Chicago rules prohibited Fletcher from using a co-authored paper in support of his Ph.D., so Fletcher was the sole author on a report of their observations of Brownian motion of the oil drops. Millikan claimed sole authorship of the paper on charge quantization and was awarded the Nobel Prize in physics in 1923 "for his work on the elementary charge of electricity and on the photoelectric effect."

measured distance. We note that the fall times are quite similar. Millikan used this observation to infer that the aerodynamic force on the droplet was consistent throughout the measurement interval.

Table 9.1. Millikan's data on drop #6

t_g (s)	t_F (s)	t_g (s)	t_F (s)	t_g (s)	t_F (s)
11.848	80.708	11.816	34.762	11.912	22.268
11.890	22.366	11.776	34.846	11.910	500.100
11.908	22.390	11.840	29.286	11.918	19.704
11.904	22.368	11.904	29.236	11.870	19.668
11.882	140.565	11.870	137.308	11.888	77.630
11.906	79.600	11.952	34.638	11.894	77.806
11.838	34.748	11.846	22.104	11.878	42.302

The presumption in the experimental design is that the droplets will reach a terminal velocity due to the resistive force (Stokes's law) opposing their motion. As we recall from Chapter 6, according to Stokes's formula, the terminal velocity of the drop would be $v_g = mg/\alpha$, where $\alpha = 6\pi\eta R$. That is, the terminal velocity of the falling drop, which is proportional to $1/t_g$, does not depend upon the charge. For the rising drop, the force on the drop is $F = qE - mg$, so the terminal velocity $v_F = (qE - mg)/\alpha$ will now depend linearly on the charge.

Exercise 9.36. Compute the mean and standard deviation of the t_g values from Table 9.1. Plot the data and horizontal lines at the values of the mean and the mean plus and minus the standard deviation. Is there a significant trend over time?

We note from Table 9.1, that the value of t_F changes over the course of the measurements. This reflects the fact that the ionization state of the droplet changes over time due to collisions with molecules in the sample chamber. Millikan employed a somewhat intricate methodology to demonstrate that the charge changes were always an integral multiple of some fundamental charge. First, he computed the values of $1/t_F$ for each of the values found in the table, using averages when there are several runs in a row. (For example $1/t_F = 0.02875$ s^{-1} for the three runs where $t_F \approx 34.75$ s^{-1}.) Next, he calculated the differences between each of those values, which he called $1/t_F - 1/t'_F$. When he divided all of the $1/t_F - 1/t'_F$ values by the smallest of the $1/t_F - 1'_F$ values, what he discovered were that all of the results were nearly integers. The result of this process Millikan called n', the number of electrons that were transferred between the two runs.

EXERCISE 9.37.   For the data in Table 9.1, compute the n' values as described. If we assume an uncertainty of 0.5 % in the value of n', how close are the computed values to the nearest integers?

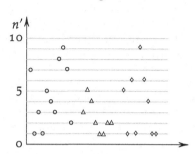

FIGURE 9.15. The changes in electric charge for three different oil drops are always integral multiples of a smallest charge

The n' values for three droplets cited in Millikan's paper are illustrated in figure 9.14. We see that all of the charge changes are by integral amounts. This was precisely the result that Millikan was seeking and enabled him to provide the first precision measurement of the electric charge. Current values of the elementary charge have a precision of two parts in 10^8 and the measurements do not rely on the rather error-prone strategy that Millikan employed. Nevertheless, Millikan's experiments provided significant support for the nascent atomic theory of matter.

Millikan's work was not without controversy in his own time. The Austrian physicist Felix Ehrenhaft was also measuring the value the electron charge, using colloidal metals. Ehrenhaft, though, found values of the charge that were significantly smaller than that obtained by Millikan, which he termed subelectrons. Millikan and Ehrenhaft engaged in a scientific debate that ran for over two decades and was not even silenced by Millikan's award of the Nobel Prize in Physics in 1923. Each continued to refine his measurements and, in each subsequent publication, provided not only an account of their own recent results but also a critique of the other's experiments. Reportedly, their strident debate persuaded the Nobel Prize committee to step back from awarding the 1920 Prize to Millikan.

EXERCISE 9.38.   Derive the expressions for the terminal velocities for falling and rising droplets. Show that the total number of electrons on the drop is proportional to the quantity $1/t_g + 1/t_F$. This result establishes that the change in number n' will be proportional to the quantity $1/t_F - 1/t_F'$.

EXERCISE 9.39.   Millikan's experiments relied on obtaining the terminal velocity by means of measuring the time the droplet required to travel a prescribed distance (the distance between two marks on the microscope reticule). A potential source of error is an ioniz-ing event that occurs during the timing interval. Suppose that the droplet loses one electron a third of the distance along the path. If the droplet begins with five, eight or ten electrons initially, how does that affect the measured t_F? (Assume that the droplet reaches its new terminal velocity instantaneously.) If the subsequent run has a droplet with six electrons, what would be the n' values for these cases?

The debate over quantization of the electron charge has long been set-tled, with Millikan on the prevailing side, but his work remains con-troversial. The ethics of omitting Fletcher's name from the 1913 paper, even in the acknowledgments, has been widely discussed. Additionally, in 1978, the American physicist and historian of science Gerald Holton pub-lished an analysis of Millikan's laboratory notebooks, which he found in the archives at the California Institute of Technology. In Millikan's note-books, it is clear that he measured many more droplets than he included in his publications. The criteria by which Millikan selected his data have been criticized, with some proposing that Millikan's methodology rises to the level of scientific fraud. Millikan certainly found droplets where the analysis demonstrated a fractional charge and excluded those from his publications. In his defense, we can see from Millikan's notebooks that he invariably had some sort of objections to the experimental conditions that caused him to question the reliability of those measurements. Millikan was, of course, painfully aware of how difficult it was to acquire valid data and used his own personal experiences to apply a personal quality-control filter on the data.

9.6. Unfinished Business

More recent experiments with improved measurement technologies have demonstrated that the electron does have a quantized charge, a quantized mass and a quantized intrinsic magnetic field; these values are among the most precisely known quantities in all of science. This result is not necessarily to be interpreted as vindication of Millikan's methodology or ethics but, even in science, it is frequently better to be lucky than good.

So at this point in the chapter, we have arrived at a rather dubious out-come. We have constructed a classical representation of the electron that is in direct conflict with experiment: the electron displays quantization effects that cannot be derived from a classical model. In a real sense, it is very discomforting to slog through a string of formidable calculations only to get the wrong answer. That is not to say that we made any alge-braic errors on the path to Equations 9.38 and 9.39. Indeed, those results stand as written but, alas, do not describe the results of the experiment illustrated in figure 9.15. It was just this sort of difficulty that confronted physicists in the 1920s: classical models of microscopic entities did not conform to experimental results. A new physical model of microscopic phenomena was required.

Indeed, a number of other, alarming results were obtained when trying to formulate classical models of the electron. As we mentioned previously, the electromagnetic fields carry momentum. If we are interested in the kinematics of charged particles, we must somehow incorporate the momentum of the fields. Students will encounter this topic in an advanced course on electrodynamics but there are curious results that arise when one attempts to deal with this issue. For example, when a charged mass is subjected to a force for a specific, finite time interval, the theory predicts that the mass will either fly away to infinity or begin accelerating before the force is applied! This behavior certainly does not seem feasible from a physical standpoint.

The subsequent course in electrodynamics will introduce the electromagnetic potentials V and $\mathbf{A}$ from which we can obtain the fields by differentiating:

$$\mathbf{E} = -\nabla V - \frac{\partial \mathbf{A}}{\partial t} \quad \text{and} \quad \mathbf{B} = \nabla \times \mathbf{A},$$

where the symbol ∇ is a shorthand for taking spatial derivatives:

$$\nabla = \left(\frac{\partial}{\partial x}, \frac{\partial}{\partial y}, \frac{\partial}{\partial z} \right).$$

For a charge distribution ρ and a current density $\mathbf{J}$, the potentials have the following integral representations:

(9.40)

$$V(\mathbf{r}_2, t_2) = \int d^3 \mathbf{r}_1 \, \frac{\rho(\mathbf{r}_1, t_1)}{|\mathbf{r}_2(t_2) - \mathbf{r}_1(t_1)|}$$

$$\mathbf{A}(\mathbf{r}_2, t_2) = \int d^3 \mathbf{r}_1 \, \frac{\mathbf{J}(\mathbf{r}_1, t_1)}{|\mathbf{r}_2(t_2) - \mathbf{r}_1(t_1)|}.$$

We shall not try to solve these equations but need to point out a particular subtlety that the equations embody. The potentials (and by inference the fields) at some point in space $\mathbf{r}_2$ at some time t_2 reflect the charge distribution ρ and current density $\mathbf{J}$ at some remote point $\mathbf{r}_1$ at the specific time t_1 where:

(9.41)

$$t_1 = t_2 - \frac{|\mathbf{r}_2 - \mathbf{r}_1|}{c},$$

where c is the velocity of light. This is known as the *retarded* time.[20] It is precisely the amount of time that a signal will take to travel the distance $|\mathbf{r}_2 - \mathbf{r}_1|$ when propagating at the velocity of light.

For an extended charge or current distribution ($R > 0$), points within the distribution are spatially separated, meaning that they are described by space-like vectors and do not share a common time. Now for an electron, which we presume to be very small, we could guess that this won't matter very much but, in fact, it does. A number of physicists have written on the subject. Notably, the English physicist Paul Dirac tried to demonstrate in 1938 that classical electrodynamics of extended objects was a self-consistent theory, albeit one that did not agree with experiment. As most physicists at that time, and subsequently, were more intent on understanding quantum mechanics, further development of the unresolved theoretical issues was abandoned. As appealing as the classical model of the electron might be, it does not describe the nature of the electron.

As a final note, we should point out that the French physicist Pierre-Simon marquis de Laplace tried applying the idea of retarded time to Newton's gravitational theory. In our derivations in the early chapters we didn't bother to address an important point: "How does mass M_1 know where M_2 is?" If we define v_g to be the velocity of gravity, then we should modify all of the earlier equations for the gravitational field to include a factor of $t_1 = t_2 - |\mathbf{r}_2 - \mathbf{r}_1|/v_g$ that represents the finite propagation velocity of gravity. In 1805, Laplace developed a model in which the earth was attracted to the point where the sun was at the retarded time and discovered that he no longer recovered elliptical trajectories.

This was, of course, an horrific result. Elliptical orbits for planets were well-established experimentally. What we can infer from Laplace's calculations, of course, is that the velocity of gravity v_g must be very large compared to the orbital velocities of the planets. As a result, we can neglect it and utilize Newton's equations of motion. In developing his own gravitational theory, Einstein made the presumption that $v_g = c$, which is, of course, very large compared to the orbital velocities of the planets. As no one has, as yet, detected gravity waves and measured their propagation velocity directly, it is commonly assumed that $v_g = c$. Proving otherwise will be left as an exercise for the reader.

---

[20]Advanced time solutions are also possible but are excluded for violating causality.

X

Modern Technology

The fall of Rome and rapid depopulation of the city of Rome itself was due, in large measure to the severing of the large aqueducts that supplied the city with water. Civilization, at least insofar as measured by the ability of peoples to live in high densities, can be measured then by its plumbing. Indeed, after any natural disaster in which essential services are disrupted, restoration of the water distribution and sewage treatment facilities are specific points of emphasis.

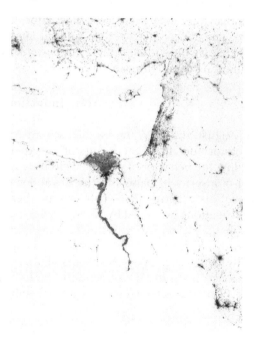

FIGURE 10.1. The image depicts the eastern Mediterranean and the Nile River valley at night. It is printed here as a negative (Image provided courtesy of NASA Earth Observatory/NOAA GFDC)

With the title Modern Technology, though, we are thinking of the transformative technologies that arose in the early 1900s that converted the theoretical ideas of Maxwell, Biot and Ampère, *et cetera*, into practical

© Mark A. Cunningham 2015
M.A. Cunningham, *Neoclassical Physics*, Undergraduate Lecture Notes in Physics, DOI 10.1007/978-3-319-10647-2_10

devices. Indeed, the exploitation of electromagnetic phenomena significantly overlaps the theoretical development of the subject. While physicists were concerned with questions like "What is the nature of matter?" others were trying to harness electrical power to industrial applications. The success of the enterprise is illustrated in figure 10.1 that depicts the eastern Mediterranean at night. With the exception of a few historical districts that retain gaslights, most cities and towns now utilize electrical energy to illuminate their streets at night.

From the figure, one can easily discern the islands of Crete and Cyprus and the Mediterranean coastline. Clearly, humans prefer to dwell near water, as is particularly evidenced by the continuous development along the Nile. Cairo is the dark spot below the Nile delta and Luxor is located at the bottom of the U-shaped bend in the river near the bottom of the figure.

The fact that one can now illuminate great spaces with the flick of a switch is now accepted without question yet cathode ray tubes were instruments of great public interest just a relatively few years earlier. What captured the imagination of many of those early researchers, as well as the public, was that light emanated from those tubes without any form of combustion. The advantages over gas and oil lamps were immediately obvious and creative individuals quickly undertook the mission of making electrical lighting commonplace.

10.1. Induction

On August 29, 1831, the English scientist Michael Faraday noted in his laboratory notebook the results of an experiment he was conducting regarding the behavior of electrical circuits. In November, Faraday read the first of a series of papers to the Royal Society in which he provided details of his experiments.[1] This was another of an extraordinary series of investigations produced by Faraday, who is widely considered the best experimentalist in history. While Faraday is claimed by physicists as one of their own, Faraday also produced significant results in chemistry, eventually being appointed as the first Fullerian Professor of Chemistry at the Royal Institute. His election to Fellow of the Royal Society in 1824 was a singular achievement for a blacksmith's son in the class-conscious

---

[1] Faraday's "Experimental Researches in Electricity" were published in the *Philosophical Transactions* of the Royal Society of London in 1831, establishing his precedence over the American Joseph Henry who conducted similar experiments in New York at approximately the same time.

England of the 1800s.[2] Faraday was even subsequently offered appointment as President of the Royal Society, a post held previously by Isaac Newton, among others, but declined to focus his remaining years on his experimental investigations.

FIGURE 10.2. Faraday wrapped two lengths of copper wire around an iron ring (*dark gray*). A battery connected across A and B produced a momentary voltage across C and D

In the summer of 1831, Faraday had fabricated an iron ring, around which he wrapped two coils of insulated copper wire, as indicated in figure 10.2. We recognize that the copper coil AB wrapped around the iron ring will generate a magnetic field in its interior when attached to a battery, i.e., when it is carrying a current. From our previous experiences with wire loops and solenoids, we can assume that the field within the coil will be relatively uniform and directed along the axis of the coil. We can also assume that computing it exactly will be a relatively formidable task, so we shall be content with a crude description for now. What we haven't discussed previously is that the magnetic permeability of iron is very large compared to that of free space, or air. As a result, the magnetic field lines emanating from the solenoid do not disperse like those in figure 9.5. Instead, the magnetic field lines remain almost entirely confined within the iron. The iron ring serves as a conduit for the magnetic field generated in the coil AB.

What Faraday observed on that fateful day in August was that, when coil AB was connected to the battery, a galvanometer attached to coil CD deflected momentarily. When Faraday disconnected the battery, the galvanometer deflected in the opposite direction. Reversing the direction of the connections to the coil AB resulted in a reversal of the effects on the galvanometer. That is, a current arose only when the magnetic field passing through the coil *changed* and the direction of the induced current depended upon the direction of the initial magnetic field.

Faraday had no formal mathematical education but possessed an extraordinary physical insight. While he might not have been able to express

---

[2] Faraday was apprenticed to a book binder at age thirteen but read voraciously and attended public lectures to better himself. He had the audacity to ask Humphrey Davy, a leading member of the Royal Society, for a position. Davy politely declined initially but shortly thereafter sacked one of his assistants for fighting. Faraday's interview had impressed Davy sufficiently that he offered Faraday the vacant position.

his ideas with sophisticated mathematical treatments, he possessed an uncanny ability to conduct experiments that exposed the detailed workings of complex systems. Additionally, Faraday possessed the ability to describe his ideas in a clear and insightful manner that made his writings accessible to a broad audience. For example, the concept of magnetic lines of force, or magnetic field lines, was introduced by Faraday.

Faraday's observations that August demonstrated that a changing magnetic field induces an electromagnetic potential. We can summarize Faraday's experimental results with the following mathematical representation:

$$(10.1) \qquad \oint_{\partial S} d\mathbf{s} \cdot \mathbf{E} = -\frac{d}{dt} \int_S d\mathbf{A} \cdot \mathbf{B}.$$

If the magnetic field through some surface S changes over time, then an electromotive force is induced along the boundary of S. Hence, the path integral on the left hand side is taken over the boundary ∂S of the surface S used in the surface integral on the right hand side. Faraday's law, as Equation 10.1 is known, was not written in this form by Faraday himself, who might have considered it ironic that a sophisticated mathematical expression now bears his name.

Nevertheless, in large measure, Faraday's law provides the underpinnings for our modern electrical world. Consider that electrical fields (and consequently currents $\mathbf{J} = \sigma\mathbf{E}$ in conductors) can be generated by a changing magnetic flux. This can be achieved, as Faraday showed initially in August of 1831, by changing the current that produces the magnetic field. He soon recognized that a current can be generated by changing the area $d\mathbf{A}$, or more efficiently, by changing the orientation of the surface with respect to the magnetic field. This principle is illustrated in figure 10.3, where a rectangular loop is positioned in a magnetic field $\mathbf{B}$ that is nominally vertically oriented. The magnetic flux through the surface will depend upon the angle θ between the vectors $\mathbf{A}$ and $\mathbf{B}$. If the loop is rotated about the y-axis with an angular velocity $d\theta/dt = \omega$, then the product $\mathbf{A} \cdot \mathbf{B}$ will vary in time sinusoidally, like $\sin \omega t$.[3] This, in turn, will generate a current that varies like $\cos \omega t$.

> EXERCISE 10.1.   For a constant field $\mathbf{B}$, the integral over the loop surface is just $\mathbf{A} \cdot \mathbf{B}$. When the orientation of $\mathbf{A}$ changes sinusoidally with an angular velocity ω, show that the voltage in the loop will vary sinusoidally.

As a result, all one has to do to generate currents is to rotate a loop in a magnetic field. This provides an extraordinarily simple means for converting mechanical energy into electrical energy that can be transmitted

---

[3]There is an arbitrary phase that we have suppressed. Generally, the time variation would be $\sin(\omega t + \phi)$, where the phase ϕ depends upon the orientation of the loop at time $t = 0$.

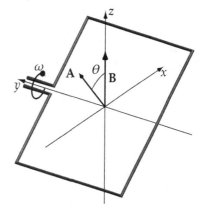

FIGURE 10.3. A loop rotating in a magnetic field **B** will generate a current that varies sinusoidally

over wires and used for whatever purposes one wants. Instead of an industrial revolution that demands local steam engines to power equipment, one can simply convert the energy of a steam engine into electrical currents. Thus the end users do not require noisy, dirty engines on their premises. The ability to cleanly deliver energy not only to industrial facilities but also to residential areas dramatically reshaped civilization.

Notice that figure 10.3 also depicts a design of a motor. A wire carrying a current in a magnetic field feels a torque. So, if on the one hand we forcibly spin a loop of wire in a magnetic field, then a current will be generated in the wire. If, on the other hand, we run a current through the wire in a magnetic field, then the magnetic torque will cause it to rotate. This is somewhat miraculous: the mechanical energy of a rotating turbine in a distant power plant is converted into currents that travel many kilometers over wires and then are converted back into mechanical rotation of vacuum cleaner motors and refrigerator compressors. This is a hugely more user-friendly use of technology than powering one's refrigerator with a gasoline engine like one finds in a lawn mower.

In addition, Faraday's initial experiment in 1831 showed the way for electrical distribution systems. The electrical power in a system is determined by the product of current and voltage: $P = IV$. Ohm's law indicates that there will be a drop in voltage due to electrical resistance, so there is a transmission power loss proportional to the square of the current: $P_{\text{loss}} = I^2 R$. Obviously, to minimize transmission losses, one wants to operate the electrical distribution system at high voltages and low currents but one does not want the end users to deal with high voltages. Faraday's ring is an example of what is termed a *transformer*. The voltage in the coil CD will be proportional to the ratio of the number of turns in the two coils and the original voltage in the coil *AB*. Just by changing the number of turns of wire, one can change one voltage into another. Thus one can readily transmit power from the generating station at very high voltages and then step it down to more modest voltages for consumers with apparatus that is no more complicated than coils of wire.[4]

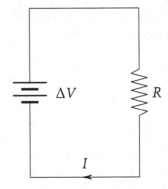

FIGURE 10.4. A simple electrical circuit can be represented by symbols that reflect the original embodiment of the elements

10.2. Circuits

Georg Ohm demonstrated the fundamental principle that the current *I* in a circuit is proportional to an electromotive force that we now term the potential. This is displayed schematically in figure 10.4. The potential source (ΔV) is depicted as a series of short and long lines that reflect the original voltaic piles that were constructed from alternating disks of zinc and copper. The resistance of the circuit (*R*) is reflected in the zigzag line that idealizes the tortuous path that electrons follow (see the Drude model). Figure 10.4 is, of course, an idealization of a real circuit. As Ohm demonstrated, the wires themselves have resistance, whereas the figure implies that all of the resistance is embedded in some discrete element. As a practical matter, the resistive element illustrated in the figure generally represents some electrical load that consumes vastly more electrical energy to do work than is lost to the resistance of the connecting wires. As a result, we can often neglect the resistance of the connecting wires; including it will make only a minor correction.

Ohm's initial investigations were extended to the problem of multiply-connected circuits by the German physicist Gustav Kirchhoff. Kirchhoff

---

[4]The American inventor Thomas Edison envisioned utilizing constant current dynamos for power distribution. He installed several of his bulky devices in American cities, with commercial operation of the Pearl Street Station in New York City beginning on September 4, 1882. Edison was rapidly supplanted in the marketplace by the superior alternating current technology of Russian emigré Nikolai Tesla and his industrial partner George Westinghouse.

recognized that coupled circuits gave rise to coupled mathematical equations and demonstrated that a set of linearly-independent equations could be developed for any number of circuits.[5] Kirchhoff's first principle we now recognize as simply another statement of the conservation of charge:

I. At any node in the circuit, the sum of the currents entering the node must equal the sum of the currents exiting the node.

Kirchhoff's second principle can be stated as follows:

II. The sum of the voltages around any closed loop in the circuit is zero.

In figure 10.4, there is an increase in voltage across the battery of ΔV and a subsequent drop in voltage across the resistor such that the following relation holds:

$$\Delta V - IR = 0.$$

For a single loop, this is just Ohm's law, albeit written in terms more familiar to engineers and circuit designers. As a practical matter, it is easier to measure the total current I flowing through a wire than the current density $\mathbf{J}$ within the wire. It is also easier to measure potential differences ΔV than electric fields $\mathbf{E}$. Note also that in figure 10.4 we associate the potential to a discrete element like a battery. Kirchhoff's second principle also holds for induced potentials like those discovered by Faraday.

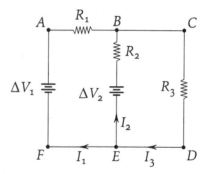

FIGURE 10.5. A more complex circuit, with multiple components

Consider the somewhat more complex situation illustrated in figure 10.5. Here there are two coupled circuits and we must utilized Kirchhoff's principles to obtain the equations that govern the behavior of the system. A typical problem involves solving for unknown currents I_i given values for

---

[5]Kirchhoff was a student of the eminent mathematician Franz Ernst Neumann when his "Über den Durchgang eines elektrischen Stromes durch eine Ebene, insbesondere durch eine kreisförmige," stating his two circuit principles was published in the *Annalen der Physik* in 1845. His demonstration that a unique solution existed was provided in "Über die Auflösung der Gleichungen, auf welche man bei der Untersuchung der linearen Vertheilung galvanischer Ströme geführt wird," also published in the *Annalen* in 1847.

the potentials ΔV_i and resistances R_i. In this situation, an initial choice for the direction of each current must be made but once those directions are chosen, the resulting system of equations is then self-consistent and can be solved uniquely. The final currents may be negative, reflecting that the actual current direction is opposite to that initially chosen.

For each element in the circuit, one considers the potential difference across the element. For a battery, this is just ΔV and for a resistor the potential difference is IR, where I is the current. The signs are obtained by following a directed path around the loop. The potential increases when traversing a battery from negative to positive and decreases when traversing a resistor in the direction of the current flow. The signs are reversed if the direction of the path is reversed.

EXERCISE 10.2.　Consider the circuit drawn in figure 10.5, which has two connected loops. For the current directions specified in the figure, write the equations that result from Kirchhoff's first principle for all of the nodes in the system A–F. (Most of these will be degenerate.)

Write the three loop equations utilizing Kirchhoff's second principle for the loops $ACDF$, $ABEF$ and $BCDE$. What is the loop equation for the loop $AFDC$ (direction is important)?

EXERCISE 10.3.　Use the Solve function to solve the system of equations generated in the previous Exercise for the currents I_1, I_2 and I_3. Show that using only the loop equations does not lead to a soluble system and that one of the current equations must be utilized. Under what conditions will I_2 be negative?

EXERCISE 10.4.　Consider the arrangement of resistors at right. What are the currents in each of the resistors? What is the total current?

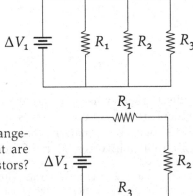

EXERCISE 10.5.　Consider the arrangement of resistors at right. What are the currents in each of the resistors? What is the total current?

Of course, in Ohm's initial investigations the resistive elements were just wires of varying lengths and thicknesses. It is a an approximation, as we have done in the past few examples, to consider the wires to be resistance-less and to attribute all of the resistance to a single element. In practice, this approach proves to be a reasonable one, at least at low frequencies. We can imagine the electrical appliances of a house, for example, as a collection of resistive elements and depict refrigerators, lamps and ovens equally as the zigzag line forms that we have used to this point. Opera-tionally, of course, these appliances have rather distinct uses but, electri-cally, they can be represented as resistors.

The tidy picture of electrical circuits that we have developed is compli-cated when the voltage source is time-varying. As we have mentioned, electrical distribution by alternating currents was preferred technically, so we must incorporate Faraday's law of induction into our discussions. Referring to figure 10.4, if the constant voltage source is replaced with a sinusoidally-varying source, then the current will be sinusoidally-varying as well. The current generates a time-varying magnetic field and the changing magnetic flux induces an additional potential in the loop, as per Equation 10.1.

In this instance, the area of the loop is unchanging, so Equation 10.1, can be rewritten as follows:

$$\oint ds \cdot \mathbf{E} = - \int d\mathbf{A} \cdot \frac{d\mathbf{B}}{dt} = -L \frac{dI}{dt}.$$

Here, we have defined the **inductance** L of the system to be the (compli-cated) result that remains after we factor out the time derivative of the current.[6] Again, at low frequencies, it is appropriate to treat the induc-tance as a lumped element, like resistance.

We can retain Kirchhoff's second principle in its original form if we also include any inductance terms. This, of course, complicates matters be-cause the potential due to inductive effects is proportional to the time derivative of the current, not the current itself. We are now faced with solving differential equations.

Consider the simple circuit illustrated in figure 10.6, where we have fol-lowed the practice of lumping all of the resistance into a single resistor and all of the inductance into a single inductor, which is identified by the coil symbol. We do this even though the inductance arises from the loop

---

[6]Note that here the symbol L does not represent the magnitude of the angular momentum vector. It would stand to reason that we use the symbol I for inductance but we are already using that symbol to denote current. To avoid confusion, it has become standard practice to use L to denote inductance in honor of the Russian physicist Heinrich Friedrich Emil Lenz, who also contributed to the understanding of electrical circuits.

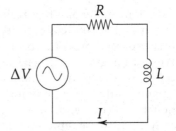

FIGURE 10.6. A time-varying voltage source is indicated by the *circle* with an embedded sine curve. The electrical potential arising from the inductance *L*, identified by the coil, depends upon the time derivative of the current *I*

as a whole. Again, our lumped element approximation is generally a good one at low frequencies. The circuit equation that arises from this example is given by the following:

$$\Delta V - IR - L\frac{dI}{dt} = 0.$$

If we assume that the driving voltage is sinusoidal, $\Delta V = V_1 \sin\omega t$, where ω is the angular frequency, we can obtain a solution if we assume that the current is of the form $I(t) = A\sin\omega t + B\cos\omega t$.[7] Substituting into the loop equation, we find the following:

(10.2) $V_1 \sin\omega t - R(A\sin\omega t + B\cos\omega t) - L(A\omega\cos\omega t - B\omega\sin\omega t) = 0.$

The functions $\sin\omega t$ and $\cos\omega t$ are linearly independent, so we collect terms containing each function:

$$\sin\omega t[V_1 - RA + L\omega B] + \cos\omega t[-RB - LA\omega] = 0.$$

The functions $\sin\omega t$ and $\cos\omega t$ do not vanish for arbitrary values of ωt, so the terms in the brackets must vanish if the equation is to hold. We now have two separate equations for the two unknowns A and B.

We can write the solution as follows:

(10.3) $$I(t) = \frac{V_1}{R} \frac{\sin\omega t - (\omega L/R)\cos\omega t}{1 + (\omega L/R)^2}.$$

From dimensional considerations, it is apparent that the quantity R/L has the dimension of a frequency (T^{-1}). We can thus surmise that the quantity V_1/R has the dimension (Q/T). Further, from Equation 10.3, we can see that the current does not oscillate in phase with the driving voltage. At high frequencies $(\omega \gg R/L)$, the current is dominated by the cosine term, as can be seen in figure 10.7. Engineers will frequently describe this phenomena by a phase angle ϕ: $I(t) = I_1 \sin(\omega t + \phi)$. At high frequencies, when the current is essentially a cosine function, the phase angle becomes $\phi = -\pi/2$.

---

[7]It is somewhat cleaner mathematically to use complex numbers in our derivations but we shall defer that discussion to subsequent courses.

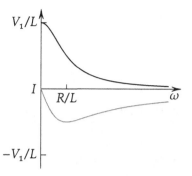

FIGURE 10.7. The current has a component that varies like $\sin \omega t$ (*black*) and a component that varies like $\cos \omega t$ (*gray*). Beyond the characteristic frequency R/L, the current is dominated by the cosine term

EXERCISE 10.6.   Use the Solve function to solve the circuit equation that arises from Equation 10.2. Show that they can be put into the form displayed in Equation 10.3.

EXERCISE 10.7.   Plot the voltage ($\sin \omega t$) and current (Equation 10.3) over the range $0 \le t \le 20$. (Let $\omega = 1$.) Use the Manipulate function to vary the inductance over the range $0 \le L/R \le 5$. What is the phase ϕ when $\omega = R/L$? Can you verify the claim that at higher frequencies (when $\omega > L/R$) the voltage leads the current?

EXERCISE 10.8.   Repeat the previous analysis for a source $\Delta V = V_2 \cos \omega t$. Solve for the current and plot the results. Show that this model provides expected results at zero frequency (DC).

In addition to the inductive effects that arise in circuits with time-varying sources, there is an additional phenomena termed **capacitance** that was discovered by early researchers. Prior to Volta's development of the voltaic pile, a number of devices that were capable of generating sizable electrostatic charges were invented. In 1745, a means for storing charge independent of the generating machine was independently developed by the German bishop Ewald Georg von Kleist and the Dutch scientist Pieter van Musschenbroek and his student Andreas Cuneaus. The apparatus has subsequently become known as a Leyden jar after the town in which van Musschenbroek and Cuneaus conducted their experiments.[8]

The Leyden jar, as sketched in figure 10.8, is simply a bottle with a conductive inner liner and, in a refinement suggested by van Musschenbroek after his initial experiment, a conductive outer sheath. (Van Musschenbroek's hand served as the conductor when the bottle was originally

---

[8]Van Musschenbroek's experimental results were reported to the Paris Academy of Science in 1746 by his correspondent René Antoine Ferchault de Réaumur. Van Musschenbroek's report of a (literally) breathtaking shock delivered by the device spurred widespread interest.

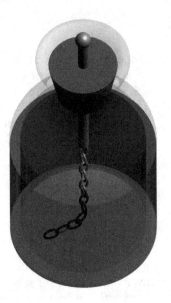

FIGURE 10.8. The Leyden jar consists of a glass bottle where the lower part of the interior surface is covered with a conducting material like gold leaf. Similarly, the outer surface is coated with a conductive material. The electrical connection to the inner conductor is usually made via a metal rod and a metal chain. Early investigators often filled the bottle with water

charged, resulting in a painful shock.) When a source of electric charge contacts the central conductor, charge is conveyed to the inner liner. Today we understand this process to involve the transport of electrons onto the inner liner. This establishes an electric field in the vicinity of the inner liner. The field acts on electrons in the outer liner, pushing them away, leaving a (relatively) positive charge on the surface of the outer sheath. At equilibrium, the charge redistributes such that the field is essentially contained in the space between the two conductors.

Somewhat surprisingly, at least to the early researchers, when the source of charge is removed, the Leyden jar retains its charge. The electrostatic potential V produced by the charge Q turns out to be proportional:

$$(10.4) \qquad\qquad Q = CV,$$

where the constant of proportionality C is called the *capacitance* of the Leyden jar. The capacitance reflects the capacity of the device to store charge.

These early experiments gave credence to the idea that charge was a material substance that could be manipulated. The simplicity of the Leyden jar provoked experiments by a host of individuals not usually associated with science. The American statesman Benjamin Franklin, for example, conducted a number of studies and discerned that the bottle and its contents were not essential to the process of storing charge. What was essential was two conductors separated by an insulator. As a result, Franklin noted that one could electrify pictures of people by coating the back of the picture with a conductor, such as gold leaf, placing the picture under glass and adding a metal crown to the front. After charging, any attempt to remove the crown would result in an electric shock to the unwary.[9]

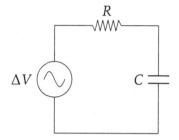

FIGURE 10.9. A capacitor C is denoted by the *two parallel lines* perpendicular to the direction of current flow. This is a stylistic representation of the parallel plates of early capacitor designs

If we consider an electric circuit containing a capacitor, such as depicted in figure 10.9, the Kirchhoff equation describing the system can be written as follows:

$$\Delta V - IR - \frac{Q}{C} = \Delta V - R\frac{dQ}{dt} - \frac{Q}{C} = 0,$$

where we have used the fact that the current I is the time derivative of the charge Q. If we again assume that the voltage is varying sinusoidally, $\Delta V = V_1 \sin \omega t$, we can look for solutions for the charge that are functions of sin and cos:

$$Q(t) = A \sin \omega t + B \cos \omega t.$$

Substituting into the loop equation, we find the following:

(10.5) $V_1 \sin \omega t - R\omega[A \cos \omega t - B \sin \omega t] - \frac{1}{C}[A \sin \omega t + B \cos \omega t] = 0.$

If we again collect terms multiplying sin and cos separately, we obtain two equations for the two unknown coefficients A and B:

(10.6) $\sin \omega t[V_1 + BR\omega - A/C] + \cos \omega t[-AR\omega - B/C] = 0.$

Each of the terms in brackets in Equation 10.6 must vanish separately if the equation is to hold.

We find then that the charge has the following solution:

(10.7) $Q(t) = \frac{V_1}{R} \frac{(1/RC)\sin \omega t - \omega \cos \omega t}{(1/RC)^2 + \omega^2}.$

---

[9]Franklin communicated his ideas in a letter to his colleague Peter Collinson in April of 1749. Franklin suggested using a portrait of the King of England and soliciting unwary subjects by declaring that the device was designed to test loyalty to the Crown.

The quantity $1/RC$ has the dimension of a frequency (T^{-1}), whereupon the product RC has the dimension of time (T). The current is just the time derivative of charge:

$$(10.8) \qquad I(t) = \frac{V_1}{R} \frac{\omega^2 R^2 C^2 \sin\omega t + (\omega RC)\cos\omega t}{1+\omega^2 R^2 C^2}.$$

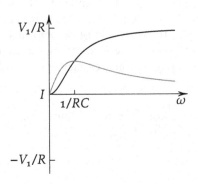

FIGURE 10.10. The current in a circuit with a resistor and capacitor has a component that oscillates in phase with the drive voltage (*black curve*) and a component that oscillates with a phase angle of $\pi/2$ (*gray curve*). At high frequencies, the in-phase component dominates

A plot of the two components of the current is illustrated in figure 10.10. At high frequencies, the current is in phase with the driving voltage. This differs from our previous results with inductors.

EXERCISE 10.9. Use the Solve function to solve the circuit equation and demonstrate that Equation 10.8 is the correct description of the current as a function of frequency. Plot the amplitudes of the terms involving sin and cos over the range $0 \le \omega RC \le 5$.

EXERCISE 10.10. Repeat the analysis leading up to Equation 10.8 but with a source term $\Delta V = V_2 \cos\omega t$.

EXERCISE 10.11. Use the Manipulate function to vary the capacitance over the range $0 \le RC \le 5$. Plot the drive voltage and current for Equation 10.8 for $\omega = 1$ over the domain $0 \le t \le 20$. What happens to the phase angle ϕ over the range?

EXERCISE 10.12. Van Musschenbroek and his followers did not have access to time-varying voltage sources. Instead, in their circuits, the source in figure 10.9 is replaced with a switch. Assume that the capacitor holds a charge Q_1 and that at time $t = 0$, the switch is closed. Solve for the charge and current and plot the results. (Hint: the resulting circuit equation can be integrated directly to solve for the charge. Any constants of integration can be defined by asserting the initial condition that $Q(0) = Q_1$.)

10.3. Resonance

If one combines all of the components that we have discovered to this point into a single circuit, as depicted in figure 10.11, we find an unexpected behavior. The circuit equation can be written as follows:

$$(10.9) \qquad \Delta V - IR - L\frac{dI}{dt} - \frac{Q}{C} = \Delta V - R\frac{dQ}{dt} - L\frac{d^2Q}{dt^2} - \frac{Q}{C} = 0,$$

where we now are confronted with a second-order differential equation for the charge Q.

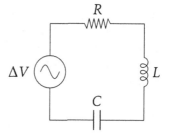

FIGURE 10.11. A circuit containing resistive R, inductive L and capacitive C elements displays resonant behavior

If we again assume that the driving voltage is sinusoidal, $\Delta V = V_1 \sin \omega t$, we can try to find solutions as before. We guess that $Q(t) = A \sin \omega t + B \cos \omega t$ and derive the following auxiliary equation:

$$V_1 \sin \omega t - R\omega[A\cos \omega t - B\sin \omega t] + L\omega^2[A\sin \omega t + B\cos \omega t]$$

$$(10.10) \qquad -\frac{1}{C}[A\sin \omega t + B\cos \omega t] = 0.$$

If we now separate the terms involving sin and cos, we find the following:

$$(10.11) \quad \sin \omega t[V_1 + BR\omega + AL\omega^2 - A/C] + \cos \omega t[-AR\omega + BL\omega^2 - B/C] = 0.$$

We again demand that the terms in brackets in Equation 10.11 vanish separately. This provides us with the following solution for the charge:

$$(10.12) \qquad Q(t) = \frac{V_1}{L}\frac{(1/LC - \omega^2)\sin \omega t - (R\omega/L)\cos \omega t}{(1/LC - \omega^2)^2 + (R\omega/L)^2}.$$

We note that the term $(1/LC)$ has the dimension of frequency squared (T^{-2}) so the product LC must have dimension of time squared (T^2).

The current in the circuit is simply the time derivative of the charge:

$$(10.13) \qquad I(t) = \frac{V_1}{L}\frac{(R\omega^2/L)\sin \omega t + \omega(1/LC - \omega^2)\cos \omega t}{(1/LC - \omega^2)^2 + (R\omega/L)^2}.$$

There are now two characteristic frequencies that govern the system. The first is the resonant frequency $\omega_0 = 1/\sqrt{LC}$. The current will be maximal at the resonant frequency, as the term $(1/LC - \omega^2)$ in the denominator vanishes. The width of the resonance is controlled by the dissipative term $\omega_d = R/L$. In the absence of resistance, the current diverges at the resonant frequency.

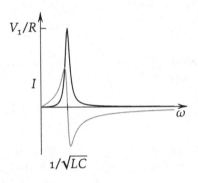

FIGURE 10.12. The LRC circuit displays resonant behavior. The current is maximal at the frequency $\omega_0 = 1/\sqrt{LC}$. At that point the current is in phase with the drive voltage. The in-phase component (*black curve*) vanishes away from the resonant frequency and the out-of-phase component (*gray curve*) changes sign at the resonant frequency

EXERCISE 10.13.   Use the Solve function to verify Equation 10.13. Plot the current components over the range $0 \le \omega \le 5$. Use the values $L = 1, C = 1$ and $R = 0.2$. What happens if you make the resistance R larger or smaller?

EXERCISE 10.14.   Plot the driving voltage and current over the domain $0 \le t \le 20$ for a frequency of $\omega = 1$. Use the Manipulate function to study the phase angle of the current and driving voltage as a function of inductance L and resistance R for a constant capacitance $C = 1$. Vary the inductance over the range $0 \le L \le 5$ and the resistance over the range $0 \le R \le 2$.

EXERCISE 10.15.   What is the limit of Equation 10.13 for $L \to 0$? Do your results seem reasonable? (Hint: what were the results for RC circuits?) What is the current at the resonant frequency $\omega^2 = 1/LC$?

Resonant phenomena are encountered in many areas of physics. Students are most likely familiar with resonances in mechanical systems. Guitar strings and piano wires, for example, support oscillations at particular frequencies determined by the wire length, wire diameter and tension in the wire. Such oscillations are often called *standing waves* due to the fact that the lateral displacement along the wire is sinusoidally varying in time and space. The end points of the wires are fixed in guitars and pianos, so the waves are fixed in place, with locations of zero amplitude (nodes) occurring at the ends.

Skyscraper designers have to contend with the fact that their buildings also display resonant behavior. Unlike guitar strings that are pinned at both ends, buildings have one free end. This modifies the resonant frequency (lowering it by a factor of two) but it means that buildings would experience large lateral displacements at the free end (roof) if systems to dampen those oscillations were not in place. Real skyscrapers have much more complex construction than simple wires but have low frequency bending modes that resemble the simple modes of guitar strings. Here, we will approximate the complex behavior of a building with a simple model but structural engineers develop much more complex models before committing to spend hundreds of millions of dollars on construction.

The Taipei 101 tower (height 509 m) uses a passive system to provide motion dampening. In essence, a large mass suspended from the top of the tower acts like a pendulum. Illustrated in figure 10.13, the spherical mass (660 mg) was too large to be lifted in one piece. Instead, the sphere was constructed on site by welding together forty-one steel plates, each of which has a thickness of 12.5 cm. The sphere hangs from cables attached to the building frame. If the building sways in the wind or is subjected to surface motion due to seismic events, the sphere does not feel those forces directly. This results in relative motion between the building frame and the sphere. Large hydraulic shock absorbers dissipate that relative motion. As a result, the tower's lateral motion during earthquakes and typhoons will be vastly reduced from what would occur without the dampening system in place, thereby increasing the likelihood that the tower will not suffer major structural damage during such potentially hazardous events.

EXERCISE 10.16. Consider a small mass M hanging from a thin wire of length L. If the mass is displaced by an angle θ from the vertical, show that the magnitude of the torque acting on the mass is given by the following expression:

$$\tau = MLg\sin\theta,$$

where g is the gravitational acceleration. Without damping, the equation of motion for this simple pendulum is given by the following equation:

$$ML^2\frac{d^2\theta}{dt^2} + MLg\sin\theta = 0.$$

For small angles, $\sin\theta \approx \theta$. Using the small angle constraint, we obtain a second-order differential equation for θ, like that for the charge in Equation 10.9. Consider an angular-velocity dependent dissipative term: $\alpha d\theta/dt$. Write the equation of motion for θ, including the dissipative term and a sinusoidal driving term: $C_1\sin\omega t$. Compare to Equation 10.9. What is the resonant frequency of the system? What is the dissipation frequency?

FIGURE 10.13. The mass damper in Taipei 101 is constructed of 41 circular steel plates, forming an approximately 5 m diameter sphere. The (660 mg) mass is suspended by four cables that attach to the building frame at the 91st floor. Hydraulic shock absorbers (*bottom* of the figure) attached to the building frame at the 87th floor serve to dampen any relative motion between the building framework and the sphere

As observed in the previous exercise, the equations of motion of mechanical systems can often be found to be described by a second-order differential equation like Equation 10.9. Such systems will display the same sort of resonant behavior illustrated in figure 10.12. The physical parameter under consideration: angular displacement, for example, will achieve maximal amplitudes when the system is excited at the resonant frequency.

10.4. Optical Molasses

We have mentioned earlier that atoms emit light at particular frequencies; each set of frequencies for a given element is known as the atomic spectrum. The atoms can both absorb and emit light at these characteristic frequencies. A mechanism for producing very cold atoms involves using a laser beam that is tuned to a frequency that is just below one of the resonant frequencies of a particular element. If cold atoms are illuminated by the beam, the atoms will "see" a laser field that is Doppler-shifted due to the motion of the atoms. Those atoms that are moving in the same direction as the laser is propagating will see the laser shifted to lower frequencies (red-shifted). Those atoms moving in the opposite direction will see the laser shifted to higher frequencies (blue-shifted). Once the laser is blue-shifted, the photons can be resonantly captured by the atoms (provoking a transition to an internal state of higher energy). After a time, the atom will re-emit the photon.

> EXERCISE 10.17. To see the effect of the Doppler shift, consider transforming to a coordinate system in which the atom is at rest. The four-momentum of a photon travelling in the x-direction in the laboratory frame is given by the following:
>
> $$\mathbf{p}_\gamma = \begin{bmatrix} h\nu/c & h\nu/c & 0 & 0 \end{bmatrix}^T.$$
>
> Apply a boost in the x-direction to obtain the four-momentum in the laboratory frame $\mathbf{p}'_\gamma$. Recall that $\cosh \zeta = [1 - (u/c)^2]^{-1/2}$ and $\sinh \zeta = (u/c)[1 - (u/c)]^{-1/2}$, where u is the velocity of the atom in the laboratory frame. Show that the frequency of the photon in the rest frame of the atom is given by the following relation:
>
> (10.14)
> $$\nu' = \nu \left[\frac{1 + u/c}{1 - u/c} \right]^{1/2}.$$
>
> Plot ν'/ν over the domain $-1 \le u/c \le 1$.

Cooling the atoms relies the fact that absorption and emission of the photons changes the momentum of the atom by the momentum of the photon. At first glance, this would seem to have no net effect but actually produces a resistive force on atoms of the form $\mathbf{F} = -\alpha \mathbf{v}$, as we have encountered before. To see how this resistive force arises, we first note that all of the photons in the original beam have the same momentum, say along the x-direction, whereas the emitted photons have random directions. As the processes of absorption and emission are repeated many times, the average momentum of the absorbed photons is just the momentum of any beam photon, but the average momentum of the emitted photons is zero. Consequently, there is a net momentum change along the x-direction.

If we direct laser beams in both directions along the x-direction (counter-propagating beams), the momenta and, hence, velocities of the atoms in the x-direction can be reduced. With lasers aligned in all three spatial directions, the atoms can be resonantly cooled to extraordinarily low temperatures. Of course, putting this relatively simple idea into practice was a non-trivial exercise and won a Nobel Prize in Physics for Stephen Chu, Claude Cohen-Tannoudji and William D. Phillips.[10] It is now possible to obtain diffuse gases with temperatures in the microKelvin range using

---

[10]Chu, Cohen-Tannoudji and Phillips were awarded the 1997 Prize "for development of methods to cool and trap atoms with laser light."

these techniques. The availability of ultracold atoms has enabled development of a number of other applications, notably precision atomic clocks.

> EXERCISE 10.18. The x-component of the momentum of the laser beam is just $p = h\nu/c = h/\lambda$, where h is Planck's constant, ν is the frequency and c is the velocity of light. Alternatively, λ is the wavelength. The absorbed beam photon will always have the momentum $p\hat{x}$ whereas the emitted (recoil) photon is randomly oriented in three dimensions. As a result, the x-component of the recoil photon can have any value in the range $-p$ to p. Use the RandomReal function to generate a large number of real numbers in the range -1 to 1 and then find their average. What is the average of 100 trials? How about 1000 trials and 10,000 trials?

10.5. Telecommunications

The discovery of resonance in electrical circuits was instrumental in developing the first radio receivers capable of detecting voices or music. The desired broadcast signal was obtained by modulating the amplitude of a high-frequency carrier signal. Human voices have a frequency content that generally lies below a few thousand Hertz (Hz).[11] The standard (U.S.) frequency band for AM broadcasting is 540–1610 kHz; each broadcast channel is separated by 10 kHz. Use of AM broadcasting permitted signals to be transmitted by multiple, independent stations that could then be separated into individual components at the receiver.

A radio receiver can be constructed to be resonant at different carrier frequencies by employing either variable capacitors or variable inductors, or both. In early receivers, often the antenna itself provided significant capacitance, so tuning was accomplished by sliding a contact along a long coil (inductor). (The inductance of a coil is proportional to its length.) When the receiver was resonant with the carrier frequency, the voltage across the inductor was a maximum, as seen in figure 10.12. A depiction of this strategy is provided in figure 10.14.

The receiver was tuned to the carrier frequency of the broadcast signal by adjusting the variable inductor L_1. Headphones were required to listen to the broadcast. The headphones were constructed from a small piece of iron attached to a paper disk and a coil of wire, as sketched in figure 10.15. A fluctuating current in the coil of wire creates a time-varying magnetic field that vibrates the iron. The paper disk was generally fixed at the outer

---

[11]The original Bell Telephone system had a frequency cutoff of 3 kHz. People in those days had a different "telephone voice." AM radio broadcasts have a bandwidth limit of 10 kHz and the effects of eliminating the high frequency content of the signal is much less noticeable.

FIGURE 10.14. Early radio receivers used an *LRC* circuit
(*left*) for tuning. Conversion
of the broadcast signal ΔV
into sound was accomplished
by rectifying the current with
the device D_1 and filtering out
the carrier signal with the
capacitor C_2. The headphones
behave like an inductor L_2 in
series with a resistor R_2

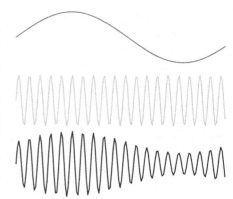

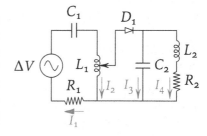

FIGURE 10.15. Early transducers
consisted of an iron cylinder (*dark
gray*) attached to a flexible (paper)
disk that was fixed at the outer
edge. A time-varying current in
the coil moves the iron, causing the
larger disk to vibrate and produce
sound waves

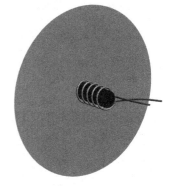

edge. Motion of the iron driver creates sound waves and the larger area
of the disk increases the intensity of the sound. As a result of the design,
the headphones behave electrically as an inductor but also possess a non-
trivial resistance because the coil is formed from many turns of fine wire.

FIGURE 10.16. A signal (*top, gray*) can be
combined with a carrier (*center, lightgray*)
frequency to produce an amplitude-
modulated broadcast signal (*bottom,
black*)

A key element in the radio receiver circuit is a new, nonlinear circuit element known as a diode, represented by the barred triangle D_1. The
diode is a *semiconductor* material that passes current in only one direction. Hence, a sinusoidal current is *rectified* such that only the positive

lobes of the sine function are passed; the negative lobes (oppositely directed currents) are suppressed. Early receivers used the mineral galena (lead sulfide) and a spring-loaded point contact known as a cat's whisker to form the diode.

To understand the function of the diode, let us assume that the driving signal can be written as $\Delta V = V_1(1 + \alpha \sin \omega_1 t)\sin \omega_2 t$, where ω_2 is the carrier frequency and ω_1 is a low-frequency amplitude modulation. The factor α is generally small compared to unity, as depicted in figure 10.16. The product of two sine functions can be written in terms of a sum of cosines: $\sin x \sin y = (\cos(x-y)-\cos(x+y))/2$, so the driving voltage will be a sum of sine and cosine terms:

$$\Delta V = V_1 \left\{ \sin \omega_2 t + \frac{\alpha}{2} \left[\cos(\omega_2 - \omega_1)t - \cos(\omega_1 + \omega_2)t \right] \right\}.$$

As a result, the broadcast signal actually consists of the carrier frequency ω_2 and two sidebands that are found at the sum and difference frequencies $\omega_2 - \omega_1$ and $\omega_2 + \omega_1$. There is no frequency component at the modulation frequency ω_1.

Rectifying the signal produces harmonics of the signals, with the net result that a component with frequency ω_1 is restored. Harmonic generation can be quantified by use of ideas introduced by the French mathematician Jean Baptiste Joseph Fourier, who described a means for decomposing uniquely any waveform into a series of sine or cosine functions of different frequencies.[12] We shall not endeavor to pursue Fourier analysis in detail here.

EXERCISE 10.19. Plot the rectified sine function.

EXERCISE 10.20. Consider the function $F_1(t)$ defined by the sum

$$F_1(t) = \sum_{j=1}^{N} \frac{\sin j\pi t}{j}$$

on the domain $0 \leq t \leq 1$. Plot $F_1(t)$ for $N = 5, 50$ and 500.

Suppose that we define $F_2(t)$ analogously to $F_1(t)$ but where only odd values of j are included in the sum. Plot $F_2(t)$. How do the functions compare?

---

[12]Fourier's *Théorie analytique de la chaleur* was published in 1822. The work provided the foundations for what is now called Fourier analysis and also introduced the concept of dimensional analysis that we have used throughout the text.

EXERCISE 10.21. Consider the function $G_1(t)$ defined by the sum

$$G_1(t) = \frac{1}{\pi} + \frac{1}{2}\sin t - \frac{2}{\pi}\sum_{j=1}^{N}\frac{\cos 2jt}{4j^2 - 1}.$$

on the domain $0 \le t \le 4\pi$. Plot $G_1(t)$ for $N = 5, 10$ and 50.

So, we can approximate that the result of incorporating the diode in the receiver circuit is this: the current leaving the diode in figure 10.14 $(I_3 + I_4)$ contains a component at a frequency of ω_1 and several other terms at frequencies of ω_2 or higher. Let us then consider a simplified circuit of just C_2 and the headphones with a driving current $I = a\sin\omega_1 t + \sin\omega_2 t$. We can utilize Kirchhoff's rules to write a series of linear equations defining the system: We find the following:

$$\frac{Q_3}{C_2} - L_2\frac{dI_4}{dt} - R_2 I_4 = 0,$$

(10.15)
$$a\sin\omega_1 t + \sin\omega_2 t - I_3 - I_4 = 0.$$

EXERCISE 10.22. Show that the Equations 10.15 with a single source term $I_0\sin\omega t$ lead to the following system of linear equations for the charges:

$$I_0 + \omega B_3 + \omega B_4 = 0 \qquad A_3/C_2 + A_4 L_2\omega^2 + B_4 R_2\omega = 0$$
$$A_3 + A_4 = 0 \qquad B_3/C_2 + B_4 L_2\omega^2 - A_4 R_2\omega = 0$$

Use the Solve function to solve for the unknown coefficients.

EXERCISE 10.23. Use the coefficients from the previous Exercise to define functions for the currents I_3 and I_4. Choose $\omega_1 = 3140\,\mathrm{s}^{-1}$ (500 Hz signal frequency) and that $\omega_2 = 3.14 \times 10^6\,\mathrm{s}^{-1}$ (500 kHz carrier frequency). Use the values $L_2 = 2.4\,\mathrm{mH}$, $R_2 = 2\,\Omega$ and $C_2 = 10\,\mathrm{nF}$. Let the input signal be $I = 0.1\sin\omega_1 t + \sin\omega_2 t$.

Plot the currents I_3 and I_4. Note that the period of oscillation of the carrier is $T_2 = 2\,\mu s$ and the period of oscillation of the signal is $T_1 = 2\,\mathrm{ms}$. What is the signal across the headphones (I_4)?

The capacitor C_2 is known as a bypass capacitor because at high frequencies, the current will preferentially flow through C_2 and not the headphones. To see how this works, we can define the **impedance** Z of a circuit element in analogy to Ohm's original equation $V = IR$. Formally, we state that for any element, the impedance is defined as follows:

(10.16)
$$V = IZ.$$

So, obviously, resistors have an impedance that is just their resistance: $Z = R$. For inductors and capacitors, we have the difficulty that the potentials associated with those elements depend on the derivative or integral of the current and are, consequently, frequency dependent. Additionally, there is a phase that arises when we take derivatives of sine functions. Dealing with these phases is, again, most neatly solved by employing complex numbers but we shall avoid introducing that subject here.

Instead, note that if the current is $I = A \sin \omega t + B \cos \omega t$ then taking the derivative with respect to time produces a factor of ω:

$$dI/dt = \omega(A \cos \omega t - B \sin \omega t).$$

Integrating the current, produces a factor of ω^{-1}:

$$\int dt\, I = (-A \cos \omega t + B \sin \omega t)/\omega.$$

Ignoring the phases for the time being, it appears that the magnitude of the impedance due to an inductor will be $|Z_L| = \omega L$. Similarly, the magnitude of the impedance of a capacitor will be $|Z_C| = -1/\omega C$.

The idea behind the bypass capacitor is that, at high frequencies, the impedance across C_2 becomes smaller and smaller. As a result, the high-frequency carrier signal will predominantly flow through the capacitor C_2. With an appropriate choice of value for the bypass capacitance, the current associated with the lower-frequency modulation of the carrier will flow predominantly through the headphones.

10.6. Vacuum Tubes

Widespread public acceptance of radio was limited by technical issues afflicting the original crystal radios. The essential cat's whisker diode element was notoriously tricky. Operators would have to find sweet spots on the crystal surface and apply just the right pressure to the contact for the semiconductor properties to be manifest. Additionally, only a single person could listen to the broadcast at any one time. The headphones were powered by the broadcast signal itself.

These problems were overcome by a series of inventions that relied on the properties of free electrons in an electric field; in these devices, the current is quite literally moving charges. In 1873, the British physicist Frederick Guthrie observed that a red-hot iron sphere held close to a positively-charged electroscope would discharge the electroscope. If the electroscope were negatively charged, the sphere had no effect. What Guthrie discovered in 1873 was thermal electron emission from the surface of the iron. A voltage-regulating device utilizing this effect was patented by the

American inventor Thomas Edison in 1884.[13] for use as a voltage regulating device. In 1904, the English physicist John Ambrose Fleming made use of the "Edison effect" to produce what he called a thermionic valve; one is sketched in figure 10.17.

FIGURE 10.17. Fleming's thermionic valve used two electrodes, called the anode and cathode. A heater coil (*dark gray*) served to create thermionic electron emissions from the cathode (inner cylinder). (Early versions of the device used the heater coil as the cathode.) Currents were controlled with the voltage difference $V_a - V_c$. The assembly was enclosed in an evacuated glass bulb to improve the sensitivity

The principle behind Fleming's thermionic valve is that electrons emitted from the cathode will only migrate to the anode if the potential difference $V_a - V_c$ is positive. If the potential difference is reversed, the electrons will be repelled from the anode. Consider now placing an oscillatory signal, like $V_1 \sin \omega t$, on the cathode. If we set $V_a = 0$, then current will flow across the gap when $V_1 \sin \omega t$ is less than zero. The thermionic valve rectifies the input signal! Fleming recognized that this would solve the problem of the unreliability of the cat's whiskers.[14] His invention required a power source to heat the cathode but greatly improved the stability of radio receivers.

> EXERCISE 10.24. Use Gauss's law and Ampère's law to solve for the electric field between the anode and cathode. Assume that the charge per unit length on the cathode is constant: $Q/L = CV_c/L$, where C is the capacitance. What is the direction of the force on an electron if $V_a - V_c$ is positive?

A key development in the history of electronics came shortly after Fleming's invention of the vacuum tube diode. The American inventor Lee de Forest devised a vacuum tube with three electrodes that he called the Audion.[15] The Audion was intended for use as a radio receiver detector but it

---

[13]Edison's U.S. Patent 307030 "Electrical meter" was the first issued in the United States for an electrical device.

[14]Fleming's device consists of two electrodes. Hence, it has subsequently been denoted as a diode, from the Greek δίς, meaning twice.

[15]De Forest's U.S. Patent 879532 "Space telegraphy" was issued in 1908. His earlier patents on a two-electrode version of the Audion provoked a patent suit by Fleming. In his testimony for the suit, de Forest admitted he did not understand how the Audion worked.

was quickly recognized that the additional electrode provided the ability to amplify signals. The triodes, like the one sketched in figure 10.18, that evolved from the original Audion were pivotal elements in the commercial success of electronic devices. Radios and, later, televisions[16] relied heavily on vacuum tubes to detect and amplify small broadcast signals.

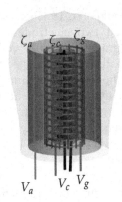

FIGURE 10.18. A triode contains another electrode, known as the grid electrode, between the cathode and anode. The grid electrode was formed by spirals of wire, or in this case, a wire mesh

To understand the operation of the triode, let us consider a simplified model in which the electrodes are infinite cylinders. The electric field in the region between the cathode and grid electrodes will be radially directed: $\mathbf{E} = \hat{\zeta} E$, where we utilize a cylindrical coordinate system. By symmetry, the field will be constant on the cylinder of radius ζ, where $\zeta_c \leq \zeta \leq \zeta_g$. The electric flux through a cylinder of length $z = L$, is just $\Phi_E = 2\pi\zeta L E$.

From Gauss's law, we know that Φ_E is a constant and, from the definition of potential, we can deduce the following relation:

$$(10.17) \qquad \int_{\zeta_c}^{\zeta_g} d\zeta\, \frac{\Phi_E}{2\pi\zeta L} = V_g - V_c.$$

We can solve this readily to show that the electric field in the region between the cathode and grid is given by the following:

$$(10.18) \qquad \mathbf{E} = \hat{\zeta}\, \frac{V_g - V_c}{\ln(\zeta_g/\zeta_c)}\, \frac{1}{\zeta}.$$

Hence, the force on an electron (charge $-e$) in this region $\mathbf{F} = -e\mathbf{E}$ will be positively directed only when $V_g - V_c$ is negative.

---

[16]The Austrian physicist Robert von Lieben procured a patent for a three-electrode cathode ray tube in 1906 that included magnetic deflection of the cathode rays. Credit for invention of what we call triodes (from the Greek τρία meaning three) is usually given to de Forest.

EXERCISE 10.25.   Fill in the details of the derivation of Equation 10.18. Plot the (normalized) field over the domain $0.1 \leq \zeta \leq 1$.

The work done on an electron as it moves from cathode to grid is obtained by integrating the force along the path. This is just $W = -e(V_g - V_c)$. Hence, if we assume the thermionic electrons have small kinetic energies at the cathode, electrons at the grid have a kinetic energy $T = -e(V_g - V_c)$. Recall that the current density can be written as $J = \rho v$, where ρ is the charge density. Hence, the current through a cylindrical surface of length L at ζ_g will be given by the following:

$$(10.19) \qquad I_g = JA = \rho_g \left[\frac{-2e(V_g - V_c)}{m_e} \right]^{1/2} 2\pi \zeta_g L.$$

EXERCISE 10.26.   Fill in the details of the derivation of Equation 10.19.

The electric field in the region between the grid and cathode can be written analogously. We find the following:

$$(10.20) \qquad \mathbf{E} = \hat{\zeta} \frac{V_a - V_g}{\ln(\zeta_a/\zeta_g)} \frac{1}{\zeta}.$$

Similarly, the current at the anode can be written as follows:

$$(10.21) \qquad I_a = JA = \rho_a \left[\frac{-2e(V_a - V_g)}{m_e} \right]^{1/2} 2\pi \zeta_a L.$$

If we now take the ratio of the currents, we have:

$$\frac{I_a}{I_g} = \frac{\rho_a \zeta_a}{\rho_g \zeta_g} \left[\frac{V_a - V_g}{V_g - V_c} \right]^{1/2}.$$

We know that charge is conserved and, thus, that the charge densities must be related.

Consider a thin, cylindrical shell. It has a volume $V = 2\pi \zeta L \, d\zeta$. An amount of charge Q distributed throughout the volume has a charge density of $\rho = Q/2\pi \zeta L \, d\zeta$. Hence, we must have that $\rho_a/\rho_g = \zeta_g/\zeta_a$ and that the current ratio from grid to anode is just given by the following expression:

$$(10.22) \qquad \frac{I_a}{I_g} = \left[\frac{V_a - V_g}{V_g - V_c} \right]^{1/2}.$$

If one makes the anode voltage much larger than the grid voltage, the current at the anode is also much larger than that at the grid. This is the definition of amplification.

In our simple model, we have not accounted for the fact that the presence of the wire mesh will block some of the electron flow from the cathode and that the field of a wire mesh is not exactly constant at the radius ζ_g. Constructing refined models of triode behavior became something of an industry in the electrical engineering world, eventually becoming quite sophisticated. Additionally a number of refinements of triode technology improved their technical capabilities. Four- and five-electrode devices ensued.[17] For our purposes, though, we have demonstrated that triodes can amplify signals.

The electronic age blossomed as a result. While early practitioners often did not understand the workings of their devices theoretically, they developed empirical models that enabled radio, and later television, become commercial successes. At the foundation of it all were the trajectories of electrons in electric and magnetic fields, governed by the Lorentz force law.

Vacuum tubes have now been supplanted from most commercial devices by solid state components. Unlike the unreliable cat's whisker diodes, modern semiconductor manufacturing techniques have developed to the point where solid state equivalents of the tubes are cheaper, more reliable and use less power to operate. One might imagine that, inside the semiconductors, there are electrons being pushed about by the Lorentz force law. If so, much of what we have already discussed will apply. That issue, however, will be the subject for future courses.

---

[17]Not surprisingly, these are called tetrodes and pentodes from the Greek roots τέτταρες (four) and πέντε (five).

XI

Emergent Phenomena

We have concentrated thus far on producing simple representations of complicated phenomena. This is precisely what is meant by the physicist's reductionist methodology. This approach has proven remarkably successful in describing a number of physical phenomena but would appear to have little value in treating truly complex systems.

In this final chapter, we shall introduce the process by which we can begin to understand truly difficult systems. This was the focus of the research by the first American PhD in Physics: J. W. Gibbs, who was interested in the problem of translating our knowledge of the microscopic behavior of systems into a description of their macroscopic behavior.

In a gas, for example, we fully expect that each collision between gas molecules will follow the rules of behavior that we have already established. Momentum and energy will be conserved. As a practical matter, we could not hope to deduce the properties of a volume of gas by studying the individual collisions of the vast number of molecules in the volume. As we have seen previously, the number of atoms in a macroscopic volume is quite large: the interatomic spacing in crystals is of the order of 10^{-10} m. As a result, we would expect roughly 10^{30} atoms in a cubic meter of material. It is unthinkable to compute 10^{30} trajectories of individual atoms.

What Gibbs demonstrated is quite remarkable. We can abandon our strategy of determining the trajectories of individual constituents but still retain our ability to make precise predictions about the behavior of systems. This is accomplished through statistical analysis and has led to the field of study that physicists call **statistical mechanics**. Predictability hinges on the key fact that, in the limit of a large number of trials, the probability distribution becomes the physical distribution. Let us try now to clarify what this statement means.

© Mark A. Cunningham 2015
M.A. Cunningham, *Neoclassical Physics*, Undergraduate Lecture
Notes in Physics, DOI 10.1007/978-3-319-10647-2_11

11.1. Probability

We have used elements of probability previously; statistical inference allowed us to interpret experimental data in terms of various models. As we mentioned at the time, laboratory experiments cannot be used to prove some hypothesis. They can certainly disprove hypotheses by contradiction but experimental results that agree with an hypothesis do not constitute proof in the mathematical sense. We have suggested that momentum is conserved and all experiments to date are consistent with this assertion. Yet it is possible that someone might conduct an experiment in which momentum is not conserved. It is, of course, unlikely but the ultimate laws of nature may not forbid such behavior.

To quantify this idea of likeliness, let us begin with the simple model of coin tossing. A coin toss has two possible outcomes: heads H or tails T.[1] If the coin is well-balanced, the probability of either outcome is one half. If the coin is biased, then if the probability of heads is p, the probability of tails is $1 - p$. That is, all tosses result in an outcome. Separate coin tosses are considered completely separate events; the probability of an individual toss is not affected by history.

We use the coin toss model because it is simple to understand and we can enumerate the outcomes. For example, tossing the coin three times produces one of the eight following possible outcomes:

HHH   HHT   HTH   THH   HTT   THT   TTH   TTT

There is only one way to obtain all three tosses as heads and one way to obtain three tails but there are three ways to arrive at either two heads and a tail or two tails and a head.

EXERCISE 11.1.   What are the possible outcomes of tossing a coin four times?

The general problem of tossing a coin N times can be formulated mathematically. Indeed, the Swiss mathematician Jacob Bernoulli discussed the problem in his landmark treatise *Ars Conjectandi*.[2] In the work, which is

---

[1] The author once managed to drop a penny that landed on its edge but this was quite literally a once in a lifetime event. Here we are considering ideal coins that do not admit a third state.
[2] Bernoulli began working on the *Ars Conjectandi* in 1684 but failed to complete the work before his death in 1705. Jacob's younger brother Johann, also an accomplished mathematician, could have finished the text but their intense sibling rivalry led to a severing of all ties in 1697. It fell to Jacob's nephew Niklaus Bernoulli to arrange for publication of *Ars Conjectandi*, printed finally in 1713.

taken as the foundation of modern statistical analysis, Bernoulli demonstrates that N coin tosses leads to the following distribution of outcomes:

$$(11.1) \qquad P(N,p,k) = \frac{N!}{k!(N-k)!} p^k (1-p)^{N-k}.$$

Here $P(N,p,k)$ is known as the binomial distribution and k is the number of times in N tosses that one obtains heads.[3]

> EXERCISE 11.2. Plot the binomial distributions for $N = 20$ and $p = 0.5, 0.6$ and 0.7 as a function of k. How does the distribution change when the probability p of an individual toss changes?

The distribution function is normalized:

$$\sum_{k=0}^{N} P(N,p,k) = 1.$$

That is a mathematical expression of the fact that N tosses of the coin will lead to one of the possible outcomes, independent of the probability p. Indeed, all that we can say about a random process is that the outcome will be selected from the list of possible outcomes; we cannot predict which of those will occur.

In the *Ars Conjectandi*, Bernoulli stated his Law of Large Numbers. This asserts that, when many experimental trials are performed, the average of the experimental results will tend toward the probability distribution. For example, we have stated that the probability of obtaining heads in a coin toss is one half. If we toss the coin once, it will be either heads or tails, with a resulting experimental probability of either one or zero. If we toss the coin 10 times, we might observe six occurrences of heads, from which we deduce a probability of six-tenths. The Law of Large Numbers states that, if we were to toss the coin millions of times, we should see a probability close to the expected one.

This is an important result. We can (and will) utilize a large number of experimental trials to establish the probability distribution for systems. Comparing the results of such experiments to models allows us to impute some level of validity to our models. There is a subtlety here that is also important to understand. In our coin tossing experiment, there is no law of nature preventing the coin from coming up heads ten thousand times in a row. Each trial is independent of all the others. The binomial distribution suggests that the probability of such an occurrence is given by $P(10,000, \frac{1}{2}, 10,000) = 2^{-10,000} \approx 5 \times 10^{-3011}$. This is highly unlikely. As a

---

[3]We are using the somewhat nonstandard notation P to represent the probability distribution in order to forestall upcoming difficulties in distinguishing amongst the variable names for pressure, probability, probability distribution and momentum.

result, were we to observe a coin toss that resulted in heads ten thousand times in a row, we could rightly infer that the probability of heads was very close to unity $p \approx 1$. This does not constitute a proof that $p = 1$ but we can infer that $p = \frac{1}{2}$ is very unlikely.

If we have some parameter x that is governed by a probability distribution like the binomial distribution, then the **expected value** (weighted average) of the parameter $\bar{x}$ will be obtained as follows:

$$(11.2) \qquad \bar{x} = \sum_{k=0}^{N} x \, P(N, p, k).$$

As an example, we can compute the expected value of k itself:

$$(11.3) \qquad \bar{k} = \sum_{k=0}^{N_{\exp}} k \, P(N_{\exp}, p, k) = N_{\exp} \, p.$$

Suppose that we flip a coin a large number $N_{\exp}$ of times and observe that heads occurs N_H times. We would naïvely expect that the probability of obtaining heads is given by $p_{\exp} = N_H / N_{\exp}$. Our personal expectations would be that in a thousand or so coin flips, we ought to see about five hundred heads. From Equation 11.3, our quantitative estimate of the value of $\bar{k}$ is precisely the expected value of the number of heads we should expect in N trials.

> EXERCISE 11.3.   Show that the result depicted in Equation 11.3 is correct. Hint: Note that the $k = 0$ term in the summation vanishes. Rewrite the sum in terms of a new variable $j = k - 1$.

The second moment of the binomial distribution also has a simple form:

$$(11.4) \qquad \mathrm{var}(k) = \sum_{k=0}^{N} (k - \bar{k})^2 \, P(N, p, k) = N p (1 - p).$$

We utilize the standard deviation $\sigma(k) \equiv \sqrt{\mathrm{var}(k)}$ frequently in physics applications. This is due to the fact that the standard deviation has the same dimensionality as the expected value. The standard deviation defines our precision.

> EXERCISE 11.4.   Demonstrate that the result in Equation 11.4 is correct. Derive a formula for the variance in terms of the quantities $\bar{k}$ and $\overline{k^2}$.

11.2. Brownian Motion

Consider now coupling our coin tossing with movement. If we obtain heads, take a step in some direction, call it east. If we obtain tails, take a step in the opposite direction. The question is, after twenty coin tosses, where will you be? The answer, of course, is "I don't know."

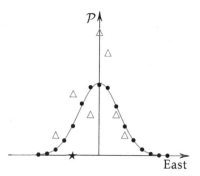

FIGURE 11.1. Random walks based on coin tossing are distributed according to the binomial distribution. For large numbers ($N = 10000$, *black dots*), the physical distribution is quite close to the probability distribution (*gray curve*). For small numbers, like one (*star*) or twenty (*triangles*) this is not true

You could be anywhere in the range from twenty paces east to twenty paces west, albeit only on even-numbered spaces because we tossed the coin an even number of times. Consider the results illustrated in figure 11.1. In this instance, you wind up several paces to the west of your starting point. If we also consider sending off the twenty members of your class, we might find them distributed like the triangles in the figure. (Note that we have scaled the distribution by the number of walkers to produce an estimate of the probability distribution $\mathcal{P}$.) Six of them wind up where they started and the remainder are scattered. There is no obvious pattern to their distribution.

If we instead consider sending off all ten thousand students in your University, then we find them distributed in a manner very close to that predicted by the binomial distribution. In this instance, we do not observe any students either 20 paces east or west of the starting point. The probability of such an occurrence is given by $\mathcal{P}(20,0,20)$ or $\mathcal{P}(20,20,20) = 2^{-20} \approx 10^{-6}$. It would be unlikely for one of ten thousand students to achieve a one in a million result. We would need to conduct vastly more trials to observe such a rare event.

EXERCISE 11.5. Use the RandomReal function to generate random numbers in the range (0,1). If the number is less than one half, take a step in the positive direction. Otherwise, take a step in the negative direction. Take 20 steps. Where do you land? Repeat 20 times. Where do those 20 land? Repeat 10000 times. How are those 10000 distributed?

In real physical systems, outcomes of events are often best described by continuous distributions not by the simple binomial distribution. It is possible to generalize the notion of distributions to continuous functions. The most common distribution is the Gaussian distribution, which we have encountered previously:

$$(11.5) \qquad \mathcal{P}(x, x_0, \sigma) = \frac{1}{\sqrt{2\pi\sigma^2}} e^{-(x-x_0)^2/2\sigma^2}.$$

The Gaussian distribution is completely specified by its first moment x_0 and standard deviation σ; higher moments are zero. The prefactor ensures that the distribution is normalized:

$$\int_{-\infty}^{\infty} dx\, \mathcal{P}(x, x_0, \sigma) = 1.$$

In physical systems in more than a single dimension, we can generally assume that the behavior in each dimension is independent. As a result, the probability for a three-dimensional system can be obtained from the product of probabilities: $\mathcal{P}(x, y, z) = \mathcal{P}(x)\mathcal{P}(y)\mathcal{P}(z)$. The proposed electron trajectory in the Drude model (See figure 9.3) exhibits just such behavior.

> EXERCISE 11.6. Use the RandomReal function on the domain $(-1, 1)$ to generate random walks in two dimensions. Plot the trajectories followed by five different walkers, where each walker takes twenty steps.

Indeed, this is just the behavior observed by the Scottish botanist Robert Brown in the summer of 1827 when observing pollen grains in water under a microscope.[4] Brown noted that the pollen emitted small particles, whose motion was "vivid." Similar observations had been reported by the Dutch scientist Jan Ingenhousz in 1785, who noted that the motion of coal dust particles on the surface of alcohol was quite erratic. Nevertheless, the random motion of small particles has come to be known as **Brownian motion**.

An explanation of Brownian motion was finally provided by Einstein in 1905 and, independently, by the Polish physicist Marian Smoluchowski in 1906. A series of detailed experiments conducted by the French physicist Jean Baptiste Perrin provided substantial support for the Einstein-Smoluchowski theory.[5] In particular, Brownian motion was determined

---

[4]Brown's "A brief account of microscopical observations on the particles contained in the pollen of plants; and on the general existence of active molecules" was communicated in the *Philosophical Magazine* in 1828.

[5]Perrin published *Les Atomes* in 1913. He was awarded the Nobel Prize in Physics in 1926 "for his work on the discontinuous structure of matter, and especially for his discovery of sedimentation equilibrium."

FIGURE 11.2. Perrin observed the motion of small particles under a microscope and recorded their positions every thirty seconds. The positions of three separate particles at each time step are depicted by the *dots*. Sequential dots are connected by lines

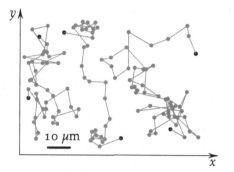

to be a macroscopic phenomenon that results from the atomic nature of matter. Perrin's experiments provided indirect proof of the discrete nature of matter at the microscopic level. Recall that in the early 1900s, Rutherford had not yet conducted his experiments demonstrating the nuclear structure of the atom and the question of the microscopic nature of matter was still widely debated.

A few of Perrin's observations are depicted in figure 11.2, taken from his book *Les Atomes*. Here Perrin observed the motion of 0.53 μm-radius spheres under a microscope and then recorded the positions of individual spheres every thirty seconds. Sequential points are connected by lines. In his text, Perrin remarks that, if one were able to shorten the time step, the apparently random behavior would continue. This property of **scale** invariance has subsequently been recognized to be a key characteristic of random processes. The subject of fractal geometry championed by the Polish mathematician Benoît Mandelbrot has its roots in self-similar curves and other geometrical objects that display scale invariance.[6]

Perrin was undoubtedly troubled by the fact that the curves displayed in figure 11.2 do not possess the attributes that we have previously ascribed to trajectories. In particular, Perrin expresses doubt that one could define a tangent to the curve in any meaningful way. We can understand his discomfort, given that the Newtonian concept of trajectories is well established. What we can now argue, given our understanding of atomic structure, is that the apparent random motion is an artifact of the time sampling. To the human eye, bats fly in erratic paths but this can be resolved with high-speed photography. Similarly, Brown's particles followed random pathways but, at an appropriate time scale, we would see that they possess Newtonian trajectories. As we shall see, the time scale

[6]Here we use the term *invariance* to mean that rescaling the x- and y-axes will not affect the results. Previously, we have used invariant to mean that a mathematical structure does not change under some transformation.

is exceptionally short (on the order of 10^{-21} s). Nevertheless, we do not need to invoke any new physics; there is no new force that causes Brownian motion.

> EXERCISE 11.7. Use the RandomReal function to generate two-dimensional random walks of fifty steps. Assuming that all walks started from the same point, plot the endpoints of one thousand walks. Consider now the distribution of the walkers. Use the Bin-Counts function to create histograms of the distribution along the x- and y-axes. (It may be useful to first compute the standard deviation of the points and to use bins that have widths of a fraction of the standard deviation.) Repeat for ten thousand walks.

What Einstein and Smoluchowski recognized was that, while the position of a single walker was not predictable using the tools of Newtonian kinematics, the *distribution* of walkers was eminently predictable. A single particle immersed in water undergoes numerous collisions with the surrounding water molecules. On average, there is no net momentum transfer to the particle but the collisions do not occur simultaneously, so each collision will impart momentum in some particular direction. The particle may accumulate significant momentum in that direction by chance, just as it is possible to toss a coin and obtain heads ten times in a row. This generates apparent randomness to the trajectory of an individual particle but let us consider what happens to many particles.

> EXERCISE 11.8. Use the RandomInteger function to generate ten thousand coin tosses. In this sample, what is the largest number of times that heads occurs consecutively? What is the largest number of times that tails occurs consecutively?

As Perrin confirmed experimentally by tracking the trajectories of hundreds of individual particles independently, the distribution of the random walkers is described by a Gaussian. Moreover, the width of the Gaussian increases with the square root of time. We can see this immediately from our simple coin-toss model of the random walk. The time will be proportional to the number N of coin tosses. The variance of the binomial distribution is just $Np(1-p)$ and the standard deviation is the square root of that. As a result, we should expect the width of the distribution to increase as the square root of time.

> EXERCISE 11.9. Use the RandomReal function to generate one-dimensional random walks of two hundred steps. Generate ten thousand trajectories and use the BinCounts function to generate an histogram of the distribution. What happens if you increase the number of steps to five hundred?

Einstein produced a differential equation for the density of walkers $\rho(x,t)$ that we can write as follows:

$$(11.6) \qquad \frac{\partial \rho(\mathbf{r},t)}{\partial t} = D \frac{\partial^2 \rho(\mathbf{r},t)}{\partial x^2},$$

where D is a constant known as the diffusion coefficient. We have introduced the notation of partial differential equations previously, when discussing the wave equation. Equation 11.6 is known as the one-dimensional **diffusion** equation. Unlike the wave equation, where there is a second derivative of time, the diffusion equation has only a first derivative of time. This drastically changes the nature of the solutions.

In particular, Einstein demonstrated that, if the walker density is normalized

$$\int_{-\infty}^{\infty} dx\, \rho(x,t) = N,$$

then a solution to the diffusion equation can be written as follows:

$$(11.7) \qquad \rho(x,t) = \frac{N}{\sqrt{4\pi D t}} e^{-x^2/4Dt}.$$

That is, the solution is a Gaussian distribution, with a width $\sigma^2 = 2Dt$. Consequently, the width of the distribution evolves as the square root of time.

> EXERCISE 11.10. Plot the function $\rho(x,t)$ as a function of x over the domain $-10 \le x \le 10$. Choose $N = 1$ and $D = 1$. Use the `Manipulate` function to vary the time over the domain $1 \le t \le 20$. How does the width of the distribution change over time?

FIGURE 11.3. Perrin tracked five hundred particles over thirty-second intervals. Displacements are indicated by the *dots*. The *dark gray circle* represents the magnitude of the standard deviation σ. Other circles are fractional divisions of σ

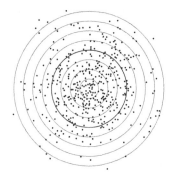

Perrin tracked the displacement of five hundred 0.367 µm-radius particles at thirty-second intervals. The results are depicted in figure 11.3. From the data, Perrin computed a standard deviation of $\sigma = 7.84$ µm.[7] The dark

---

[7]Perrin's definition of the deviation differs by a factor of $\sqrt{2}$ from the one we utilize. This is noted in the table.

TABLE 11.1. Perrin's displacement data for 0.367 μm-radius particles

Radius ($\sqrt{2}\sigma/4$)		
Inner	Outer	Number
0	1	34
1	2	78
2	3	106
3	4	103
4	5	75
5	6	49
6	7	30
7	8	17
8	9	9

circle in the figure has a radius of σ. Perrin drew a series of concentric circles of radius $n\sigma/4$, where $n = 1, 2, \ldots, 8$ and then counted the dots within each annular region. His results are listed in Table 11.1.

To compare Perrin's results with what Einstein's model predicts, we must perform an integration of the probability distribution over finite intervals. For Cartesian coordinates, the result is not expressible in terms of simple functions but occurs so often that it has been given the name of **error function**. Formally, we can write define the error function as follows:

$$(11.8) \qquad \mathrm{erf}(x) = \frac{2}{\sqrt{\pi}} \int_0^x d\xi\, e^{-\xi^2}.$$

With this definition, integration of the Gaussian probability distribution over the interval $[a, b]$ produces the resulting expression:

$$(11.9) \qquad \frac{1}{\sqrt{2\pi\sigma^2}} \int_a^b dx\, e^{-x^2/2\sigma^2} = \frac{1}{2}\left[\mathrm{erf}\left(\frac{b}{\sqrt{2}\sigma}\right) - \mathrm{erf}\left(\frac{a}{\sqrt{2}\sigma}\right)\right].$$

Because the error function occurs so frequently in statistical analysis, it was widely tabulated and is, of course, available in the *Mathematica* software.

EXERCISE 11.11. Plot the error function erf(x) over the domain $0 \le x \le 5$.

EXERCISE 11.12. Use Equation 11.9 to analyze the results of Exercise 11.9. Plot the random (experimental) data from the exercise and theoretical predictions. Do they agree?

In two dimensions, Perrin utilized polar coordinates where one can actually perform the integration in terms of known functions. Note that the probability is assumed to be the product of independent probabilities in the x- and y-directions:

$$P(x,y) = P(x)P(y) = \frac{1}{2\pi\sigma^2} \int dx\, e^{-x^2/2\sigma^2} \int dy\, e^{-y^2/2\sigma^2}$$

and that the standard deviations in each direction are the same. If we convert to polar coordinates, then we have that the probability of finding a particle in the range $\zeta_a < \zeta < \zeta_b$ is given by the following:

$$P(\zeta_a, \zeta_b) = \frac{1}{2\pi\sigma^2} \int_{\zeta_a}^{\zeta_b} d\zeta \int_0^{2\pi} d\varphi\, \zeta\, e^{-\zeta^2/2\sigma^2}$$

$$(11.10) \qquad\qquad = e^{-\zeta_a^2/2\sigma^2} - e^{-\zeta_b^2/2\sigma^2},$$

where we have used the fact that the distribution is independent of the azimuthal direction φ.

EXERCISE 11.13. Use Equation 11.10 to compute the predictions of the Einstein model for Perrin's data (Table 11.1). Recall that $N = 500$. Plot the results. Are they in agreement?

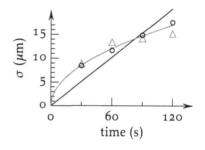

FIGURE 11.4. The standard deviations of the displacements of 0.212 µm-radius particles demonstrate a $\sqrt{t}$ dependence (*gray curve*) and not a linear dependence (*black curve*). Two separate time series with slightly different experimental conditions (*triangles* and *open circles*) behave similarly

Perrin and his students measured the displacements of fifty 0.212 µm-radius particles at thirty-second intervals for two minutes. The results of two series of their experimental results are indicated in figure 11.4. Also plotted in the figure are two theoretical curves: $a_1 \sqrt{t}$ (gray) and $b_1 t$ (black). From the data, it would appear that the linear (black) fit would be improved if we added an offset term ($b_1 t + b_0$) but Perrin measured the difference in position of the particles at each time interval. Thus, all the particles start from the origin and this constraint requires that $b_0 = 0$. It is apparent from the figure that the data are better represented by the square-root of time curve, in agreement with Einstein's model for Brownian motion.

The importance of Einstein's model for diffusion is that it relates microscopic behavior to macroscopic observations. Brownian motion can be interpreted as the result of the apparently random behavior of macroscopic particles interacting with (unobserved) microscopic particles; it is an **emergent** phenomenon. Moreover, we can obtain such behavior without resort to new forces of nature. Collisions of two molecules are governed primarily by the electromagnetic force. We have already studied the relatively simple case of α particle scattering through the Coulomb force; there are complications, to be sure, when considering the interactions of extended (non-point) molecules but there is no experimental evidence for new forces.

Some fifty years earlier than Einstein's publication, the German physiologist Adolf Fick had determined that the diffusion of salt in water could be described by two mathematical formulas that have come to be known as Fick's laws of diffusion:

$$J = -D\frac{\partial \phi}{\partial x}$$

$$(11.11) \qquad \frac{\partial \phi}{\partial t} = D\frac{\partial^2 \phi}{\partial x^2},$$

where J is the mass flux and ϕ is the (salt) concentration. We can recognize the second law as Einstein's model for Brownian motion with a minor change of notation.

Chemistry students have long been taught that chemical species diffuse along the concentration gradient, which is a statement of Fick's first law. Without further explanation, one might then infer that high concentrations of chemicals generate some form of motive force that drives the diffusion process. Students might also ponder the question of how does a single molecule in the system "know" the direction of the concentration gradient. Einstein's model provides answers to both questions. The behavior of the macroscopic particles observed by Brown and Perrin can be seen as indicative of the behavior of microscopic (unseen) particles. Diffusion is simply the result of individual molecules conducting random walks and the width of the ensemble spreading as the square root of time. There is no additional motive force that arises from the accumulation of chemical species and individual molecules do not need to somehow sense the direction of the concentration gradient in order for diffusion to occur.

11.3. Statistical Thermodynamics

The industrial revolution and the development of the steam engine provided a number of technical problems that were addressed by scientists and engineers of the day. One might wonder, for example, how to improve upon the efficiency of or power produced by steam engines. How does operating temperature or size of the boiler affect these quantities? A series of principles, today known as the Laws of Thermodynamics, were developed that explained the experimental observations and provided a guide for engineering improvements in machinery.

The first or fundamental law of thermodynamics is simply the statement that energy is conserved, with the additional caveat that work and energy are equivalent entities. This latter principle was discovered by Benjamin Thompson (later Count Rumford) in 1798 who demonstrated that boring cannon barrels could boil water indefinitely, as long as the horses driving the boring apparatus could continue their efforts. His experiments stood in direct conflict with the *caloric* theory of matter that suggested that matter was imbued with a caloric fluid and that the heat[8] released during machining operations originated within the matter itself. Thompson's experiments demonstrated that there was no mass difference between the original cannon barrel and the machined barrel and the metal shavings. Caloric fluid could not have mass and could not therefore be a material property.

In this regard, the fundamental law of thermodynamics is not a new principle. We have already shown that energy is conserved in mechanical systems. In modern terminology, we can state the first law as follows:

$$(11.12) \qquad \Delta Q = \Delta E + \Delta W.$$

If a small (infinitesimal) amount of energy ΔQ is added to a system, it will be partitioned into a corresponding change in the internal energy ΔE of the system and, potentially, work ΔW performed by the system on the external universe. The study of thermodynamics is, in some measure, detailed accounting of the flow of energy through systems. That promises to be a reasonably straightforward exercise, on its face.

Other aspects of the behavior of thermodynamical systems do not appear to be so simple. For example, it was recognized that temperature is somehow a measure of the internal energy of a system. If you heat[9] a system, by placing a flask over a fire for example, there is generally an increase in temperature.

Consider the experimental results sketched in figure 11.5. If we place a pan of water over an open flame and measure the temperature at regular intervals, we produce results like those indicated by the black dots.

---

[8]Here we use the word *heat* to mean thermal energy. Historically, the definition of heat is muddled. We should use the more precise terminology *thermal energy*. Unfortunately, heat is succinct and is used ubiquitously as a noun to mean thermal energy.
[9]We use the word *heat* here as a verb. This is the modern preferred usage.

The water temperature increases linearly for a time and then approaches a constant temperature asymptotically. We call this the boiling point of water. If the flame is maintained appropriately, we can assume that the energy being added to the pan is approximately constant in time. So, how can we account for the fact that the temperature increase diverges from linear?

Similarly, if we cooled a pan of water, or heated a block of ice, we would produce a set of results like those indicated by the gray dots in the figure. There are regions where the temperature appears to change in a linear fashion but, at what is termed the freezing point of water, the temperature does not change as the energy of the water changes.

FIGURE 11.5. The temperature of liquid water will rise linearly with time when heated (*black points*) until it reaches the boiling point. Similarly, it would decrease linearly when cooled (*gray points*) until it reaches the freezing point

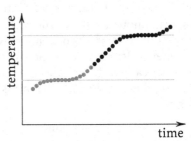

To accommodate the observations that energy and temperature have a more complex relationship than simple linear behavior, the concept of **entropy** was introduced. While having a precise mathematical rôle in the equations that define the thermodynamic behavior of systems, entropy was usually likened to the disorder of the system. The explanation of the experiments depicted in figure 11.5 is this: while the temperature of the system stabilizes around the boiling and freezing points, the entropy of the system continues to increase. Hence, energy is conserved, the energy input into the system has just been converted into entropy and does not therefore manifest itself by a rise in temperature.

The second law of thermodynamics is then formulated in terms of the observed property of this new quantity of entropy: as systems evolve, entropy increases. Formally, we can write the following relation:

$$(11.13) \qquad\qquad \Delta Q = T\Delta S,$$

where ΔS is the change in entropy. While this formulation of thermodynamical principles is effective in explaining the behavior of systems and serving as a guide for engineering improved forms of machinery, the concept of entropy is rather nebulous, at least in this author's experience. What exactly is disorder? Why should entropy increase? There are numerous questions that arise for which there are no physical insights that would clarify the mathematical expressions.

This situation changed significantly with a series of insightful discoveries made by the American physicist Josiah Willard Gibbs, who developed a rigorous mathematical framework for determining the macroscopic consequences of microscopic behavior[10] and the Austrian physicist Ludwig Eduard Boltzmann, whose work on the kinetic theory of gases was conducted independently of Gibbs.[11] This new, atomistic methodology begins with the definition of the microscopic state of the system. For concreteness, let us define our system to be a cubical box of volume V that contains N atoms. At some point in time, each atom has a location x_i and a momentum p_i. The state of the system is defined by the collection of variables: $\Psi = \{x_1, \ldots, x_N, p_1, \ldots, p_N\}$. The energy of the system $\mathcal{E}$ we have seen can be divided into two parts, kinetic $\mathcal{T}$ and potential $\mathcal{U}$. We have inevitably found that the kinetic energy depends quadratically on the momenta:

$$\mathcal{T} = \sum_{i=1}^{N} \frac{p_i^2}{2M},$$

where we have made the simplification that all of the atoms possess mass M. The potential energy can be a complicated function of all of the positions: $\mathcal{U} = \mathcal{U}(x_1, \ldots, x_N)$.

The probability of finding the system in this state is proportional to the so-called Boltzmann factor:

$$\mathcal{P} \propto e^{-\mathcal{E}/k_B T},$$

where k_B is now known as the Boltzmann constant and T is the absolute temperature of the system.[12]

To obtain the normalized probability, we integrate over all of the possible configurations of position and momentum:

$$(11.14) \qquad \mathcal{Z} = \int d^3 x_1 \cdots \int d^3 x_N \int d^3 p_1 \cdots \int d^3 p_N \, e^{-\mathcal{E}/k_B T}.$$

The use of the letter $\mathcal{Z}$ for the function defined in Equation 11.14 arises from the German *Zustandssumme* or "sum over states." As the momenta and positions are continuous variables, it is appropriate to integrate over

---

[10]Gibbs shared the same name with his father, who went by Josiah. The younger Gibbs was known as Willard to his family. His monograph *On the Equilibrium of Heterogeneous Substances* was published in two parts in 1875 and 1878 by the Connecticut Academy of Sciences.

[11]Boltzmann wrote several papers on the kinetic theory of gases that were published in the *Wiener Berichte* in 1871 and 1872. He revisited the subject in two further papers in 1877.

[12]The absolute temperature scale inevitably utilizes the SI Kelvin scale, where a 1 K temperature difference is equivalent to a 1 C temperature difference. The two scales differ in their origins, where the Celsius scale defines the freezing point of water to be 0 C.

these variables instead of summing. The function $\mathcal{Z}$ is known as the **partition function**. More than just a simple normalization factor, we shall see that essentially all thermodynamic properties of a system can be obtained, at least formally, from a knowledge of the partition function.

We note that the Boltzmann probability of finding a system in the state with energy $\mathcal{E}$ is then given by the normalized quantity:

$$(11.15) \qquad \mathcal{P}(\mathcal{E}) = \frac{e^{-\mathcal{E}/k_B T}}{\mathcal{Z}}.$$

From this, we can deduce that the expected value of the energy will be defined by the following relation:

$$\overline{\mathcal{E}} = \frac{\int d^3 \mathbf{x}_1 \cdots \int d^3 \mathbf{x}_N \int d^3 \mathbf{p}_1 \cdots \int d^3 \mathbf{p}_N \, \mathcal{E} \, e^{-\mathcal{E}/k_B T}}{\mathcal{Z}}.$$

The integrals define the **phase space** of the system. The phase space is composed of the dynamically accessible regions in position and momentum spaces. In our example of particles in a box, the position integrals only extend over the interior of the box.

Rather than trying to attempt the integrations, let us first note that if we take the derivative of the exponential with respect to temperature, that we obtain the following:

$$\frac{\partial}{\partial T} e^{-\mathcal{E}/k_B T} = \frac{\mathcal{E}}{k_B T^2} e^{-\mathcal{E}/k_B T}.$$

Using this result, we can recast the expression for the expected value of energy as follows:

$$(11.16) \qquad \overline{\mathcal{E}} = k_B T^2 \frac{\partial \ln \mathcal{Z}}{\partial T}.$$

We are, of course, assuming that the orders of differentiation and integration can be interchanged and that other mathematical issues can be clarified. Nevertheless, Equation 11.16 provides a statement that the expected energy of a system can be obtained from the temperature dependence of the partition function.

> EXERCISE 11.14.　　Fill in the details of the derivation of Equation 11.16.

The entropy can also be defined in terms of the partition function:

$$(11.17) \qquad S = \frac{\partial}{\partial T} (k_B T \ln \mathcal{Z}) = k_B \ln \mathcal{Z} + \frac{\overline{\mathcal{E}}}{T}.$$

The partition function represents the sum of all of the (weighted) probabilities of configurations of the dynamical system. The entropy is proportional (through the factor k_B) to the natural logarithm of the number of these configurations. In this sense, entropy is not the (vague) disorder of the system; it represents the (generally large but definable) number of different configurations that are accessible to the system.

Time evolution of dynamical systems can be seen as the evolution of the system toward the most probable configurations. Consequently, entropy will generally increase but there are no external forces at work that coerce such behavior. At the microscopic level, molecules are effectively conducting random walks and diffusing with a square root of time behavior. If one adds a drop of food coloring to a glass of water, over time and without stirring, the water will become uniformly colored. One never observes the dye molecules to coalesce back into a small space within the glass. That is because the number of configurations in which all of the dye molecules are located within a small fraction of the volume is vastly less than the number of configurations in which they are randomly scattered about the volume.

Consider requiring only that the molecules be on the left side of the glass. There is a probability of $1/2$ of that occurring for any one of the N molecules, so the cumulative probability is $\mathcal{P} = (1/2)^N$. A drop of dye molecules probably contains 10^{19} or so molecules, so the probability of finding them all on the left side of the glass is $\mathcal{P} \approx 10^{-7 \times 10^{18}}$, which is a fantastically small number. As a result, we never observe the dye molecules to even congregate in half of the glass.

EXERCISE 11.15.   Consider excluding just a small volume of the glass, say 1 % of the volume. The probability of finding one atom in this volume is $\mathcal{P} = 0.99$. What is the probability of finding all $N = 10^{19}$ dye molecules in this region? What can you infer about the distribution of dye molecules?

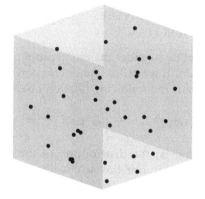

FIGURE 11.6. A simple dynamical system consists of N (weakly interacting) particles confined inside a box of volume V

In practice, evaluating the partition function can be quite challenging and, consequently, the simple expression in Equation 11.16 for the expected energy may also be difficult to evaluate. Nevertheless, the framework is quite rigorous and does not depend on the nature of the interactions $\mathcal{U}$. In simple cases, the evaluation can be performed. Consider, for example, the case of N weakly interacting atoms in a box of volume V, as illustrated in figure 11.6. We take $\mathcal{U} = 0$ or at least negligible compared to the kinetic energies. The partition function then separates into the product of N terms:

$$\mathcal{Z} = \prod_{i=1}^{N} \int d^3 x_i \int d^3 p_i e^{-p_i^2/2Mk_BT} = \left(\int d^3 x \int d^3 p \, e^{-p^2/2Mk_BT} \right)^N,$$

where in the last step we utilize the fact that, if the atoms are the same, each of the product terms will be equivalent.

Here there is no dependence upon position, so the integrals over position yield just the volume of the box V. The momentum integrals are integrals over Gaussian functions, so we obtain the following result:

$$(11.18) \qquad \mathcal{Z} = \left[(2\pi M k_B T)^{3/2} V \right]^N.$$

Using this result, we can compute the expected energy to be given by the following:

$$(11.19) \qquad \overline{\mathcal{E}} = \frac{3N k_B T}{2}.$$

Equation 11.19 is a statement of the *equipartition theorem*, which holds that every degree of freedom contributes $k_B T/2$ to the average energy. In our example, we had N atoms and three dimensions. Hence, the expected energy is $k_B T/2$ multiplied by $3N$. This result also confirms that the energy is linearly dependent upon the temperature.

EXERCISE 11.16.   Fill in the details of the derivation of Equation 11.19.

Let us consider the momentum portion of the single atom partition function. If we utilize spherical coordinates in momentum space, then we can show the following relation holds:

$$(11.20) \qquad \int d^3 p \, e^{-p^2/2Mk_BT} = 4\pi \int_0^\infty dp \, p^2 e^{-p^2/2Mk_BT}.$$

That is, the distribution of the magnitude of momentum amongst the atoms in our box scales like the square of momentum times the Boltzmann exponential factor. As the momentum is just the product of mass and velocity, we can infer that the magnitude of velocity distribution inside the box has the following form:

(11.21)
$$P(v) = P_0 v^2 e^{-Mv^2/2k_B T},$$

where P_0 is a normalization factor. This particular form of the velocity distribution was derived initially by Maxwell and is often called the Maxwell velocity distribution. It is also known as the Maxwell-Boltzmann distribution, in an effort to acknowledge the later contributions of Boltzmann.

It is possible to sample this distribution experimentally. If we have an oven that is heated to some temperature T, then the atoms within should have a distribution like that defined in Equation 11.21. If we drill a small hole in the side of the oven, then atoms that would normally strike the wall and rebound back into the volume will, instead, emerge as a narrow beam. If the hole is small enough, the escape of a relatively few atoms does not affect the internal state of the atoms within the box appreciably. This process is known as *effusion*.

The beam is allowed to strike a rotating velocity selector, which can be simplified to consist of two circular plates with holes drilled in their faces. If the holes are separated by an azimuthal angle ϕ_1, then only atoms with a velocity $v = L\omega/\phi_1$ will emerge through the second plate and strike a detector. Such an apparatus is illustrated in figure 11.7.

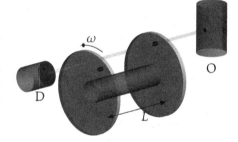

FIGURE 11.7. An atomic beam effuses from a small hole in an oven O. The beam passes through a rotating velocity selector and impinges on the detector D

EXERCISE 11.17. Show that the rotating plates ideally select atoms with velocities $v = L\omega/\phi_1$. As a practical matter, slower atoms may also be admitted. What would be the velocity of atoms that emerge from the apparatus after it has rotated by the angle $\phi_1 + 2\pi$?

The German-American physicist Polykarp Kusch and his student R. C. Miller constructed a more sophisticated apparatus than that depicted in the figure but utilized the same principle of operation.[13] Their velocity

---

[13]Miller and Kusch published their results in 1955, the same year in which Kusch won the Nobel Prize in Physics "for his precision determination of the magnetic moment of the electron." Kusch shared the Prize with the American physicist Willis Lamb, who won "for his discoveries concerning the fine structure of the hydrogen spectrum."

TABLE 11.2. Velocity distribution of 944 K thallium atoms

v (m/s)	$P(v)$	v (m/s)	$P(v)$	v (m/s)	$P(v)$
59.9	3.90	206.1	17.32	408.4	13.25
91.3	6.29	224.8	18.45	429.2	12.01
103.1	7.10	246.0	19.53	455.4	9.80
116.7	8.30	267.4	19.98	471.8	8.96
132.7	10.11	283.7	19.98	494.4	7.20
141.0	11.46	309.5	19.43	520.5	5.45
156.2	12.81	330.2	18.77	530.8	4.90
177.4	14.58	356.2	17.33	554.2	3.93
191.7	15.97	379.7	15.61	588.7	2.65

selector utilized a rotating cylinder; helical slots had been machined into the surface to improve the ability to reject velocities outside of a small window and noise was further suppressed by cryogenically cooling the apparatus. Some of their results utilizing thallium metal vapor are depicted in figure 11.8. From Equation 11.21, we expect that the velocity distribution will be maximal at the velocity $v_{max} = (2k_BT/M)^{1/2}$. For the oven temperature at which these data were acquired (870 K), the distribution should be maximal at $v = 266$ m/s. This is precisely where Kusch and Miller observe the peak in the experimental velocity distribution.

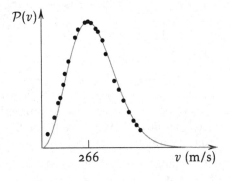

FIGURE 11.8. The experimental velocity distribution for thallium metal atoms at a temperature of 870 K (*black dots*) agrees well with the predicted Maxwell-Boltzmann distribution (*gray curve*). An arbitrary normalization constant was used to match the distributions at the peak ($v = 266$ m/s)

EXERCISE 11.18.   The maximum of the Maxwell-Boltzmann velocity distribution will occur when the derivative vanishes. Find the velocity at which this occurs. Plot the distribution.

EXERCISE 11.19. The data in Table 11.2 were measured for thallium atoms at a temperature of 944 K. What is the velocity at the maximum of the distribution for this temperature? Plot the Maxwell-Boltzmann distribution and the (suitably scaled) experimental results. Are they in agreement?

The experimental evidence provides significant support to the statistical framework developed by Boltzmann and Gibbs. This has somewhat astonishing implications for the nature of the microscopic world around us. For example, the atmosphere is composed principally of nitrogen and oxygen molecules. According to the Maxwell-Boltzmann distribution, these molecules will have velocity distributions that peak at roughly 400 m/s. (This is higher than the velocity of sound in air and comparable to the muzzle velocities of .22 caliber rifle rounds.) As the inter-molecular spacing in air is a few nanometers, this suggests that the characteristic time for collisions between gas molecules is of the order of 10^{-10}–10^{-11} s. What emerges is a microscopic picture of utter chaos: individual gas molecules fly through space at macroscopically large velocities and endure 10^{10} or more collisions per second. This is certainly not what we perceive at the macroscopic level.

EXERCISE 11.20. In weakly interacting systems, the partition function of N particles is just $Z = (Z_1)^N$, where Z_1 is the partition function of a single particle. Hence, it suffices to consider just a single particle: $\ln Z = N \ln Z_1$. Consider a system of magnetic particles that can have energies $\mathcal{E} = \pm \mu B$ depending upon whether the magnetic moment μ of the particle is aligned (−) or anti-aligned (+) with the magnetic field B. Here the partition function is just a sum over the possible states.

Write the partition function explicitly. Show that the expected value of the magnetization is given by $\bar{\mu} = \mu \tanh(\mu B / k_B T)$. Plot the function $\tanh(1/x)$ over the domain $0 \le x \le 20$. What happens to the magnetization at low temperatures? What happens to the magnetization at high temperatures?

11.4. Applications

On April 27, 1661, the English natural philosopher Richard Townley and his colleague Henry Power conducted an experiment in which they measured the air pressure with a barometer at several different altitudes on Pendle Hill in Lancashire. When examining the results, the two noted that there was a correlation between the air density and pressure. Townley discussed his results with the English natural philosopher Robert Boyle, who conducted further experiments and published an analysis in which he stated that the product of pressure and volume is a constant: $pV = c$.[14] This relationship is now known as Boyle's Law.

In 1802, the French natural philosopher Joseph Louis Gay-Lussac published the results of his investigations into gases and concluded that if one holds the pressure constant, then the volume of a gas is linearly related to the temperature: $V = cT$.[15] Finally, in 1834 the French physicist Benoît Paul Émile Clapeyron combined the two results into what is now called the ideal gas law. In modern notation, we can write this as follows:[16]

$$(11.22) \qquad\qquad pV = Nk_B T.$$

This fundamental, experimentally-determined relationship, along with Gay-Lussac's recognition that gases form in ratios such that the volumes of reactants and products can be expressed in simple whole numbers, form part of the founding pillars of modern chemistry.

This fundamental relation can also be derived from the microscopic theory of Gibbs and Boltzmann. The expected value of the pressure is given by the following relation:

$$(11.23) \qquad\qquad \overline{p} = k_B T \frac{\partial \ln \mathcal{Z}}{\partial V}.$$

Using the partition function for N particles in a box, we recover Equation 11.22. This is a remarkable result. The observed relationship between the macroscopic parameters that define the thermodynamic state of gases emerges from the microscopic consideration of particles in a box.

EXERCISE 11.21. Use Equations 11.18 and 11.23 to derive the ideal gas law.

One of the macroscopic thermodynamic properties that characterizes materials is the somewhat unfortunately named *heat capacity*.[17] Formally, if we add a small amount of energy ΔQ to a system, then we could well expect a proportional change in the temperature:

$$\Delta Q = C \Delta T,$$

---

[14]Boyle referred to the proposal as "Townley's hypothesis" in a 1662 appendix to his 1660 monograph *New Experiments Physio-Mechanicall, Touching the Spring of Air and its Effects*.

[15]In his "Recherches sur la dilatation des gaz et des vapeurs" published in *Annales de Chimie*, Gay-Lussac cited previously unpublished work by the French scientist Jacques Charles. The relation $p/T = c$ is now known as Charles' Law.

[16]In chemistry texts, this is usually written as $pV = nRT$ where n is the number of moles of gas and R is known as the gas constant. Note also that here we are using the symbol p to mean pressure and not probability or momentum.

[17]The language reflects the then-prevailing attitude that *heat* was an intrinsic property of matter, a caloric fluid. One could envision that a given amount of matter could only hold so much heat before it overflowed, in some sense.

where the constant of proportionality C is termed the heat capacity. We see such behavior sketched in figure 11.5. It was recognized by early investigators that the heat capacity is actually a complex function of the other thermodynamic variables. For example, if one measured the temperature rise of a quantity of material where the volume was constrained to be constant then those results would be different from measurements where the pressure was held constant. In gases, this is not particularly surprising, given that the thermodynamic variables are related through the ideal gas equation. The heat capacity will also, of course, depend upon the amount of material present, so one can define a **specific heat** capacity for a particular amount of material. With the advent of the atomic model of matter, the molar mass has become the most common standard. The lower case c is, at times, utilized to reference the specific heat capacity per mole of material.

Formally, we propose that the heat capacity is related to the derivative:

$$\Delta Q = \frac{\partial Q}{\partial T} \Delta T,$$

whereby $C = \partial Q/\partial T$. There are different definitions of the specific heat that depend upon the nature of the thermodynamic experiment being conducted. For example, suppose that we conduct an experiment in which we hold the volume of the box constant. The specific heat at constant volume is then defined as follows:

(11.24)
$$C_V \equiv \left(\frac{\partial Q}{\partial T}\right)_V = \left(\frac{\partial \bar{\mathcal{E}}}{\partial T}\right)_V,$$

where the subscript V indicates that we are holding the volume constant. Because the incremental work done on the external universe is $\Delta W = p\,\Delta V$, then the energy added to the system will only increase the internal energy of the system $\bar{\mathcal{E}}$. On the other hand, if we conduct an experiment in which the pressure is held constant, then the relevant specific heat will be defined as follows:

(11.25)
$$C_p \equiv \left(\frac{\partial Q}{\partial T}\right)_p = \left(\frac{\partial [\bar{\mathcal{E}} + \bar{p}\,\Delta V]}{\partial T}\right)_p.$$

From Equation 11.19, we see that our model of weakly-interacting atoms in a box will be characterized by a specific heat at constant volume of $C_V = 3Nk_B/2$, whereas the specific heat at constant pressure will be $C_p =$

$5Nk_B/2$. In SI units, we find $C_V = 12.47$ J/mol-K and $C_p = 20.78$ J/mol-K, which are precisely the measured values for specific heats of the noble gases helium, neon, argon, krypton and xenon.[18]

EXERCISE 11.22. Use the ideal gas law and Equations 11.24 and 11.25 to demonstrate that $C_V = 3Nk_B/2$ and $C_p = 5Nk_B/2$.

TABLE 11.3. Molar specific heats of diatomic molecules

Molecule	C_p (J/mol-K)	Molecule	C_p (J/mol-K)
H_2	28.836	CO	29.142
N_2	29.124	NO	29.845
O_2	29.376	Br_2	36.048

EXERCISE 11.23. The specific heats at constant pressure for several diatomic gases measured at 25 °C (298.15 K) are listed in Table 11.3. Use the equipartition theorem to count the number of degrees of freedom. If you think of the molecules as classical particles, what might be the extra internal degrees of freedom?

Consider now a simple model of solids. The electrons in a solid interact in a complex fashion with the lattice formed by the nuclear centers and other electrons. Suppose instead that we replace the complex interactions with simple harmonic potentials. That is, each atom in the lattice sees an effective quadratic potential:

$$(11.26) \qquad \mathcal{U}(\mathbf{x}_1, \ldots \mathbf{x}_N) = \tfrac{1}{2} \sum_{i=1}^{N} \kappa_i (\mathbf{x}_i - \mathbf{r}_i)^2,$$

where κ_i is the effective spring constant, the $\mathbf{r}_i$ represent the lattice sites and the $\mathbf{x}_i$ represent the nuclear (atomic) positions. In this case, the position integrals in the partition function are integrals over Gaussians. One can evoke the Equipartition Theorem to immediately conclude that the expected energy of the system (with six degrees of freedom for each of N atoms) will be given by the following:

$$\bar{\mathcal{E}} = 6N(k_B T/2) = 3N k_B T.$$

EXERCISE 11.24. Using the definition of $\mathcal{U}$ in Equation 11.26, compute the partition function $\mathcal{Z}$. Compute the expected energy $\bar{\mathcal{E}}$ from Equation 11.16.

---

[18]Thermodynamic values can be found in databases managed by the U.S. National Institute of Standards and Technology (www.nist.gov). See, for example, the NIST-JANAF thermochemical tables.

In 1819, the French physicists Alexis-Thérèse Petit and Pierre-Louis Dulong noted the specific heat capacities of elemental materials was nearly a constant.[19] In Table 11.4, we list some of the molar specific heats of metals measured at 298 K, where $3Nk_B = 24.9$ J/mol-K.

TABLE 11.4. Molar specific heats of metals. Units are J/mol-K

Solid	C_p	Solid	C_p
Aluminum	24.3	Lead	26.4
Bismuth	25.7	Silver	24.9
Copper	24.5	Tungsten	24.8
Gold	25.6	Zinc	25.2

The heat capacities for metals are remarkably close to the simple theoretical prediction. We have not included any sort of quantum mechanics or details of the interactions between atoms yet have still managed to produce a result that is in general agreement with the experimental observations. What we might infer from this situation is that macroscopic observables are not particularly sensitive to the details of the underlying microscopic interactions. At least in the case of metals, it appears that the assumption that atoms form a regular lattice in the solid form is enough to explain the observed heat capacity.

This is not the case for diamond, where the specific heat at 298 K is approximately 6 J/mol-K. As diamond is a hard, crystalline material, it is somewhat surprising that the model appears to fail so badly. The discrepancy caught the attention of Einstein, who utilized an idea from the (then) new theory of quantum mechanics. In 1901, the German physicist Max Planck provided a theoretical explanation for the observed spectral intensity of black-body radiation.[20] Planck's seminal idea was that the energy of electromagnetic radiation is quantized. That is, energy is not a continuous function but there is a smallest amount of energy.[21] Note that this idea echoes our earlier discussion on treating electric charge as a continuous function even though we know that charge comes in integral multiples of the fundamental charge. Planck's suggestion can be expressed mathematically by the following equation:

$$(11.27) \qquad \mathcal{E} = h\nu = \frac{hc}{\lambda},$$

---

[19]Petit and Dulong published their "Recherches sur quelques point importants de la Théorie de la Chaleur" in the *Annales de Chimie et de Physique*.

[20]Planck's "Über das Gesetz der Energieverteilung im Normalspectrum" was published in the *Annalen der Physik*.

[21]Planck was awarded the 1918 Nobel Prize in Physics "in recognition of the services he rendered to the advancement of Physics by his discovery of energy quanta."

where ν is the frequency of oscillation and λ is the wavelength of the radiation. The proportionality constant h is now known as Planck's constant.

Einstein decided to apply Planck's approach to the problem of the specific heat of diamond. The consequence of using a discrete energy step is that the integrals that define the partition function become sums, as shown in the following expression:

$$(11.28) \qquad \mathcal{Z} = \sum_{n=0}^{\infty} e^{-nh\nu/k_B T} = \left[1 - e^{-h\nu/k_B T}\right]^{-1}.$$

This is a geometric series that has a finite sum.

EXERCISE 11.25. Define the finite sum S_N as follows:

$$S_N = \sum_{n=0}^{N} x^n.$$

Compute the value of $(1-x)S_N$ explicitly. Show that if $|x| < 1$, then the infinite sum has the finite result indicated in Equation 11.28.

The expected value of the energy in Einstein's model can be obtained from the following:

$$(11.29) \qquad \overline{\mathcal{E}} = 3Nk_B T^2 \frac{\partial \ln \mathcal{Z}}{\partial T} = 3N \frac{h\nu \, e^{-h\nu/k_B T}}{1 - e^{-h\nu/k_B T}}.$$

It is customary to define the Einstein temperature $\Theta_E = h\nu/k_B$, whereupon we can rewrite Equation 11.29 in the following fashion:

$$(11.30) \qquad \overline{\mathcal{E}} = 3Nk_B \frac{\Theta_E/T \, e^{-\Theta_E/T}}{1 - e^{-\Theta_E/T}}.$$

From this last result, we can now compute the specific heat:

$$(11.31) \qquad C_V = 3Nk_B \frac{(\Theta_E/T)^2 \, e^{-\Theta_E/T}}{[1 - e^{-\Theta_E/T}]^2}.$$

Einstein's model makes a new prediction for the specific heat that depends on a new parameter, the Einstein temperature.

EXERCISE 11.26. Fill in the details of the derivation of Equation 11.31. Plot the function $(1/x)^2 e^{(-1/x)}/(1 - e^{(-1/x)})^2$ for x in the range $0 \le x \le 5$.

As one can see from figure 11.9, Einstein's modification to the formula for specific heats greatly improves the correspondence between theory and experiment. The failure of the Dulong-Petit rule for diamond can now

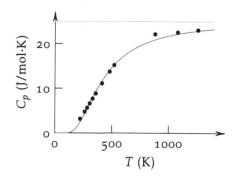

FIGURE 11.9. The specific heat of diamond was measured over a wide temperature range (*black dots*). The prediction of the Einstein model with a temperature $\Theta_E = 1296\,\text{K}$ is shown as the *gray curve*. The asymptotic value $3Nk_B$ is indicated by the *light gray line*

be attributed to the fact that the Einstein temperature for diamond is approximately 1300 K. Thus, room temperature is effectively low temperature for diamond and the Dulong-Petit rule does not work at low temperatures. Gratifyingly, at higher temperatures, the specific heat for diamond approaches the value $3Nk_B$.

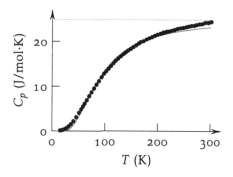

FIGURE 11.10. The measured specific heat of aluminum (*black dots*) is compared to the Einstein model with a temperature of $\Theta_E = 278\,\text{K}$ (*gray curve*)

It turns out that metals also possess temperature-dependent specific heats. In figure 11.10, we show the fit of the Einstein model to data for aluminum. At room temperature, aluminum has a heat capacity of $3Nk_B$ but this is due to the fact that the Einstein temperature is approximately $\Theta_E = 278\,\text{K}$. For metals, it would appear that the success of the Dulong-Petit rule can be attributed to the fact that their Einstein temperatures are comparable to room temperature. As a result, the specific heats are close to the asymptotic value of $3Nk_B$. For diamond, this is clearly not the case; the Einstein temperature is 1000 K higher. This difference in behaviors between diamond and metals undoubtedly reflects the difference in the microscopic interactions in the different systems. At present, we have not yet addressed quantum mechanical effects in any real sense but we have devised a one-parameter model based on quantum insights.

Any further progress, like predicting the value of Θ_E based on some sort of first-principles calculations, will have to be deferred to subsequent courses.

TABLE 11.5. Specific heat for copper in units of J/mol·K

T (K)	C_p	T (K)	C_p	T (K)	C_p
14.82	0.17	101.24	16.29	201.39	22.73
17.63	0.30	106.72	16.98	207.07	22.94
19.75	0.45	112.25	17.61	213.03	23.15
23.35	0.78	117.86	18.10	218.9	23.17
28.21	1.42	123.4	18.65	224.23	23.33
33.52	2.38	128.99	19.05	229.66	23.43
38.86	3.51	134.54	19.47	235.25	23.58
44.21	4.76	140.2	19.88	240.71	23.73
48.17	5.78	146.02	20.26	245.98	23.86
53.34	7.10	151.01	20.60	251.58	23.96
59.08	8.46	156.91	20.92	256.6	23.99
65.12	9.81	162.77	21.19	261.34	24.08
70.12	10.90	168.28	21.57	266.61	24.18
75.36	12.03	174.05	21.70	272.18	24.22
80.60	13.04	179.36	21.92	277.69	24.35
85.62	13.91	184.72	22.18	283.59	24.39
90.73	14.70	190.18	22.35	289.51	24.38
95.78	15.49	195.81	22.57	300.15	24.43

EXERCISE 11.27. Use the specific heats for copper listed in Table 11.5 to estimate the Einstein temperature. Use the FindFit function and plot the results. How does the value of Θ_E compare with that of aluminum?

If one looks closely at figure 11.10, Einstein's model of the specific heat still does not quite describe the data, particularly at low temperatures. A further refinement was produced by the Dutch physicist Peter Debye, who reasoned that excitations of the lattice had a high-frequency cutoff. If we think about the modes of oscillation of a guitar string, then the fundamental mode is represented by a displacement of $\sin(\pi x/L)$. Higher order modes are represented by displacements of the form $\sin(n\pi x/L)$. At some point, the distance L/n becomes comparable to the lattice spacing and that mode has each atom oscillating in the direction opposite to its neighbor. Higher frequencies are not supported.

When Debye included this constraint in his model, he derived the following formula for the specific heat:

$$(11.32) \qquad C_V = 9Nk_B\left(\frac{T}{\Theta_D}\right)^3 \int_0^{\Theta_D/T} dx\, \frac{x^4 e^x}{(e^x - 1)^2},$$

where we have introduced the Debye temperature $\Theta_D = h\nu_D/k_B$ for some maximum oscillation frequency ν_D.

EXERCISE 11.28.   The integral in Equation 11.32 cannot be performed analytically. The Debye function $D_n(x)$ is defined as follows:

$$D_n(x) = \int_0^x dt\, \frac{t^n}{e^t - 1}.$$

Plot $D_3(x)$ over the domain $0 \leq x \leq 5$ in steps of $dx = 0.5$.

EXERCISE 11.29.   Plot C_V as defined in Equation 11.32 for $0 \leq T \leq 300$ in steps of $T = 10$ Use the Einstein temperature found in the previous exercise for the copper data and plot the Einstein model results as well. How do the two models compare?

EXERCISE 11.30.   Find the value of the Debye temperature that best fits the copper data listed in Table 11.5. (The FindFit function will likely be too time consuming. Adjust the value of Θ_D by hand.) Does the Debye model fit the data better than the Einstein model? What is the difference in the models in the low temperature region?

11.5. Equilibrium

The statistical approach to thermodynamics has had profound influence on other fields, notably chemistry. A simple model of a chemical reaction begins with reactants on one side of the equation and products on the other:

$$2H_2 + O_2 \leftrightarrows 2H_2O,$$

where the arrows indicate that the reaction might proceed in either direction. This particular reaction is one often studied in high-school chemistry labs, where the hydrolysis of water yields hydrogen and oxygen gas. The reverse reaction, combustion of the hydrogen gas with atmospheric oxygen, is not often sanctioned but is nevertheless often conducted by curious students. The resulting explosion is a clear demonstration that the product state is energetically favored.

One might well then wonder why hydrogen and oxygen gases exist at all? If water molecules are energetically favored, then how can molecules of oxygen and hydrogen exist? The answer is that there is generally

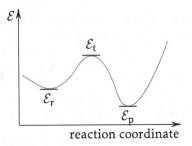

FIGURE 11.11. The energy surface (*gray curve*) for a chemical reaction includes reactant ($\mathcal{E}_r$), product ($\mathcal{E}_p$) and transition ($\mathcal{E}_t$) states

an energy barrier to the chemical pathway from reactant to product. Microscopically, as one atom approaches another, the electron clouds surrounding the nucleus repel one another. It isn't until the nuclear centers are much closer that an energetically-favorable molecular bonding arises. In a vastly simplified sense, a chemical process can be thought of as proceeding along some reaction coordinate from reactant to product, as depicted in figure 11.11. (Actually, a complex reaction like water hydrolysis will have several alternative pathways and a multidimensional coordinate but that does not alter the main point.) The peak of the barrier is known as the *transition state* although there is not a single transition state in a quantum mechanical sense. Instead, there is an ensemble of transition states, all characterized by an energy $\mathcal{E}_t$.

Suppose that the system is initially in the reactant state. Then the probability that the system can surmount the energy barrier and reach the transition state is proportional to the Boltzmann factor:

$$(11.33) \qquad \mathcal{P} \propto e^{-(\mathcal{E}_t - \mathcal{E}_r)/k_B T}.$$

Similarly, the probability of the system beginning in the product state and reaching the transition state is given by the following:

$$(11.34) \qquad \mathcal{P} \propto e^{-(\mathcal{E}_t - \mathcal{E}_p)/k_B T}.$$

Equations 11.33 and 11.34 explain a number of observations about chemical reactions. First, the existence of the barrier explains why hydrogen and oxygen do not spontaneously combust and form water. The rate of reaction (governed by the probabilities) is not proportional to the difference in energies of product and reactant states but is proportional to the difference in energies between the transition state and reaction end points. Additionally, the exponential dependence on the energies means that the rate will be a strong function of temperature. This explains why chemists rely so heavily on heating their reaction vessels. Bunsen burners were once standard fixtures in chemistry labs; safety concerns have greatly restricted their use.

EXERCISE 11.31. In SI units, the value of $k_B T$ at 300 K is approximately 2.5 kJ/mol. Compute the probability of occupying each of two states given an energy difference of $\Delta \mathcal{E} = 0, 5, 10, 15, 20, 25, 30$ and 50 kJ/mol between the two. Suppose that you raise the temperature to 400 K. How does that change the probability table?

The transition state model of chemical reactions implies that reactants reach the transition state with the probability shown in Equation 11.33. At that point, the system may evolve into the product state or revert to the initial, reactant state. Similarly, products can also return to the transition state with the probability defined in Equation 11.34. Over time, we expect to see a distribution arise that depends on the energy difference between reactants and products. If, as depicted in figure 11.11, the product state is energetically lower, we expect the distribution to be dominated by the product state.

Gibbs recognized that if we want to be quantitative about these ideas, we need to consider the case where the number of particles is not constant. In our water example, if there are N_{H_2} molecules of hydrogen, N_{O_2} molecules of oxygen and N_{H_2O}, then the values of those will change as the reaction proceeds in either direction. Gibbs suggested that there must be a thermodynamic variable that corresponds to this change of number and called it the potential. It is now generally called the chemical potential and identified by the symbol μ. The second law of thermodynamics is expanded to include the situations where the particle number is not constant:

$$(11.35) \qquad d\mathcal{E} = T\,dS - p\,dV + \sum_{i=1}^{n} \mu_i \, dN_i,$$

where the sum extends over n different species. Formally, we can define the chemical potential as follows:

$$(11.36) \qquad \mu_i = \frac{\partial \mathcal{E}}{\partial N_i}.$$

The astute student may recognize that N_i is an integer and, thus, might wonder how the partial derivative can be defined. (There can be no limit of $N_i + \delta$ as $\delta \to 0$ when N_i can only have integral values.) In practice, this quantity can be defined as $\mu_i = \mathcal{E}(N_i + 1) - \mathcal{E}(N_i)$, where we compute the difference in energies when the particle number changes by one. We utilize the partial derivative notation for consistency with our previous mathematical notation.[22]

---

[22]This is a primary example of the need to understand what the equations mean at a fundamental level. One could well argue that, particularly in this instance, our notational choice is exceptionally poor. Nevertheless, Equation 11.36 has become the standard for expressing the concept, at least in physics texts.

Gibbs proposed that chemical equilibrium occurs when the system occupies the most probable state and that condition is satisfied when the derivative of the **free energy** of the system vanishes.[23] Thus far, we have avoided dealing with many of the details of thermodynamic systems but the concept of free energy is actually relatively simple. If we think about defining the energy of a system $\mathcal{E}$, it is obviously a function of many variables $\mathcal{E}(T, p, V, S, N, \ldots)$ which are related through an equation of state like the ideal gas law. So, when we talk about taking partial derivatives, we need to be explicit about what variables are actually being controlled. For example, if we have a number of particles in a box, then we can control the number N, the volume of the box V and the temperature T. The pressure p cannot be separately controlled, its value is a consequence of the ideal gas law. The energy in this case is a function of three variables $\mathcal{E}(N, V, T)$.

Alternatively, we might instead want to control the pressure in the box. In this case, we will find that the volume will be defined via the ideal gas law. The energy here is then a function of three variables $\mathcal{E}(N, p, T)$. As a notational aid and in order to differentiate the different combinations of independent variables, each of the energies associated with a particular set of independent variables has been given a specific name.[24] The standard internal energy that we have been considering, as per Equation 11.35, is a function of volume and entropy: $\mathcal{E} = \mathcal{E}(N, V, S)$. The Helmholtz free energy is defined as follows:

$$\mathcal{F} = \mathcal{E} + TS.$$

The equivalent form of the second law is written as follows:

$$(11.37) \qquad d\mathcal{F} = d\mathcal{E} + S\,dT + T\,dS = p\,dV + S\,dT + \sum_{i=1}^{n} \mu_i dN_i.$$

Hence, the Helmholtz free energy is a function of the variables N, V and T, $\mathcal{F} = \mathcal{F}(N, V, T)$. Similarly, one can define the Gibbs free energy $\mathcal{G}(N, p, T)$ and the enthalpy $\mathcal{H}(N, p, S)$.

> EXERCISE 11.32. Using the example of Equation 11.37, what are the definitions of the Gibbs free energy and enthalpy in terms of the standard internal energy $\mathcal{E}$ and the other thermodynamic properties?

---

[23]A vanishing derivative marks an extreme point: a maximum or minimum. In this instance, equilibrium occurs at the minimum of the free energy.

[24]Imagine the difficulties that would ensue if we expected students to remember that $\mathcal{E}(N, V, T)$ and $\mathcal{E}(N, p, T)$ were entirely different quantities. We could decorate them as $\mathcal{E}^{NVT}$ but this still leaves us with rather ugly notation.

The statement of chemical equilibrium is then provided by the following expression:

(11.38)
$$\sum_{i=1}^{n} \mu_i \, dN_i = 0,$$

which arises from the fact that the relevant free energy is a minimum ($d\mathcal{E} = 0$, for example) at equilibrium. Because atoms are indestructible at energies relevant to chemical processes, the numbers N_i are related. We can write a chemical reaction like water hydrolysis in an abstract form:

(11.39)
$$\sum_{i=1}^{n} \alpha_i X_i = 0,$$

where the α_i are the stoichiometric indices[25] and the X_i are the species. For the water reaction, this is just the statement:

$$-2H_2 - O_2 + 2H_2O = 0.$$

with the identification that $\alpha_{H_2} = -2$, etc.

If we now have a chemical reaction in which one of the species changes by an amount, dN_i, we recognize that the change is constrained by Equation 11.39 and it must be proportional to the stoichiometric index: $dN_i = \lambda \alpha_i$. As a result, we must also have that the following relation is true:

(11.40)
$$\sum_{i=1}^{n} \alpha_i \mu_i = 0.$$

This is a general statement of chemical equilibrium.

For ideal gases, where we maintain control over the volume and temperature, then the relevant free energy is the Helmholtz free energy $\mathcal{F}$. In this case, we have that the chemical potential is defined by the following:

$$\mu_i = \frac{\partial \mathcal{F}}{\partial N_i} = -k_B T \ln \mathcal{Z}_i / N_i,$$

where we have assumed that the total partition function $\mathcal{Z}$ factors into separate partition functions for each molecular species $\mathcal{Z}_i$.

Substituting into Equation 11.40, we find the following:

$$-k_B T \sum_{i=1}^{n} \alpha_i \ln \mathcal{Z}_i / N_i = \Delta \mathcal{F} + k_B T \sum_{i=1}^{n} \ln N_i^{\alpha_i},$$

---

[25]The Greek root στοιχεῖον means element. Hence, stoichiometry is the measure of the elements.

where we have lumped the single-particle partition functions into the free energy and used the fact that $a \ln x = \ln x^a$. Exponentiating this expression, we now find the following result:

$$(11.41) \qquad \prod_{i=1}^{n} N_i^{\alpha_i} = e^{-\Delta \mathcal{F}/k_B T} \equiv K(T).$$

The term $K(T)$ is known as the equilibrium constant and Equation 11.41 is known as the **equation of mass action**. For our example of water hydrolysis, the equation of mass action would read as follows:

$$\frac{N_{H_2O}^2}{N_{H_2}^2 N_{O_2}} = K.$$

The law of mass action was developed initially by the Norwegian chemists Peter Waage and his brother-in-law Cato Maximillian Guldberg in 1864,[26] building on work by the French chemist Claude Louis Berthollet, who first recognized that chemical reactions were reversible.

EXERCISE 11.33. At a particular temperature T, the gases

$$CO + H_2O \leftrightharpoons CO_2 + H_2$$

are in chemical equilibrium in a vessel of volume V. What is the appropriate free energy for this system? What is the statement of chemical equilibrium? If the total partition function is separable ($\mathcal{Z} = \mathcal{Z}_{CO} \mathcal{Z}_{H_2O} \mathcal{Z}_{CO_2} \mathcal{Z}_{H_2}$), does the rate constant depend upon the volume?

We can recover this fundamental behavior of chemical species from the statistical approach devised by Gibbs and Boltzmann. Remarkably, the mathematical framework devised by Gibbs can be utilized with little modification when considering quantum phenomena. There are numerous examples of such behavior but this text must end somewhere and we must leave something to be discovered in subsequent courses.

Students at this juncture should have a modest appreciation for how one might utilize mathematical language to describe physical phenomena and how the process of developing models unfolds. Students should also be rather proficient in utilizing the numerical tools in the *Mathematica* package and those skills will indeed be transferable to other courses. We have endeavored to link the mathematical results with experimental observations and students should not lose sight of the fact that science depends fundamentally on experiment. Moreover, the analytical approach that we take in physics classes can certainly be used to analyze problems in other arenas. Good fortune.

---

[26]Waage and Guldberg published their "Studier over Affiniteten," in the proceedings of the Norwegian Academy of Sciences: *Avhandlinger Norske Videnskaps-Akademi Oslo*.

Vectors and Matrices

Vector notation provides a compact representation of the equations dealing with motion in multiple dimensions. In simple linear motion, an object can be characterized by its position x in some reference system. For objects that are constrained to move in a plane, the object is characterized by two numbers x and y that specify its position in the plane. By convention, we can write the position as (x,y), where the order is important. In general, the point (y,x) is not the same as the point (x,y). For objects that move through three dimensions, we now require three numbers to specify location (x,y,z). If we had more dimensions, we could continue analogously.

So, in some basic sense, a vector is an ordered list of numbers. One might imagine a fruit vector, in which each component of the vector represents the number of apples, bananas, grapefruit, etc., that one possesses. In the pharmaceutical industry, chemists construct quantitative structure/activity relationship (QSAR) vectors for small molecules that capture their chemical properties like acidity and polarity and may have millions of components. In these examples, there is no particular relation between the components of the vector: one obviously cannot mix apples and bananas. In physics applications, however, we make a further restriction on the vectors that we utilize: the components must have the same dimensionality. This means that if x has the units of a length, then so will y and z.

In this text, and most printed matter, we utilize the notation that boldface characters represent vectors. Hence, we will write $\mathbf{r} = (x,y,z)$, which has the obvious advantage that we can replace several characters with just one. On the blackboard, or in homework submissions, it is more common to utilize an arrow atop the character: $\vec{r} = (x,y,z)$. This style is not nearly as aesthetically pleasing as the use of boldface but is a practical solution to the problem that it is quite difficult and time consuming to draw boldface characters by hand.

© Mark A. Cunningham 2015
M.A. Cunningham, *Neoclassical Physics*, Undergraduate Lecture
Notes in Physics, DOI 10.1007/978-3-319-10647-2

One can perform algebraic manipulations on vectors. The scalar product of a number and a vector is defined as

$$\alpha \mathbf{x} = (\alpha x_1, \alpha x_2, \ldots, \alpha x_n),$$

where the vector is assumed to have n components. One can add vectors:

$$\mathbf{x} + \mathbf{y} = (x_1 + y_1, x_2 + y_2, \ldots, x_n + y_n),$$

provided that the vectors have the same dimensionality. There is also an additive inverse $-\mathbf{x}$:

$$\mathbf{x} + (-\mathbf{x}) = 0,$$

where the inverse is obtained by negating all of the components of $\mathbf{x}$.

There are two means of multiplying vectors. The first is the inner product and results in a scalar:

$$\mathbf{x} \cdot \mathbf{y} = x_1 y_1 + x_2 y_2 + \cdots + x_n y_n$$

The inner product of a vector with itself produces the square of the magnitude of the vector:

$$\mathbf{x} \cdot \mathbf{x} \equiv |\mathbf{x}|^2 = x^2,$$

where we utilize the common notation that the magnitude of a vector is written in italics. In places in the text where this notation may be confusing, we utilize the first form.

The second vector product is the outer product, which results in a vector. Restricting ourselves to three dimensions for the moment, the outer product is also known as the cross product:

$$\mathbf{x} \times \mathbf{y} = \left((x_2 y_3 - x_3 y_2), (x_3 y_1 - x_1 y_3), (x_1 y_2 - x_2 y_1) \right).$$

The cross product does not commute. In fact, $\mathbf{x} \times \mathbf{y} = -\mathbf{y} \times \mathbf{x}$. The vectors $\mathbf{x}$ and $\mathbf{y}$ are orthogonal to their cross product, by which we mean that:

$$\mathbf{x} \cdot (\mathbf{x} \times \mathbf{y}) = 0 \quad \text{and} \quad \mathbf{y} \cdot (\mathbf{x} \times \mathbf{y}) = 0.$$

Higher order products can generally be reduced via one of several vector identities:

$$\mathbf{a} \cdot (\mathbf{b} \times \mathbf{c}) = \mathbf{b} \cdot (\mathbf{c} \times \mathbf{a}) = \mathbf{c} \cdot (\mathbf{a} \times \mathbf{b})$$

$$\mathbf{a} \times (\mathbf{b} \times \mathbf{c}) = (\mathbf{a} \cdot \mathbf{c})\mathbf{b} - (\mathbf{a} \cdot \mathbf{b})\mathbf{c}$$

$$(\mathbf{a} \times \mathbf{b}) \cdot (\mathbf{c} \times \mathbf{d}) = (\mathbf{a} \cdot \mathbf{c})(\mathbf{b} \cdot \mathbf{d}) - (\mathbf{a} \cdot \mathbf{d})(\mathbf{b} \cdot \mathbf{c})$$

Students are undoubtedly familiar with the Cartesian representation of vectors, which is predicated on some choice of coordinate system. As we shall see, the choice of coordinate system will not affect the physical interpretation of our equations, so any choice of coordinate system will suffice. As a practical matter, though, it will often prove highly useful to choose coordinate systems to minimize the algebraic effort in solving the resulting equations.

To define a coordinate system, we will most often define a set of normalized, orthogonal *basis vectors*. In this text, we shall define these as follows:

$$\hat{\mathbf{x}} = (1,0,0) \qquad \hat{\mathbf{y}} = (0,1,0) \qquad \hat{\mathbf{z}} = (0,0,1).$$

For an arbitrary vector $\mathbf{v} = (v_x, v_y, v_z)$, the inner product of the basis vectors with the vector produces the *components* of $\mathbf{v}$:

$$\mathbf{v} \cdot \hat{\mathbf{x}} = v_x \quad \mathbf{v} \cdot \hat{\mathbf{y}} = v_y \quad \mathbf{v} \cdot \hat{\mathbf{z}} = v_z.$$

The basis vectors serve as projection operators.

It is not necessary for the basis vectors to be orthogonal to still serve as a basis. In fact, it is common in crystallography to choose a set of basis vectors that are aligned with the crystal lattice. In this case, the basis vectors $\mathbf{e}_i$ are not orthogonal and are often not normalized to have unit magnitude. To obtain the components of a vector in such a basis set, it is necessary to define dual vectors $\tilde{\mathbf{e}}_i$:

$$(A.1) \qquad \tilde{\mathbf{e}}_1 = \frac{\mathbf{e}_2 \times \mathbf{e}_3}{\mathbf{e}_1 \cdot (\mathbf{e}_2 \times \mathbf{e}_3)} \quad \tilde{\mathbf{e}}_2 = \frac{\mathbf{e}_3 \times \mathbf{e}_1}{\mathbf{e}_1 \cdot (\mathbf{e}_2 \times \mathbf{e}_3)} \quad \tilde{\mathbf{e}}_3 = \frac{\mathbf{e}_1 \times \mathbf{e}_2}{\mathbf{e}_1 \cdot (\mathbf{e}_2 \times \mathbf{e}_3)}.$$

The components of an arbitrary vector $\mathbf{v}$ are obtained by projections with the dual vectors:

$$v_i = \mathbf{v} \cdot \tilde{\mathbf{e}}_i.$$

In the text, we shall want to differentiate and integrate vector quantities. We shall interpret the derivative of a vector as the derivative of its components:

$$\frac{d}{dt}\mathbf{x} = \left(\frac{dx_1}{dt}, \frac{dx_2}{dt}, \dots, \frac{dx_n}{dt}\right).$$

The derivative is an operator that acts on the components of the vector. Similarly, we will treat integration as an operator acting on the components of a vector:

$$\int dt\,\mathbf{x} = \left(\int dt\,x_1, \int dt\,x_2, \dots, \int dt\,x_n\right).$$

Vector calculus admits additional forms of integration. For example, one may integrate a vector function along a directed path or over a surface. This gives rise to integrals of the following forms:

$$\int d\mathbf{s} \cdot \mathbf{F} \quad \text{and} \quad \int d\mathbf{A} \cdot \mathbf{B},$$

where **F** and **B** are vector functions. Both of these integrals result in scalar values. In Cartesian coordinates the infinitesimal path element ds can be resolved into coordinates (See figure A.1):

$$d\mathbf{s} = \hat{\mathbf{x}}\,dx + \hat{\mathbf{y}}\,dy + \hat{\mathbf{z}}\,dz.$$

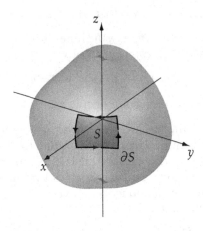

Figure A.1. Consider a volume in space (*light gray object*) and a portion of the surface of that volume S. The direction of the surface is routinely taken to be directed outward and a path around the boundary ∂S of the surface is positively directed in a right-handed sense

Differential surface elements are constructed from the cross products of the unit vectors:

$$d\mathbf{A} = \hat{\mathbf{x}}\,dx \times \hat{\mathbf{y}}\,dy + \hat{\mathbf{y}}\,dy \times \hat{\mathbf{z}}\,dz + \hat{\mathbf{z}}\,dz \times \hat{\mathbf{x}}\,dx$$
$$= \hat{\mathbf{x}}\,dy\,dz + \hat{\mathbf{y}}\,dx\,dz + \hat{\mathbf{z}}\,dx\,dy.$$

Integrals over the volume take the following form:

$$\int d^3\mathbf{r}\,\mathbf{F},$$

where the volume element is composed by the triple product:

$$d^3\mathbf{r} = \hat{\mathbf{x}}\,dx \cdot (\hat{\mathbf{y}}\,dy \times \hat{\mathbf{z}}\,dz) = dx\,dy\,dz.$$

The result of this integration will be a vector function. Note that the triple product is an invariant; one constructs the same invariant from permutations of $\hat{\mathbf{x}}$, $\hat{\mathbf{y}}$ and $\hat{\mathbf{z}}$.

For problems that have cylindrical symmetry, we shall frequently utilize a cylindrical coordinate system. In cylindrical coordinates, the z-axis remains the same as in the Cartesian system but we utilize polar coordinates

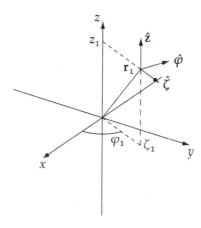

Figure A.2. A cylindrical coordinate system utilizes the same $\hat{z}$ vector as the Cartesian system but uses polar coordinates in the x-y plane

in the x-y plane. In physics, the most common notation for this is to denote the azimuthal angle by θ and the radial distance by r. We shall frequently also utilize spherical coordinates, where θ is used by physicists to mean the polar angle and φ denotes the azimuthal angle. The distance r in spherical coordinates reflects the distance to the origin, not the distance to the z-axis. To avoid these issues, we shall use the (somewhat nonstandard) symbol φ to denote the azimuthal angle as measured from the x-axis and the (completely nonstandard) symbol ζ to denote the radial distance to the z-axis, as illustrated in figure A.2.

It is somewhat more common to utilize the symbol ρ to denote the radial distance to the z-axis but, in physics, we shall often use ρ to mean a density. This results in the particularly awkward equation for computing the total charge from a charge distribution:

$$Q = \int d^3\mathbf{r}\,\rho(\mathbf{r}) = \int_{r_a}^{r_b} d\rho \int_0^{2\pi} d\varphi \int_{z_a}^{z_b} dz\,\rho\,\rho(\rho,\varphi,z),$$

where we intend to integrate the function $\rho(\rho,\varphi,z)$ over the range of the radial coordinate ρ. We note though that our choice of ζ to replace r and ρ (which is the Greek form of the Latin r) is not completely unjustified. We are using the string of Latin characters x, y, z to denote positions. The equivalent Greek characters are χ, υ and ζ. It it quite likely that handwriting-challenged faculty will have difficulty differentiating χ from x on the blackboard but it seems completely impossible for even the most diligent faculty to differentiate the Greek υ from the Latin v. It is imperative that we differentiate the radial position υ from the velocity v.

So, weighing these three choices, we are left with ζ as the radial coordinate in a cylindrical system. Another possibility would be the Greek symbol ξ, which has the advantage that it is dramatically different than any Latin character and not used commonly. In the author's mind, ζ is sometimes a pseudonym for the direction z, not orthogonal to it, but the Greek ξ will prove overly taxing to the aforementioned handwriting challenged to use routinely. So, we'll stick with ζ as the radial coordinate in cylindrical systems. Again, the choice of notation is partly an aesthetic choice. Students are ultimately free to make their own choices.

The same point $\mathbf{r}_1 = (x_1, y_1, z_1)$ can be expressed in both Cartesian and cylindrical forms. To avoid confusion, we shall only use the parenthesis form of the vector to mean Cartesian coordinates. That is,

$$\mathbf{r}_1 = (x_1, y_1, z_1) = \hat{\mathbf{x}}\, x_1 + \hat{\mathbf{y}}\, y_1 + \hat{\mathbf{z}}\, z_1$$

are different representations of the same point in space. In cylindrical coordinates, we have the following:

$$\mathbf{r}_1 = (\zeta_1 \cos\varphi_1, \zeta_1 \sin\varphi_1, z_1)$$

where $\zeta_1 = (x_1^2 + y_1^2)^{1/2}$ and $\varphi_1 = \tan^{-1}(y_1/x_1)$. This latter definition implies that the following relations hold:

$$\cos\varphi_1 = \frac{x_1}{(x_1^2 + y_1^2)^{1/2}} \quad \text{and} \quad \sin\varphi_1 = \frac{y_1}{(x_1^2 + y_1^2)^{1/2}}.$$

The unit vectors in a cylindrical coordinate system are given by the following:

$$\hat{\zeta} = (\cos\varphi, \sin\varphi, 0) \quad \hat{\varphi} = (-\sin\varphi, \cos\varphi, 0) \quad \hat{\mathbf{z}} = (0, 0, 1).$$

These are related to the Cartesian unit vectors by the following:

$$\hat{\mathbf{x}} = \hat{\zeta} \cos\varphi - \hat{\varphi} \sin\varphi \qquad\qquad \hat{\zeta} = \hat{\mathbf{x}} \cos\varphi + \hat{\mathbf{y}} \sin\varphi$$

$$\hat{\mathbf{y}} = \hat{\zeta} \sin\varphi + \hat{\varphi} \cos\varphi \qquad\qquad \hat{\varphi} = -\hat{\mathbf{x}} \sin\varphi + \hat{\mathbf{y}} \cos\varphi$$

The differential elements in cylindrical coordinates are given by the following:

$$d\mathbf{s} = \hat{\zeta}\, d\zeta + \hat{\varphi}\zeta\, d\varphi + \hat{\mathbf{z}}\, dz$$

$$d\mathbf{A} = \hat{\zeta}\zeta\, d\varphi\, dz + \hat{\varphi}\, d\zeta\, dz + \hat{\mathbf{z}}\zeta\, d\zeta\, d\varphi$$

$$d^3\mathbf{r} = \zeta\, d\zeta\, d\varphi\, dz.$$

We will also make use of spherical coordinates, as depicted in figure A.3, in situations where we can exploit the symmetry. A point in space $\mathbf{r}_1$ is characterized by the distance to the origin $r_1 = (x_1^2 + y_1^2 + z_1^2)^{1/2}$, and two angles. The azimuthal angle is measured from the x-axis and we denote it

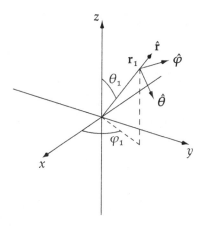

Figure A.3. In a spherical coordinate system, points are represented by the radial distance to the origin, the polar angle θ and the azimuthal angle φ

by $\varphi_1 = \tan^{-1}(y_1/x_1)$. The polar angle is measured from the z-axis and we denote it by $\theta_1 = \tan^{-1}(\sqrt{x_1^2 + y_1^2}/z_1)$.

Unit vectors in the spherical coordinate system are given by the following:

$$\hat{\mathbf{r}} = (\sin\theta_1 \cos\varphi_1, \sin\theta_1 \sin\varphi_1, \cos\theta_1)$$
$$\hat{\boldsymbol{\theta}} = (\cos\theta_1 \cos\varphi_1, \cos\theta_1 \sin\varphi_1, -\sin\theta_1)$$
$$\hat{\boldsymbol{\varphi}} = (-\sin\varphi_1, \cos\varphi_1, 0).$$

These are related to the Cartesian vectors by the following expressions:

$$\hat{\mathbf{x}} = \hat{\mathbf{r}} \sin\theta_1 \cos\varphi_1 + \hat{\boldsymbol{\theta}} \cos\theta_1 \cos\varphi_1 - \hat{\boldsymbol{\varphi}} \sin\varphi_1$$
$$\hat{\mathbf{y}} = \hat{\mathbf{r}} \sin\theta_1 \sin\varphi_1 + \hat{\boldsymbol{\theta}} \cos\theta_1 \sin\varphi_1 + \hat{\boldsymbol{\varphi}} \cos\varphi_1$$
$$\hat{\mathbf{z}} = \hat{\mathbf{r}} \cos\theta_1 - \hat{\boldsymbol{\theta}} \sin\theta_1.$$

Differential elements are listed below:

$$d\mathbf{s} = \hat{\mathbf{r}}\,dr + \hat{\boldsymbol{\theta}}\,r\,d\theta + \hat{\boldsymbol{\varphi}}\,r\sin\theta\,d\varphi$$
$$d\mathbf{A} = \hat{\mathbf{r}}\,r^2\sin\theta\,d\theta\,d\varphi + \hat{\boldsymbol{\theta}}\,r\sin\theta\,dr\,d\varphi + \hat{\boldsymbol{\varphi}}\,r\,dr\,d\theta$$
$$d^3\mathbf{r} = r^2\sin\theta\,dr\,d\theta\,d\varphi.$$

Just as vector calculus includes new forms of integration, there are additional derivative operators that can be defined. We can define the symbol ∇ to mean the following:

$$\nabla \equiv \left(\frac{\partial}{\partial x}, \frac{\partial}{\partial y}, \frac{\partial}{\partial z}\right).$$

This mathematical quantity is not a vector in the sense that we have used before but it does have ordered components. (Mathematicians will call this entity a 1-form.) We can construct a vector from the gradient of a scalar function f:

$$\nabla f \equiv \left(\frac{\partial f}{\partial x}, \frac{\partial f}{\partial y}, \frac{\partial f}{\partial z} \right).$$

We can also define the (scalar) divergence of a vector function:

$$\nabla \cdot \mathbf{F} = \frac{\partial F_x}{\partial x} + \frac{\partial F_y}{\partial y} + \frac{\partial F_z}{\partial z}.$$

Finally, we can define the curl of a vector function:

$$\nabla \times \mathbf{F} = \left(\frac{\partial F_z}{\partial y} - \frac{\partial F_y}{\partial z}, \frac{\partial F_x}{\partial z} - \frac{\partial F_z}{\partial x}, \frac{\partial F_y}{\partial x} - \frac{\partial F_x}{\partial y} \right).$$

The vector derivatives can be used to derive an alternative representation of our various equations of motion. This is done primarily because, while it is exceptionally difficult to solve the resulting partial differential equations, it is even more difficult to solve integral equations. Students may recognize that transforming between integral and differential forms of equations amounts to the mathematical equivalent of pushing one's peas about the plate in the hopes that one's parents do not notice that they have been eaten by the dog. The author cannot provide a cogent argument against such a proposal.

We have introduced matrix notation in the text primarily as a visual convenience. If we define a vector $\mathbf{x}$ to have N components x_j and a matrix $\mathbf{A}$ to have $M \cdot N$ components A_{ij}, then the product $\mathbf{y}$ of the multiplication of $\mathbf{Ax}$ is defined as follows:

(A.2)
$$y_i = \sum_{j=1}^{N} A_{ij} x_j.$$

In a more graphical display, we can write this in the following form:

(A.3)
$$\begin{bmatrix} y_1 \\ y_2 \\ y_3 \\ \vdots \\ y_M \end{bmatrix} = \begin{bmatrix} A_{11} & A_{12} & A_{13} & \cdots & A_{1N} \\ A_{21} & A_{22} & A_{23} & \cdots & A_{2N} \\ A_{31} & A_{32} & A_{33} & \cdots & A_{3N} \\ \vdots & \vdots & \vdots & \ddots & \vdots \\ A_{M1} & A_{M2} & A_{M3} & \cdots & A_{MN} \end{bmatrix} \begin{bmatrix} x_1 \\ x_2 \\ x_3 \\ \vdots \\ x_N \end{bmatrix},$$

by which we mean that, for example,

$$y_2 = A_{21} x_1 + A_{22} x_2 + A_{23} x_3 + \cdots + A_{2N} x_N.$$

Thus, Equations A.2 and A.3 are simply different representations of the same mathematical quantity.

The matrix representation of a system of linear equations provides a convenient and compact means for representing those equations. There are a number of courses on linear algebra that provide far more detail on matrices than we shall attempt to cover here. We note that one can also define multiplication of two matrices $\mathbf{A}$ and $\mathbf{B}$ in a systematic manner where the components of $\mathbf{AB}$ are obtained by taking the inner products of the columns of the right-hand matrix $\mathbf{B}$ with the rows of the left-hand matrix $\mathbf{A}$, similar to the form displayed in Equation A.2. For this to make sense, the number of columns of the right-hand matrix $\mathbf{B}$ must be the same as the number of rows of the left-hand matrix $\mathbf{A}$. Matrix algebra differs from the algebra of the real numbers in that the commutative property is generally lost, i.e., $\mathbf{AB} \neq \mathbf{BA}$.

In the graphical representation of matrices, for $\mathbf{Ax}$ to make sense, the vector $\mathbf{x}$ needs to have as many rows as the matrix $\mathbf{A}$ has columns. Consequently, $\mathbf{x}$ is represented as a column vector. The *Mathematica* syntax defines vectors as ordered lists and matrices as lists of lists. Performing a matrix-vector multiplication implicitly uses the rule stated in Equation A.2. For example the *Mathematica* script v={1,2,3} creates a list of three numbers that will be interpreted in the usual sense as a vector. Thus the script yhat={0,1,0}; yhat.v will produce the result 2, i.e., the second component of the vector v.

As a consequence, there is no need to define column vectors, although in a *Mathematica* script, we could define a single-column matrix as follows:

$$p = \{\{E/c\}, \{px\}, \{py\}, \{pz\}\}.$$

The dual vector would then be the row vector pd={E/c,-px,-py,-pz}. When using the MatrixForm function, these two entities would appear in the expected form. The dual vector can be constructed from the transpose of the product of the vector and the metric tensor. For the four-vector $\mathbf{x}$, we have the following:

$$\tilde{\mathbf{x}} = \begin{bmatrix} ct & -x & -y & -z \end{bmatrix} = \left(\begin{bmatrix} 1 & 0 & 0 & 0 \\ 0 & -1 & 0 & 0 \\ 0 & 0 & -1 & 0 \\ 0 & 0 & 0 & -1 \end{bmatrix} \begin{bmatrix} ct \\ x \\ y \\ z \end{bmatrix} \right)^T.$$

From which we can recover the invariant interval:

$$\tilde{\mathbf{x}} \cdot \mathbf{x} = \begin{bmatrix} ct & -x & -y & -z \end{bmatrix} \begin{bmatrix} ct \\ x \\ y \\ z \end{bmatrix} = (ct)^2 - x^2 - y^2 - z^2.$$

Note that the graphical representation of row vectors and column vectors is not required if we simply utilize the definition of Equation A.2. Both vectors and their duals are ordered lists of n elements. Their inner product can be defined without resort to a graphical display. Indeed, the astute student will recognize that we have refrained from illustrating vector addition in the text with pictures of arrows linked head to tail. We have used the precise definition that vector addition is defined by adding the components of the vectors. This activity can be performed exactly where sketching arrows cannot.

We referred to the metric tensor for the theory of Special Relativity, which we can write as the matrix with $(1,-1,-1,-1)$ on the diagonal and zero elsewhere. The metric tensor is a mathematical extension of the Euclidean distance between two points $A = (x_1, y_2)$ and $B = (x_2, y_2)$ on the plane: $d = [(x_2 - x_1)^2 + (y_2 - y_1)^2]^{1/2}$. The Euclidean metric tensor is the identity matrix, i.e., a matrix with ones on the diagonal and zero elsewhere. In this initial course, the mathematical structure underlying metric spaces is not necessary to provide adequate descriptions of physical systems. It was, of course, essential for the development of Einstein's General Theory of Relativity.

We have made no effort in this text to distinguish four-vectors from three-vectors with different notation. This may be something of a controversial choice but texts that make this distinction generally jump into notational choices that are difficult to understand initially. We made a conscious choice in this text to use notation that is not compact, so that students will learn to appreciate the value of compact notation as they progress. Eventually, use of the Einstein summation convention will prove useful but not for introductory students. It cannot be expected that such students will grasp the subtlety that by p_μ, with a Greek subscript, we mean the four-momentum of something, whereas by p_i, with a Latin subscript, we mean the ith component of the three-momentum of something. We shall leave such intricacies to the instructors of subsequent courses.

Noether's Theorem

We would now like to make the connection between symmetries of the equations of motion and conservation laws. To do so, we will first need to revisit a result first established by the mathematician Joseph-Louis Lagrange.[1] Lagrange suggested that one should consider the quantity $T - U$, where T is the kinetic energy and U is the potential energy that we have defined previously. Recall that T is a function only of the velocities and U is only a function of the positions. For concreteness, let us use the energies defined for two masses interacting through the gravitational force; these were defined by Equations 2.26 and 2.27.

The Lagrangian function $\mathcal{L} = T - U$ is a function of six quantities: the three components of the vector $\mathbf{r}_2(t) - \mathbf{r}_1(t)$ and the three components of the vector $\mathbf{v}_2(t) - \mathbf{v}_1(t)$. Consider taking the derivative of $\mathcal{L}$ with respect to one of the components of $\mathbf{r}_2(t) - \mathbf{r}_1(t)$. As T does not depend upon the position, we have, for the x-component:

$$\frac{\partial \mathcal{L}}{d(\mathbf{r}_2(t) - \mathbf{r}_1(t))_x} = -\frac{\partial U}{d(\mathbf{r}_2(t) - \mathbf{r}_1(t))_x}$$

(B.1)
$$= G \frac{M_1 M_2}{|\mathbf{r}_2(t) - \mathbf{r}_1(t)|^3}(\mathbf{r}_2(t) - \mathbf{r}_1(t))_x,$$

where by the $()_x$ notation we mean the x-component of the vector. Similarly, if we take the derivative of $\mathcal{L}$ with respect to the x-component of the vector $\mathbf{v}_2(t) - \mathbf{v}_1(t)$, we would obtain the following result:

$$\frac{\partial \mathcal{L}}{d(\mathbf{v}_2(t) - \mathbf{v}_1(t))_x} = \frac{\partial T}{d(\mathbf{v}_2(t) - \mathbf{v}_1(t))_x}$$

(B.2)
$$= \frac{M_1 M_2}{M_1 + M_2}(\mathbf{v}_2(t) - \mathbf{v}_1(t))_x.$$

---

[1] Lagrange was actually born in Turin, Italy and baptized as Guiseppe Lodovico Lagrangia in 1736. He spent much of his working life in Berlin before moving to Paris in 1787, where he remained for the remainder of his life. Lagrange considered himself French at heart, signing his name with the French spelling even in his youth.

© Mark A. Cunningham 2015
M.A. Cunningham, *Neoclassical Physics*, Undergraduate Lecture Notes in Physics, DOI 10.1007/978-3-319-10647-2

What Lagrange observed is that if we take the time derivative of Equation B.2 and add Equation B.1 we obtain the following:

(B.3) $\dfrac{M_1 M_2}{M_1 + M_2} \dfrac{d}{dt}(\mathbf{v}_2(t) - \mathbf{v}_1(t))_x + G \dfrac{M_1 M_2}{|\mathbf{r}_2(t) - \mathbf{r}_1(t)|^3}(\mathbf{r}_2(t) - \mathbf{r}_1(t))_x$

This is just the x-component of the equations of motion that we wrote down in Equation 2.6 and we know that it vanishes!

Lagrange's approach to mechanics seems somewhat contrived here but, in fact, offers a very general strategy for deriving the equations of motion of physical systems, which is completely equivalent to the Newtonian methodology that we have investigated previously. In using the Lagrange strategy, one must define the kinetic energy T and the potential energy U of the system in terms of whichever variables are convenient and then the equations of motion can be systematically constructed by taking derivatives of the Lagrangian $\mathcal{L}$. Suppose that we have a system defined by a set of variables $\{x_i\} = x_1, \ldots, x_n$. The equations of motion for the x_i are then obtained from the following:

(B.4) $\dfrac{\partial \mathcal{L}}{\partial x_i} - \dfrac{d}{dt} \dfrac{\partial \mathcal{L}}{\partial \left[\dfrac{\partial x_i}{dt}\right]} = 0,$

where there will be n total equations.

> EXERCISE B.1. We just stated the result in Equation B.1. Consider using our concise notation: $\mathbf{r}_2(t) - \mathbf{r}_1(t) = \mathbf{r}_{21} = (x_{21}, y_{21}, z_{21})$. Here, we would write $U = -GM_1 M_2/r_{21}$, where $r_{21} = |\mathbf{r}_{21}|$. Show that you do indeed obtain the result found in Equation B.1

> EXERCISE B.2. Likewise, we just stated the result in Equation B.2. Consider using our concise notation: $\mathbf{v}_2(t) - \mathbf{v}_1(t) = \mathbf{v}_{21} = (v_x, v_y, v_z)$. Here, we would write $T = M_1 M_2 v_{21}^2 / 2(M_1 + M_2)$, where $v_{21} = |\mathbf{v}_{21}|$. Show that you obtain the result found in Equation B.2

Our introduction of Lagrange's methodology allows us now to discuss Noether's ideas on conservation laws. Consider that the parameters that define Lagrange's function $\mathcal{L}$ are subjected to some continuous transformation, like our Lorentz transformation. Then, the parameters can be thought of as themselves functions of another parameter like ζ: $x_i = x_i(\zeta)$. That is, if $\mathcal{L}$ is a function of x and its time derivative dx/dt, we would have the following:

(B.5) $\dfrac{d\mathcal{L}}{d\zeta} = \dfrac{\partial \mathcal{L}}{\partial x} \dfrac{\partial x}{\partial \zeta} + \dfrac{\partial \mathcal{L}}{\partial dx/dt} \dfrac{\partial dx/dt}{\partial \zeta}.$

Now this quantity vanishes if Lagrange's function $\mathcal{L}$ and, consequently, the equations of motion do not depend explicitly on ζ.

Noether defined the following quantity:

(B.6)
$$\eta = \frac{\partial \mathcal{L}}{\partial(dx/dt)} \frac{\partial x}{\partial \zeta}.$$

Noether's theorem is that the time derivative of η vanishes; η is a conserved quantity. We find explicitly that:

$$\frac{d\eta}{dt} = \left[\frac{d}{dt} \frac{\partial \mathcal{L}}{\partial dx/dt}\right] \frac{\partial x}{\partial \zeta} + \frac{\partial \mathcal{L}}{\partial dx/dt} \left[\frac{d}{dt} \frac{\partial x}{\partial \zeta}\right]$$

(B.7)
$$= \left[\frac{d}{dt} \frac{\partial \mathcal{L}}{\partial dx/dt} - \frac{\partial \mathcal{L}}{\partial x}\right] \frac{\partial x}{\partial \zeta} = 0,$$

where we have used the result that Equation B.5 vanishes to replace the second term on the right-hand side of the first line. Note that the quantity in the square brackets in Equation B.7 is just the equation of motion for the parameter x. This vanishes by definition, completing the proof that the quantity η does not depend upon time and is thereby conserved. It is commonly referred to as the Noether current.

For a free particle of mass m and velocity $\mathbf{v}$, there is no potential energy, just kinetic energy. In this case, we find the Lagrangian is given by the following:

$$\mathcal{L} = m \frac{v_x^2 + v_y^2 + v_z^2}{2},$$

where the v_j are the components of velocity in a Cartesian coordinate system. The Lagrangian has no explicit dependence upon x, so here $\zeta = x$ and the corresponding Noether current is just

$$\eta_x = \frac{\partial \mathcal{L}}{\partial v_x} \frac{\partial x}{\partial x} = mv_x.$$

This is the x-component of momentum and we have just demonstrated that it is an invariant of the system, through the use of Noether's theorem.

As a second example, consider the theory of Newtonian gravitation, where the Lagrangian has the following form:

$$\mathcal{L} = \frac{M_1 M_2}{M_1 + M_2} |\mathbf{v}_2 - \mathbf{v}_1|^2 + G \frac{M_1 M_2}{|\mathbf{r}_2 - \mathbf{r}_1|}.$$

Suppose that we rotate the coordinate system by an angle ζ around the z-axis. In the new (primed) coordinate system, we have the following:

$$\mathbf{r}_2' - \mathbf{r}_1' = \begin{pmatrix} \cos\zeta(\mathbf{r}_2 - \mathbf{r}_1)_x - \sin\zeta(\mathbf{r}_2 - \mathbf{r}_1)_y \\ \sin\zeta(\mathbf{r}_2 - \mathbf{r}_1)_x + \cos\zeta(\mathbf{r}_2 - \mathbf{r}_1)_y \\ (\mathbf{r}_2 - \mathbf{r}_3)_z \end{pmatrix}$$

and we have a similar expression for the velocity vector:

$$\mathbf{v}_2' - \mathbf{v}_1' = \begin{pmatrix} \cos\zeta(\mathbf{v}_2 - \mathbf{v}_1)_x - \sin\zeta(\mathbf{v}_2 - \mathbf{v}_1)_y \\ \sin\zeta(\mathbf{v}_2 - \mathbf{v}_1)_x + \cos\zeta(\mathbf{v}_2 - \mathbf{v}_1)_y \\ (\mathbf{v}_2 - \mathbf{v}_3)_z \end{pmatrix}$$

The x-component of the Noether current is given by the following:

$$\eta_x = \frac{\partial \mathcal{L}}{\partial(\mathbf{v}_2' - \mathbf{v}_1')_x} \frac{\partial(x_2' - x_1')_x}{\partial\zeta}$$

$$= -\frac{M_1 M_2}{M_1 + M_2}(\mathbf{v}_2' - \mathbf{v}_2')_x(\mathbf{r}_2' - \mathbf{r}_1')_y.$$

Similarly, the y-component of the Noether current is given by the following:

$$\eta_y = \frac{\partial \mathcal{L}}{\partial(\mathbf{v}_2' - \mathbf{v}_1')_y} \frac{\partial(x_2' - x_1')_y}{\partial\zeta}$$

$$= \frac{M_1 M_2}{M_1 + M_2}(\mathbf{v}_2' - \mathbf{v}_2')_y(\mathbf{r}_2' - \mathbf{r}_1')_x.$$

Each of these two components is individually conserved, so their sum will be conserved as well. Thus the following quantity is conserved:

$$\eta_x + \eta_y = \frac{M_1 M_2}{M_1 + M_2}\left((\mathbf{r}_2' - \mathbf{r}_1')_x(\mathbf{v}_2' - \mathbf{v}_2')_y - (\mathbf{r}_2' - \mathbf{r}_1')_y(\mathbf{v}_2' - \mathbf{v}_2')_x\right)$$

$$= \frac{M_1 M_2}{M_1 + M_2}\left((\mathbf{r}_2' - \mathbf{r}_1') \times (\mathbf{v}_2' - \mathbf{v}_2')\right)_z$$

$$= L_z$$

This is just the z-component of the angular momentum vector defined in Equation 2.24. Hence, invariance to rotations about any axis leads to conservation of angular momentum about that axis.

It is a bit beyond most student's present mathematical abilities but one can define a Lagrangian function from which Maxwell's equations can be derived. The application of Noether's theorem results in a conservation law for electric charge:

$$\int_V d^3\mathbf{r}\,\rho(\mathbf{r}) + \int_{\partial V} d\mathbf{A} \cdot \mathbf{J}(\mathbf{r}) = 0.$$

We recognize this result as Equation 9.3. While electric charge conservation is an experimental observation, it is also the result of an important symmetry that is reflected in the Maxwell equations.

Index

abstraction, 1, 4, 164
acceleration, 22, 24, 134, 161, 177, 184,
  221, 227
  definition, 14
  relation to force, 17
accuracy, 94
  definition, 90
Jean-Baptiste le Rond d'Alembert
  wave equation, 102, 132
algebra
  matrix, 128, 360, 361
  vector, 6, 354
α particle, 64, 239
  scattering, 66, 74–85
alphabet
  Greek, 7, 64, 244, 358
  Latin, 7
André-Marie Ampère
  magnetic field, 251, 267
Ampère-Maxwell equation, 267
amplifier, 316
Anders Jonas Ångström
  solar spectrum, 157
angular momentum
  definition, 36
anode, 315
antimatter, 243
approximation, 2, 13, 253
Dominique François Jean Arago, 251
Argonne National Laboratory
  neutrino event, 249
asteroid, 214
Francis Aston
  isotopes, 230
  Nobel Prize (1922), 230
astronomical unit (AU), 43
asymptote, 79
asymptotic behavior, 61, 63, 123, 208,
  234
atom
  graphical representation, 157
  nuclear structure, 86, 239
  size, 153

axis, *see also* coordinate system
azimuthal angle, 80
  definition, 357, 359

backspin, 183
baryon, 119, 244
baseball, 173–184
  pitch, 183–184
basketball, 8–11
battery, 218, 253
beam
  α particle, 82
  canal rays, 226
  cathode rays, 222
beats, 138
Daniel Bernoulli
  vibrating string, 134
Jacob Bernoulli
  statistics, 320
Claude Louis Berthollet
  reversibility of chemical reactions, 352
William Bertozzi
  relativistic electrons, 127
β particle, *see also* electron
Jean-Baptiste Biot
  magnetic field, 251
Biot-Savart equation, 257, 260, 267, 272
Ludwig Eduard Boltzmann
  kinetic theory of gases, 333
boost, *see also* Lorentz transform
Robert Boyle
  gas law, 340
William Henry Bragg
  Nobel Prize (1915), 148
  x-ray diffraction experiments, 148
William Lawrence Bragg
  Nobel Prize (1915), 148
  x-ray diffraction theory, 148
Bertram N. Brockhouse
  neutron spectroscopy, 158
  Nobel Prize (1994), 158
Robert Brown
  random motion, 324

© Mark A. Cunningham 2015
M.A. Cunningham, *Neoclassical Physics*, Undergraduate Lecture
Notes in Physics, DOI 10.1007/978-3-319-10647-2

Brownian motion, 285, 323–330
bubble chamber
    interpretation, 244–250
    operation, 241

calculus, 40, 52, 254
    fundamental theorem, 28, 32
    integral representation, 273
    vector, 355, 359
caloric fluid, 331
canal rays, 226–232
capacitance
    definition, 302
capacitor, 302, 303
    bypass, 314
Giovanni Domenico Cassini, 88
cathode, 315
cathode rays, see also electron
causality, 113, 290
center of mass, 178
    coordinate origin, 34
    definition, 33
    motion, 33, 178, 189, 279
central limit theorem, see also statistics
CERN, 119–121
James Chadwick
    neutron, 233–238
    Nobel Prize (1935), 238
changeup, see also baseball
channel rays, see also canal rays
chaos, 215
charge
    atomic, 65
    conservation, 254, 297, 317
    current, 253
    definition, 55
    density, 253, 272, 317
    electric field, 66, 252
    electric force, 56, 217, 240
    electron, 223–225, 271
    magnetic, 252
    magnetic force, 219
    mass ratio, 228, 232
    storage, 301
    thermionic emission, 314
    transport velocity, 256
Jacques Charles
    gas law, 340
chemical reaction, 347
chemistry, 347
    reaction coordinate, 348
    transition state, 348

Stephen Chu
    Nobel Prize (1997), 309
    optical molasses, 309
circuit, 257, 292, 296
    LR, 299
    LRC, 305
    multiply-connected, 297
    RC, 303
circular orbit, 29
Benoît Paul Émile Clapeyron
    ideal gas law, 340
CODATA, 239
Claude Cohen-Tannoudji
    Nobel Prize (1997), 309
    optical molasses, 309
Peter Collinson, 303
Arthur Holly Compton
    Nobel Prize (1927), 236
    photon scattering, 236–238
conductivity, 253, 257
conductor
    definition, 255
conic section, 35, 58
    definition, 29
conservation, 11
    angular momentum, 26, 31, 36, 42,
        57, 178, 219
    definition, 10
    eccentricity, 31
    energy, 171, 210
    momentum, 34, 197, 233, 241
    Noether current, 365
coordinate system
    center of mass, 33
    definition, 6
    invariance, 9
    polar, 40
coordinates
    Cartesian, 6, 161, 356
    chemical reaction, 348
    cylindrical, 49, 260, 329, 357
    spherical, 168, 273, 358
Charles Augustin de Coulomb, 55
    electric force, 85, 86, 286
    inverse square law, 56
    torsion balance, 56
cross section, see also scattering
crystal lattice, 148, 149, 342
Andreas Cuneaus
    Leyden jar, 301

current, 252–256
  induced, 294
curve fitting, 3, 46
  least squares, 95
curveball, *see also* baseball

Clinton Davisson
  electron diffraction, 157
  Nobel Prize (1937), 158
Humphrey Davy, 293
Louis Victor Pierre Raymond duc de
    Broglie
  electron wavelength, 158
  Nobel Prize (1929), 158
Lee de Forest
  Audion, 315
René Antoine Ferchault de Réaumur,
    301
Peter Debye
  heat capacity, 346
deflection
  α particle in an electric field, 76
  moving charge in magnetic field, 219
  of a spinning baseball, 180
density, 174
diffraction, 144–154
diffusion, 327–330
dimensional analysis, 3, 22, 164, 174,
    256, 265
diode, 312
Paul Dirac
  classical electrodynamics, 290
  delta function, 258
distribution
  charge, 73
  electrical power, 295
  Gaussian probability, 91, 326
  mass, 49
  Maxwell-Boltzmann, 339
  probability, 320
  random, 82
Christian Doppler
  frequency shift of a moving source,
    119
Doppler shift, 119, 122, 309
Paul Karl Ludwig Drude
  electrons in metals, 255, 296
Pierre-Louis Dulong
  heat capacity of metals, 343

eccentricity, 29
Thomas Edison, 296

electrical meter patent, 315
effusion, 337
Felix Ehrenhaft
  subelectron, 287
Albert Einstein, 10
  Brownian motion, 324
  Eigenzeit, 121
  general theory of relativity, 129
  heat capacity, 343
  relativistic mechanics, 124–128
  respect for mathematics, 44
  special theory of relativity, 114, 116
  temperature, 344
electric field
  charged sphere, 278
  parallel plates, 224
  point charge, 66
electrodynamics, 289
electron, 64, 65, 74, 93, 156, 217, 222,
    239, 253
  cloud, 79, 86, 158
  diffraction, 158
  finite size, 74
  lepton, 119
energy, 60
  conservation, 32, 37
  definition, 32
  free, 350
  kinetic, 32, 37, 60
  potential, 32, 37, 60
  relativistic, 127
entropy, 332
ephemeris table, 88
equations of motion, 16, 25, 180
  electron in metal, 255
equilibrium, 350
equipartition theorem, 336, 342
error function, 93
  definition, 328
Leonhard Euler
  Lagrange points, 195
  wave equation, 132
Paul Peter Ewald
  graphical solution to diffraction, 152
experiment
  β decay, 240
  μ lifetime (CERN), 121
  π⁰ decay/constant light velocity
    (CERN), 120

Bertozzi relativistic electrons, 127
Davisson-Germer electron diffraction, 157
existence of the ν, 249
flag waving, 2
G. P. Thomson electron diffraction, 157
Geiger-Marsden, 66, 81, 130
Miller-Kusch, 337
Millikan oil drop, 285
Perrin, 324
statistical analysis, 90
Stern-Gerlach, 281
Thomson cathode ray, 229
exponential
mass dependence of rockets, 199

Fagnano
elliptical arc length, 262
Michael Faraday
magnetic field lines, 252, 266
magnetic induction, 292–296
fastball, see also baseball
Enrico Fermi
neutrino, 241
Adolf Fick
diffusion, 330
field
electric, 66
gravitational, 48
magnetic, 257
Hyppolite Fizeau
velocity of light, 101
John Ambrose Fleming
thermionic valve, 315
fluorescence, 66
force
aerodynamic, 177
attractive, 56
central, 20, 56, 78
conservative, 169
Coulombic, 217
definition, 16
dissipative, 159, 244, 307, 309
electromagnetic, 55
gravity, 21
inverse square, 24
magnetic, 219
repulsive, 56
resistive, 164–167, 256
strong nuclear, 85
weak nuclear, 239

Léon Foucault
velocity of light, 101
Jean Baptiste Joseph Fourier
sine and cosine series, 135, 312
Benjamin Franklin
electrical novelties, 302
free energy, 350
free-body diagram, 160
frequency
definition, 119
friction, 185–188
Walter Friedrich
x-ray diffraction, 147
function
continuous, 253
of a continuous variable, 4
piecewise continuous, 18
vector, 356

Galileo Galilei, 23, 53, 88
Jovian moons, 87
γ radiation, 64
Carl Friedrich Gauss, 67
Law of electric flux, 67
mathematical proof, 133
Gauss's law, 67–72, 315, 316
Joseph Louis Gay-Lussac
gas law, 340
geosynchronous orbit, 201
Walther Gerlach
magnetic moment of the electron, 281
Lester Germer
electron diffraction, 157
Josiah Willard Gibbs, 319
chemical potential, 349
free energy, 350
statistical mechanics, 333
Donald Glaser
bubble chamber, 241
Nobel Prize (1960), 241
glibness, 132, 133
gold, 77
Eugen Goldstein
canal rays, 226
golf, 159
gravitation
restricted three-body problem, 190–195
gravitational field, 52
Henry Grayson
ruling engine, 155
James Gregory

diffraction grating, 144
grid electrode, *see also* triode
Francesco Maria Grimaldi
    diffraction, 144
Marcel Grossmann, 44
    differential geometry, 129
Frederick Guthrie
    thermal electron emission, 314

hadron, 244
Edmund Halley, 189
William Rowan Hamilton
    dynamics, 189
Serge Haroche
    Nobel Prize (2012), 121
heat, 331
Joseph Henry
    magnetic induction, 292
David Hilbert, 10
Walter Hohmann
    transfer orbit, 201
Christiaan Huygens, 54
    wavelets, 145

idealization, 13, 27, 48, 75, 129, 135,
    162, 177, 186, 192, 200, 203, 214,
    254, 296
impact parameter, *see also* scattering
impedance
    definition, 313
inductance, 300
inductor, 300
Jan Ingenhousz
    coal dust particles, 324
insulator
    definition, 255
interference, 137, 140–144
interpolation, 46, 62
invariance, 10
    coordinate, 21
    definition, 10
    scale, 325
invariant interval, 106, 110, 125
isotope, 230, 239

Jupiter
    Galilean moons, 87

Kanalstrahlen, *see also* canal rays
Johannes Kepler
    First law of planetary motion, 24, 30
    Second law of planetary motion, 42
    Third law of planetary motion, 42

kinematic
    equation, 15, 165, 326
    relativistic theory, 124–128, 235
Gustav Kirchhoff
    circuit equations, 296–299, 313
    spectra of alkalis and earth-alkalis,
        155
Daniel Kirkwood
    orbital resonance, 212
Felix Klein, 10
Ewald Georg von Kleist
    Leyden jar, 301
Paul Knipping
    x-ray diffraction, 147
knuckleball, *see also* baseball
Polykarp Kusch
    Maxwell-Boltzmann distribution, 337
    Nobel Prize (1955), 337

Joseph-Louis Lagrange
    co-orbital points, 196
    dynamics, 189, 363
    wave equation, 134
Willis Lamb
    Nobel Prize (1955), 337
Pierre-Simon marquis de Laplace
    finite velocity of gravity, 290
large number limit, 83
Joseph Larmor
    radiation by accelerated charges, 86
latitude, 87
Max von Laue
    Nobel Prize (1914), 148
    x-ray diffraction, 147
Lawrence Berkelery Laboratory
    pair production, 242
Lawrence Berkeley Laboratory
    strange matter, 245
Leon Lederman
    Υ discovery, 96
    Nobel Prize (1988), 96
Adrien-Marie Legendre
    elliptic integrals, 262
Heinrich Friedrich Emil Lenz
    inductance, 299
lepton, 119, 243
lightning, 55
Gabriel Lippmann
    Nobel Prize (1908), 65
lodestone, 218, 269
longitude, 87
Hendrik Antoon Lorentz

electromagnetic force, 219, 279
Nobel Prize (1902), 105
wave equation invariance, 105
Lorentz force, 219, 244, 318
Lorentz transform, 105–114
physical interpretation, 122

magic, 126
magnetic field
current density, 268
current loop, 260–267, 277
equivalence, 269
line segment, 258–260
spinning electron, 272–277
spinning sphere, 277
torque, 280
magnetic permeability, 257
Heinrich Magnus
force on spinning sphere, 178
Benoît Mandelbrot
fractal geometry, 325
many body problem, 52, 333
Simon Marius
Jovian moons, 87
Mars, 8
mass, 55
definition, 17, 22
mass spectrometry, 230, 232
mathematics
differential geometry, 129
group theory, 128
need for more, 1, 44, 54, 62
representation, 1, 135, 275, 294, 361
Cato Maximillian
law of mass action, 352
James Clerk Maxwell
electromagnetic theory, 101, 218, 267
mean, see also statistics
mechanics
statistical, 319
meson, 119, 244
meteor
Chelyabinsk, 163
R. C. Miller
Maxwell-Boltzmann distribution, 337
William Hallowes Miller
crystallographic indices, 150
Robert Millikan
charge quantization, 287
electron charge, 285
Nobel Prize (1923), 285
Michael Minovitch

gravity assist, 207
modulation, 139, 310
molecular biology, 117
momentum
definition, 17
electromagnetic field, 277
photon, 235
relativistic, 126
motor, 295
Andrew Motte, 120
muon, 121
Pieter van Musschenbroek
Leyden jar, 301

Napoleon, 218
NASA, 11, 53, 122, 198
GOES satellite, 36
Franz Ernst Neumann, 297
neutrino, 249
neutron
lifetime, 93
Isaac Newton, 29
*Philosphiæ Naturalis Principia
Mathematica*, 17, 120
calculus, see also calculus
corpuscular theory of light, 146
First law of motion, 85, 185
flight of tennis balls, 178
light through a prism, 154
mechanics, 124, 128, 159
Opticks, 100
Second law of motion, 17
Third law of motion, 20
three-body problem, 189
Universal gravitation, 19, 24, 37, 52
NIST, 342
Emmy Noether
conservation theorem, 10, 103, 171,
365–366
normal distribution, see also
distribution, Gaussian
notation, 7, 15, 21, 24, 25, 34, 40, 77,
353, 358
numerical solution
baseball trajectory, 176–185
electron trajectory, 244
precession of magnetic moment, 280
three-body problem, 192

Hans Christian Ørsted
magnetic field, 251
Georg Simon Ohm

electrical circuits, 253, 296
optical molasses, 308
orbit, *see also* trajectory

partial derivative
  definition, 101
partial differential equation, 102, 128
particle, 158, 242
partition function, 334
Wolfgang Pauli
  neutrino, 241
pendulum, 307
perfect differential, 28, 50
period, 117, 142
periodic motion, 42
Jean Baptiste Perrin
  Brownian motion, 324
  Nobel Prize (1926), 324
Alexis-Thérèse Petit
  heat capacity of metals, 343
William D. Phillips
  Nobel Prize (1997), 309
  optical molasses, 309
Thomas Phipps
  magnetic moment of hydrogen, 282
physical insight, 19, 59, 78, 116, 122
physics
  as an experimental science, 3
  definition, 1
  process, 1, 2
  terminology, 2
  vocabulary, 12
pion, 119
Max Planck
  energy-frequency relationship, 156,
    343
  Nobel Prize (1918), 343
Plato
  magnetism in ancient Greece, 217
plumbing, 291
polar angle, 80
  definition, 359
polonium, 77, 237
potential
  electric, 296
potential energy surface, 190
Henry Power
  gas law, 339
power, 295
precession, 192, 280
precision, 2
  definition, 90

product, 347
propagation
  of light, 118
proper time, *see also* Einstein, Eigenzeit
Ptolemy
  geocentric model of the universe, 88

QSAR, 353
quantization
  charge, 285
  energy, 156
  spin, 285
quantum
  definition, 1
  theory, 11, 86, 156, 217

radio receiver, 310
radium, 66
radon, 66
random
  sample, 320
  walk, 323
John William Strutt, third Baron of
    Rayleigh of Terling Place, Witham
  Nobel Prize (1904), 174
  velocity-dependent force, 174
reactant, 347
rectifier, 312, 315
refraction, 100, 154
relativity, 114
representation
  atomic, 85, 157, 239
resistor, 296
resonance, 212
  guitar string, 306
  LRC circuit, 306
  radio receiver, 310
  skyscraper, 306
rings of Saturn, 54
rocket equation, 197–203
Wilhelm Conrad Röntgen
  Nobel Prize (1901), 156
  x-ray tube, 156
Ole Rømer, 88
  eclipses of Io, 88–89, 96–99, 115
Ernest Rutherford, 64
  α, β radiation, 64
  finite nuclear size, 85
  Nobel Prize (1908), 65
  nuclear model of the atom, 79, 270

satellite, 190

Saturn, 53
Félix Savart
  magnetic field, 251
scattering
  cross section, 80, 84
  elastic, 234
  impact parameter, 80
Melvin Schwartz
  Nobel Prize (1988), 96
scientific priority, 87, 196
semiconductor, 312
sequence
  time ordered, 114
Clifford Shull
  neutron diffraction, 158
  Nobel Prize (1994), 158
Le Système International d'Unités (SI),
  76
significance, 93, 98
singular, 61, 63, 67, 275
skew, *see also* statistics
Marian Smoluchowski
  Brownian motion, 324
solenoid, 267
solid angle, 80
Arnold Sommerfeld, 44
spacetime, 124
  event, 106
specific heat capacity, 341
spectroscopy, 217
spectrum
  definition, 154
  visible, 153
square root of time, 326
standard deviation, *see also* statistics
statistics, 90
  central limit theorem, 92
  inference, 94
  mean, 91
  skew, 91
  standard deviation, 91, 97
  variance, 91
Jack Steinberger
  Nobel Prize (1988), 96
Otto Stern
  magnetic moment of the electron, 281
  Nobel Prize (1943), 281
stoichiometry, 351
George Stokes
  resistive force on droplets, 165–167,
    286

Reynolds number, 173
Strategic Defense Initiative (Star Wars),
  102
Gerald Sussmann
  chaotic solar system, 215

tankage factor, 199
Brook Taylor
  vibrating string, 134
John Taylor
  magnetic moment of hydrogen, 282
Taylor, Brook
  series expansion, 12
telecommunications, 310
Nikolai Tesla, 296
thermodynamics, 331
  first law, 331
  second law, 332
Benjamin Thompson, Reichsgraf von
    Rumford
  work/energy equivalence, 331
George Paget Thomson
  electron diffraction, 157
  Nobel Prize (1937), 158
Joseph John Thomson, 64
  cathode rays, 222
  electron, 65
  Nobel Prize (1906), 223
  plum pudding model, 65, 74, 225
Richard Townley
  gas law, 339
trajectory, 8, 12, 18, 20, 232, 247
  baseball, 174–185
  bats, 5
  bounded, 192
  definition, 5
  elliptical, 30, 38–48
  gravity assist, 208–211
  helical, 221
  hyperbolic, 30, 58–64, 78
  parabolic, 30
  random walk, 325
  time on orbit, 43–48
transducer, 311
transform
  coordinate, 34, 103, 273
  coordinate rotation, 110
  rotation, 196
transformer, 295
transmutation, 64
trend, *see also* statistics
triode, 316

uncertainty, 2, 90, 91
uncorrelated, 91
unit system
  British, 177
  conversion between, 198
  le Système International d'unités(SI), 3
universe, 11, 20, 88, 243
unphysical solution, 58

vacuum tube
  diode, 315
  triode, 316
variance, *see also* statistics
vector
  algebra, 6, 354
  calculus, 13, 289
  components, 6
  cross product, 25
  definition, 6, 353
  dual, 149, 355
  identity, 29, 38, 354
  light-like, 107
  normal to a surface, 68
  space-like, 107
  time-like, 107
velocity
  asymptotic, 208
  characteristic, 101
  definition, 13
  exhaust, 198
  gravity, 290
  relativistic, 125
  relativistic addition, 123
  sound, 115
  terminal, 166
Venus, 8, 21
vibration
  damping, 307
viscosity, 174

Alessandro Volta
  battery, 218, 253
voltaic pile, *see also* battery
Robert von Lieben
  triode, 316

Peter Waage
  law of mass action, 352
wave
  intensity, 142
wave equation
  one-dimensional, 101–106, 107
  point source, 135–137
  three-dimensional, 111–113
  two-dimensional, 108–111
wave vector, 150, 153
wavelength, 150
  definition, 118
Victor Weisskopf, 247
George Westinghouse, 296
Charles Thomson Rees Wilson
  Nobel Prize (1927), 236
wind, 184
  as a vector field, 67
David J. Wineland
  Nobel Prize (2012), 121
Jack Wisdom
  chaotic solar system, 215
Ernest Wolland
  neutron diffraction, 158
work
  definition, 168
  energy equivalence, 169

Pieter Zeeman
  Nobel Prize (1902), 105
Zeus, 87
Zustandsumme, *see also* partition
  function